The Biblical Cosmos versus Modern Cosmology

Why the Bible Is Not the Word of God

David Presutta

Scripture quotations marked (ESV) are from The Holy Bible, English Standard Version, copyright © 2001 by Crossway Bibles, a division of Good News Publishers. Used by permission. All rights reserved.

Quotations designated (NIV) are from THE HOLY BIBLE: NEW INTERNATIONAL VERSION®. NIV®. Copyright © 1973, 1978, 1984 by International Bible Society. Used by permission of Zondervan Publishing House. All rights reserved.

New American Bible (NAB) © 1991, 1986, 1970, Confraternity of Christian Doctrine

Scripture taken from the NEW AMERICAN STANDARD BIBLE, (NASV), © The Lockman Foundation 1960, 1962, 1963, 1968, 1971, 1972, 1973, 1975, 1977, 1988, 1995. Used by permission.

New English Bible (NEB) © 1961, 1970, Oxford University Press and Cambridge University Press

New Jerusalem Bible (NJB) Copyright © 1985, Doubleday

New King James Version (NKJV) Copyright ©, Thomas Nelson, Inc. All rights reserved.

New Revised Standard Version (NRSV) The Scripture quotations designated (NRSV) contained herein are from the New Revised Standard Version of the Bible, Copyrighted 1989 by the Division of Christian Education of the National Council of Churches of Christ in the United States of America, and are used by permission. All rights reserved.

Jewish Publication Society Tanakh. Copyright 1985 by The Jewish Publication Society. All rights reserved.

JPS Holy Scriptures 1917. Electronic text copyright © 1995-98 by Larry Nelson.

Usage of the astronomical photographs and of the different versions of the Bible herein is by standard permission and does not necessarily imply endorsement of this book by the originating institutions or individuals.

© 2007 David Presutta

All rights reserved. No part of this publication may be reproduced or transmitted in any form or by any means electronic or mechanical, including photocopy, recording, or any information storage and retrieval system, without permission in writing from both the copyright owner and the publisher.

Requests for permission to make copies of any part of this work should be mailed to Permissions Department, Llumina Press, PO Box 772246, Coral Springs, FL 33077-2246

ISBN: 978-1-59526-829-7

Printed in the United States of America by Llumina Press

Library of Congress Control Number: 2007904261

Contents

Introduction	1
Bibles and Other Resources Used in this Book	5

Part One: Why Ask if the Bible is the Word of God?
Chapter 1	The Historical Origins of the Bible	9
Chapter 2	The Consequences of Belief in the Bible	15

Part Two: The Modern View of the Cosmos
Chapter 3	The Earth in Space	29
Chapter 4	The Evidence for the Modern View	38
Chapter 5	The Big Bang	53
Chapter 6	The Size of the Cosmos	59

Part Three: The Biblical View of the Cosmos
Chapter 7	In the Beginning	65
Chapter 8	The Firmament of Heaven	78
Chapter 9	The Circle of the Earth	98
Chapter 10	The Pillars of the Earth	127
Chapter 11	The Pit of Sheol	148
Chapter 12	The Lights in the Firmament	169

Part Four: The Cosmological Answer
Chapter 13	The Biblical Cosmos: Metaphor or Mythology?	189
Chapter 14	The Implications	201

Appendixes
A	Problems in the Old Testament	213
B	Problems in the New Testament	244
C	The Book of Enoch and the Biblical Cosmos	280
	Supplement to Appendix C	299
D	Babylonian Myths and the Bible	305

Contents

Notes	311
Bibliography	329
Subject Index	337
Bible Verse Index	348

Illustrations
- Figure 1. The center of Omega Centauri, a massive star cluster — 31
- Figure 2. M31, the nearest major galaxy to our own — 33
- Figure 3. Spiral galaxy M101 — 35
- Figure 4. Abell 1689, a galaxy cluster — 36
- Figure 5. The biblical cosmos — 190

Introduction

In the beginning God created the heaven and the earth.

Those are, of course, the opening words to the Bible. They are from the first chapter of Genesis and are an introduction to the steps of creation that are described in the rest of the chapter.

Although the opening words of the Bible make a cosmological statement and numerous passages that have a bearing on the nature of the biblical cosmos can be found throughout the Bible, surprisingly few in-depth analyses have been written about the cosmology of the Bible and how it compares to the findings of science. In contrast, numerous analytical books have been written about many other facets of the Bible in relation to various other fields of study—for example, how the history of the ancient Near East relates to the stories that are told in the Bible, how archaeological findings relate to the structures and events that are described in the Bible, and, of course, how the geological, paleontological, and biological sciences relate to the biblical accounts of the creation and of the Flood.

To be sure, some individuals have written books that attempt to square certain limited aspects of biblical cosmology with the findings of modern cosmology—particularly concerning scientific evidence indicating that the universe is billions of years old in contrast to the implication in Genesis that creation occurred relatively recently.[1] However, biblical cosmology consists of much more than just its temporal aspect; a significant number of passages in the Bible shed light on the structure and physical characteristics of the biblical cosmos, and those passages provide a more complete picture of the cosmology that can be derived from the Bible. It is remarkable, then, that biblical cosmology has not been investigated in the literature about the Bible to the extent that is warranted.

Despite the fact that it has not been the subject of as much analytical investigation as many other facets of the Bible have been, the cosmos that is revealed in the Bible is a fundamental aspect of the biblical worldview, just as the cosmos that science has revealed is a fundamental aspect of the modern worldview. The biblical cosmos is, in fact, an integral part of the narrative that unfolds in the Bible. In the Bible (specifically, the Christian Bible, consisting of both the Old Testament and the New Testament), God creates heaven and earth to provide a stage upon which to play out the drama of his program of salvation for mankind—a program that begins with the creation of the cosmos and has its completion in the portended future annihilation of the cosmos.

According to the Bible, then, as culminated in the New Testament, the origin and the fate of the cosmos are tied to events here on the earth and to God's salvation program for mankind. In that aspect, the Bible presents a geocentric view of the cosmos that is obviously and significantly different from the modern view. In the modern view of the cosmos there is nothing special about the earth; it is only one planet among an uncountable number of planets that must exist in the universe, and the earth's sun is only one star among an equally uncountable number of stars. The explicit reality of the modern view is that the earth is but an insignificant mote of no particular relevance to the rest of the cosmos.

Those differences between the biblical view of the cosmos and the modern view are so fundamental with respect to the natural order of things that they should give one pause to wonder whether they can be reconciled. Did the God of the Bible create the incredibly vast and multi-faceted universe that science has uncovered just so he could destroy it all in the "Last Days" when he administers his judgment on the earth and its inhabitants? Or, on the other hand, does that modern view of such a vast universe negate and render meaningless the cosmological descriptions of divine judgment that are put forth in the Bible?

Those questions cannot be lightly brushed aside, for—unless one totally rejects the objective reality of the cosmos that science has uncovered—they would certainly play a part in determining the credibility of the Bible. And, of course, the credibility of the Bible is crucial in any attempt to answer the question: Is the Bible the word of God?

Because some two billion people on this planet would answer it in the affirmative, that question has a great deal of relevance to present-day society. It cannot be ignored, for it deeply affects us all, regardless of whether we acknowledge so or not, and regardless of whether we, as individuals, are believers or nonbelievers. Indeed, the question of whether or not the Bible is the word of God has consequences for us and our future here on the earth regardless of whether the answer is "yes" or "no."

If the Bible is the word of God, then the future of mankind, as well as that of the earth and of the cosmos of which it is a part, is laid out in the Bible, and an acceptance of the Bible as a supernaturalistic revelation from God is a prerequisite for achieving the personal salvation it holds forth as a divine promise.

On the other hand, if the Bible is not the word of God, there is no reason to accept its pronouncements relating to the supernatural, and it must be only a collection of myths, folklore, embellished histories, and superstitious assumptions and speculations. If that is the case, the Bible gives us a false view of ourselves and our place in nature, and belief in the Bible is therefore a delusion—a delusion that can have severe and dire repercussions for us and the world on which we live.

Those consequences highlight the importance of determining the answer to the question of whether the Bible is the word of God or not. For centuries, of course, people have argued over the answer to that question, but no one has ever come up with a clear-cut, undeniable argument for one side or the other that would convince everyone. For that matter, it is likely there will always be those who will remain unconvinced by any argument that is contrary to what they believe, no matter how persuasive and undeniable that argument may be. Nevertheless, the question remains, and there are many people who are seeking an answer that can be based on evidence and reason.

There have been, of course, many investigators—biblical scholars as well as nonbelievers—who have pointed out numerous problems, contradictions, and inconsistencies in the Bible. But devout believers in the Bible tend to downplay or simply ignore all such problems. Many believers in the Bible hold the position that there are no such problems at all in the Bible and that any appearance to the contrary is a matter of faulty understanding or interpretation. Some believers in the Bible simply say that acceptance of the Bible must ultimately be a matter of faith and that the Bible is above critical analysis.

But why should that be? The question of whether to accept the Bible as a potential—even the ultimate, as some would have it—source of truth and knowledge should be subject to the same rules of evidence that are applied to any other area of study. Simply saying that the Bible is the word of God and that is good enough is *not* good enough.

And that brings us back to what the Bible says about the cosmos—that is, heaven and the earth. If the Bible is the word of God, its statements about the nature and structure of the cosmos should reflect the view of the cosmos that science has uncovered; after all, the God of the Bible is supposed to have created heaven and the earth, and that same God is supposed to be the ultimate author of the Bible. Moreover, as we noted, the cosmos that is revealed in the Bible is an integral part of the biblical worldview; in fact, it is so integrated with the biblical worldview that the credibility of the Bible is dependent upon the validity of its cosmology.

On the other hand, the findings of science through direct observations over the centuries have revealed a cosmos that, in the modern worldview, is basic to existence and to objective reality. In light of that modern cosmological worldview, the Bible must therefore stand or fall according to *its* cosmological worldview. In this book, then, we will analyze what the Bible has to say about the cosmos, and we will see how the biblical cosmos compares to the modern view of the cosmos as defined by the findings of science.

And that brings up an important point: Referencing only a few Bible verses here and there would hardly be sufficient to demonstrate to everyone's satisfaction the actual nature of the biblical cosmos. Therefore, the analysis of the biblical cosmos that is presented in this book is necessarily very extensive and detailed, and it will show that the Bible presents a consistent cosmological view that permeates the Bible from beginning to end.

Doubtless, there will be those who will remain unconvinced by anything that goes against their beliefs, but, for those who are open to the evidence, this analysis will provide a basis for arriving at a reasoned answer to the question of whether or not the Bible is the word of God.

A final note: I would like to give special thanks to Dr. Stuart Jordan, former Senior Staff Scientist in the Laboratory for Astronomy and Solar Physics at Goddard Space Flight Center, for verifying the accuracy of the astronomical material in Part Two of this book. I would also like to thank Dr. Hector Avalos, Professor of Religious Studies, Iowa State University, for his scholarly review of the manuscript and his suggestions for additional research material. Finally, I want to express my appreciation to Richard Akin, Ken Marsalek and Don Evans for their helpful comments and suggestions.

Bibles and Other Resources used in this Book

The following is a list of the different versions of the Bible that are referred to in this book, along with their abbreviations. Copyright notices of copyrighted Bibles are listed on the copyright page of this book.

American Standard Version (ASV)
English Darby Bible (Darby) 1884/1890
English Standard Version (ESV)
King James Version (KJV)
New American Bible (NAB)
New American Standard Bible (NASV)
New English Bible (NEB)
New International Version (NIV)
New Jerusalem Bible (NJB)
New King James Version (NKJV)
New Revised Standard Version (NRSV)
Jewish Publication Society Holy Scriptures 1917 (JPS 1917)
Jewish Publication Society Tanakh 1985 (JPS Tanakh)
Septuagint English Translation by Brenton

When one searches for the meaning of a passage in the Bible, the question often arises as to what a particular word means in the original Hebrew of the Old Testament or in the original Greek of the New Testament. Of value in answering such questions are concordances of the Bible and lexicons of the Hebrew and Greek languages. One of the more popular and readily available of the concordances is *Strong's Exhaustive Concordance of the Bible* (hereinafter referred to as *Strong's Concordance*), which, as well as providing a relatively complete concordance, also conveniently provides Greek and Hebrew lexicons. In *Strong's Concordance*, each English word in the King James Version of the Bible is assigned a number, and that number can be used to cross reference to the original Hebrew word for English words appearing in the Old Testament, or to the original Greek word for English words appearing in the New Testament. However, the lexicons in *Strong's Concordance* are not necessarily the best available, so we will also refer to other lexicons in order to get a consensus on the proper translation of critical Hebrew and Greek words. It should also be noted that there are several versions of *Strong's Concordance* currently available. Some versions have lengthier interpretations of, and more background information about, the original Hebrew and Greek words than other versions do. Some of these versions also have more complete lists of the Hebrew and Greek words in the main part of the concordance, whereas others relegate many of the less significant words to an appendix where they are simply listed without any interpretation. *The New Strong's Expanded Exhaustive Concordance of the Bible* is one that provides lengthier information.

The following is a list of the other lexicons that are also used in this book along with their abbreviations:

Greek-English Lexicon of the New Testament, Joseph Henry Thayer (Thayer)

Greek-English Lexicon of the New Testament and Other Early Christian Literature, Bauer, Danker, Arndt & Gingrich (BDAG)

Analytical Lexicon to the Greek New Testament, Timothy Friberg and Barbara Friberg (Friberg)

Hebrew-Aramaic and English Lexicon of the Old Testament, Francis Brown, S.R. Driver, and Charles A. Briggs (BDB)

The Hebrew and Aramaic Lexicon of the Old Testament, Ludwig Koehler and Walter Baumgartner (KB)

A Concise Hebrew and Aramaic Lexicon of the Old Testament, William Holladay, editor (Holladay)

Another important help in analyzing the Bible is a good computer software program that can be used in making searches of words, both in the original languages of the Bible and in English. One such program is BibleWorks™, which contains most of the above-referenced versions of the Bible and lexicons, as well as many other resources. It provides considerable assistance in doing word searches in Hebrew, Greek, and English, and in finding out the original meaning of biblical passages.

Part One
Why Ask if the Bible Is the Word of God?

Chapter 1
The Historical Origins of the Bible

The Bible had its origins in an age when people belonging to most cultures customarily interpreted the forces of nature and the events in their lives according to their particular mythological and supernaturalistic beliefs. Those beliefs usually included a pantheon of gods who could be propitiated by worship and sacrifice, and who would assist a supplicant in overcoming life's obstacles. Tales about those gods—how they interacted with each other and with mankind—were also usually included in the myths of the time. Even today, most educated people have at least a passing familiarity with the myths of the ancient Greeks, Romans, and Scandinavians, and, in our younger days, many of us learned stories about Zeus, Jupiter, and Odin. Less well known in our western culture are the stories of Gilgamesh and Marduk, who belonged to the myths of the Mesopotamian civilizations. And, of course, numerous other cultures had gods and supernatural heroes that most of us have never even heard about.

But what about the ancient Hebrews? Did they likewise interpret events according to an established mythological outlook and a belief in the supernatural? When they wrote the books of the Old Testament, did they embellish the narratives with mythic and supernatural overtones in the same way that those living in other early cultures embellished the narratives of their own histories? Did the ancient Hebrews, for example, add their own mythic coloration to the biblical stories about the battles for the land of Canaan in much the same way that Homer wove Greek myth into his story of the siege of Troy? Specifically, is much of the Bible, like much of the literature of the other cultures of ancient times, a collection of myths, folklore, embellished histories, and superstitious assumptions and speculations?

Or should the stories that the ancient Hebrews told, as passed on in the Old Testament, be taken as being something completely different from the fanciful tales that were told in other cultures of the time? Should the biblical stories of the wanderings and wars of the ancient Hebrews, with their accounts of interventions and judgments by Yahweh, the tribal God of Israel, be considered true in every detail? And what of the earlier stories in the Old Testament, such as the creation and the Flood? Should those stories also be considered true and supernaturally revealed in everything they relate?

And what about the Christian New Testament? Those who wrote the books of the New Testament expounded on a new series of traditions, many of which were based on the supernaturalistic beliefs and cultural traditions delineated in the Old Testament. Must the New Testament also be considered true and supernaturally revealed in every detail?

In short, must the Bible and everything in it be regarded as the true word of God?

Before we begin the process of answering that question, we should take a brief look at the process by which the Bible was produced, particularly because most people, including most Bible believers, have only a vague understanding of that process. Understanding the process by which the Bible was produced is certainly relevant in trying to find the answer to the question of whether the Bible is the word of God, and it will also help to put things in perspective as we analyze various aspects of the Bible and of belief in it.

The Bible had its beginnings in a number of scriptures that had been written, accumulated, and edited by the ancient Hebrews over a period of about eleven hundred years or so prior to the Christian era. There were many such scriptures, and they consisted of those books that eventually made it into the Hebrew scriptural canon (that is, the collection of books considered to be authoritative), as well as many other books that ultimately failed to enter the canon. Toward the latter part of this period, a particular collection of those scriptures had gained a general acceptance among the Jewish remnant of the Hebrew peoples. This collection was divided into three groups of books: the Law (also called the Pentateuch, or the Five Books of Moses), the Prophets, and the Writings. These scriptures were written primarily in Hebrew, but had a few passages in Aramaic, a Semitic dialect that came to be spoken by many Hebrews in the latter part of the period in question.

As early as the third century B.C.E.[1], a translation of some of the Hebrew Scriptures into Greek had been started. Initially, the Pentateuch was translated into the Greek, and then eventually the books of the Prophets and the Writings were translated as well. This Greek translation of the Hebrew Scriptures came to be called the Septuagint, which is derived from the Greek word for seventy because seventy Hebrew scholars were supposed to have worked on the translations. During the second and first century B.C.E., some other "scriptural" books were also added to the text of the Septuagint, including some that covered historical events that took place during this period.

Prior to the fall of Jerusalem in the year 70 C.E., there does not appear to have been an absolutely fixed and exclusive set of scriptural books comprising the Hebrew canon, even though the books of the Pentateuch, the Prophets, and the Writings were generally favored. About the year 90 C.E., the Jewish Council of Jamnia set out to resolve that situation by determining exactly which books were to be considered authoritative. They selected twenty-four books, which consisted of the Pentateuch, the Prophets, and the Writings in their final form. These books were eventually accepted by Judaism at large and became the established Hebrew scriptural canon. With the establishment of the Hebrew canon, great care was taken in the copying process, and the texts were transmitted down the years with considerable accuracy.

When Christianity came on the scene, a variety of resultant Christian scriptures brought about similar questions concerning which texts were to be considered authoritative. Those questions did not arise, however, until after several decades and with an eventual accumulation of a very large number of diverse writings. The first followers of Jesus were mostly Jews who considered themselves still to be Jews and who believed that Jesus was the Jewish Messiah—that is, "the Anointed One" ("*christos*" in Greek) of God who would establish God's kingdom here on the earth. These early Christians used the same scriptures for reading and worship that they had used before they had become followers of Jesus, and, because many of them lived in a Greek-oriented society and spoke Greek, the scriptures that they were likely to use were those of the Septuagint.

Aside from the initial Jewish believers, few Jewish converts came into the Christian church during the following years, and, as a result of the efforts of Paul, the church grew by bringing in mostly Gentiles. As additional church congregations came into existence, there was little consensus among them as to which versions of the Hebrew Scriptures were to be considered authoritative. Some congregations accepted only the twenty-four book collection that comprised the Jewish canon, some the Septuagint text with the added material, and some the Septuagint but without most of the added material, which was called the "Apocrypha."

As previously indicated, the early followers of Christianity initially had no scriptures uniquely relating to their own beliefs. However, some of the apostles wrote letters to various individuals and congregations, and many of those letters were eventually considered authoritative and were copied and frequently read during services. Several gospels and other writings were also eventually written and were read in the early church as well. These writings included a large number of other gospels, books, and letters besides those that are presently in the New Testament.

During the first few centuries of the Christian era, there was no consensus by all believers on which of these writings were authoritative. Indeed, Christianity itself was composed of numerous groups that based their beliefs on different collections of writings. Many of these groups considered other Christian groups to be "heretical" and became antagonistic toward those who did not accept their view of the "true" faith.[2]

In the year 325 C.E. the emperor Constantine convened the Council of Nicaea to establish a consensus on what would be the major tenets of the faith. At the time, there were two major Christian groups. One was led by Arius, the presbyter of Alexandria. This group believed that Jesus was the begotten son of God, and, as such, came after and was lesser than God. The other group was led by Athanasius, the bishop of Alexandria. This group believed that Jesus was unbegotten and was co-equal with God as part of a triune Godhead. The group led by Athanasius managed to hold sway at the council proceedings and got the triune view established as the correct view of the nature of Jesus. However, ten years later a council at Tyre rejected the triune view and established the Arian view as the correct one. Subsequently, other councils went back and forth in declaring which was the correct view. Eventually the triune view won out and became the established belief, while all other views were considered heretical.

A parallel dispute occurred concerning which of the many collections of Christian writings was considered authoritative and reflected the true nature of the ministry of Jesus. Some of the early church leaders made up lists of those writings that they considered to be inspired, but there continued to be many disagreements. Some of the list makers rejected some of the writings that are now considered canonical, while others accepted writings that are not now in the canon. By the end of the fourth century C.E., the twenty-seven books of the present New Testament canon were generally accepted, but by no means unanimously. Some Christian churches still do not accept some of the New Testament books that are accepted by others; some branches of the Syriac church, for example, do not accept 2 Peter, 2 and 3 John, Jude, and the Book of Revelation.

There was also disagreement about the inspiration of the books of the Apocrypha. The Eastern and the Roman Catholic churches accepted the Septuagint, including most of the additions constituting the Apocrypha, as canonical, and their translations of the Old Testament still contain their

1. The Historical Origins of the Bible

versions of the Apocrypha. After the Protestants split from the Catholic Church centuries later, the early editions of their versions of the Bible, including the King James Version, often contained the books of the Apocrypha, but in a separate section and usually with an indication that they were not considered inspired. Eventually, however, the Apocrypha were dropped from most Protestant Bibles.

In the Christian Bibles, both Protestant and Catholic, the order of the Old Testament books was revised, and two of the original books in the Hebrew canon were each divided into two separate books. These were the Book of Kings, which was divided into I Kings and II Kings, and the Book of Chronicles which was divided into I Chronicles and II Chronicles. Thus, Christian Bibles have a larger number of individual books in the Old Testament than does the Hebrew canon covering the same material.

It is generally acknowledged by New Testament scholars that the books of the New Testament were written during the period from around 50 C.E. to the early part of the second century. Certain letters ascribed to Paul—specifically, I Thessalonians or Galatians—were probably the earliest of these writings. Some other letters that were supposedly written by Paul—such as Ephesians, 1 and 2 Timothy, and Titus—are in dispute and some scholars consider them to have been written by someone who used Paul's name to give them a measure of authority. The earliest gospel is considered to be that of Mark, which was probably written around 70 C.E. The Gospel of John is generally thought to have been the latest of the gospels to have been written, and is usually dated around 100 C.E. The authors of the gospels of Matthew and Luke are believed to have used the Gospel of Mark as a source of material for their own gospels and used additional material from other sources to fill out the story of the ministry of Jesus. Thus, none of the New Testament writings were written during the time of the ministry of Jesus, but were written decades later. With the possible exception of Luke, the authors of the gospels are essentially unknown, and the names ascribed to the gospels came about through tradition rather than from a clear designation of authorship. For that matter, the author of the Gospel of Luke, by his own admission in the preamble to his book, was not an eyewitness to the events described in his gospel, but merely retold stories about Jesus that had been handed down.

The Bible, then, was not put together as a unique revelation from God at a single point in time. The origin of the Bible encompassed a period of well over one thousand years and its final composition came about only after many disputes over which of the many writings that were produced during that period were considered authoritative and which were not. For the most part, the process of selecting the writings that were considered authoritative depended on which parties were dominant or which parties were the most effective in espousing their views.

Because the books of the Bible had such varied origins in both place and time and came together as an established canon only after centuries of disputes over which of the many potential source documents were authoritative, one should not be surprised to find many problems, inconsistencies, and contradictions in the Bible. That, of, course, raises the question as to whether those who put the Bible together were "inspired" in their selection of the books they included in the canon. Many Bible believers hold that they were and that the Bible is the complete word of God. Yet, several of the books of the Bible refer to other books, such as the Book of Jasher and the Book of Enoch, as if they were inspired scripture, but which are not included in the canon. If the "inspired" books of the canon cite other books that are not in the canon as being authoritative, should not those other books also be in the canon? Can the Bible be considered complete without them?

We shall take a more detailed look at that question elsewhere, but first, we should understand why it is important to determine the answer to the question: Is the Bible the word of God?

Chapter 2

The Consequences of Belief in the Bible

A large number of people—certainly a majority in the western world—have faith that the Bible is the word of God. But that a particular faith is believed by a large number of people is no assurance that that faith is based on truth. In some countries, the majority of people have faith that the Koran is the word of God, yet the followers of other religions reject that faith. In one of the American states, the majority of people have faith that the Book of Mormon, as well as the Bible, is the word of God, but most Bible believers in other states have faith that the Mormon scripture is not the word of God. The Jews have faith that the books of their scriptures are the word of God and that the books that the Christians have added to those scriptures are not the word of God. Christians, on the other hand, have faith that both sets of books are the word of God. The Christian religion itself is divided into hundreds of denominations. The followers of some of those denominations have faith that they alone have the true Bible-based gospel and that those who belong to different denominations are following false gospels.

Faith, then, does not produce a universal belief in a single creed. And having faith, no matter how sincere or how intense it may be, is no guarantee that one has a faith that is based on truth, nor is it a guaranteed means of convincing others that one's beliefs are true.

So that brings us back to the previously asked question: Must the Bible and everything in it be regarded as the true word of God? For that matter, is it really all that important to determine the answer to that question?

In fact, the answer to those questions is of considerable importance. For most of the time following the inception of Christianity, belief in the Bible—specifically, belief in the Christian version of the Bible with its New

Testament—was a force that drove many people to try to foster that belief upon others and upon the institutions of society. Though that force often expressed itself benevolently and with a concern for alleviating suffering (as many present-day Bible believers frequently emphasize), such has not always been the case. Belief in the Bible, and in the religion that is based on it, has a darker side, and oftentimes over the centuries that belief expressed itself through violent means. Religious wars, crusades, and inquisitions played a dominant role in the history of western civilization and were the result of efforts by Bible-believing Christians to spread their faith and to prevent the spread of "heresies." During those times there was no separation of church and state, and the leaders of the Church could use the power of the state to further their religious goals.

Over time, however, and especially during the Enlightenment, the power of the Church waned and the state became more secular in its governance. Currently, in most western nations, church and state are no longer enmeshed to the extent they were in former times, and religious leaders no longer have the absolute means to use the power of the state to enforce religious conformity.

Nevertheless, even today, there are those who seek to impose their Bible-based religious beliefs upon society, and, in fact, many groups in the American Religious Right are working toward the complete elimination of the principle of separation of church and state. Despite historical evidence to the contrary, these "anti-separationist" groups maintain that the United States of America was founded as a Christian nation and was based on biblical principles. According to these groups, a liberal court system has, over the years, subverted the original intent of the country's founders and has moved the nation away from its religious roots. These groups include the Christian Coalition, Coral Ridge Ministries, the Center for Christian Statesmanship, the National Association of Evangelicals, the American Family Association, the Campaign for Working Families PAC, Concerned Women for America, the Family Research Council, Focus on the Family, and others. These groups are frequently aided by Christian legal entities, such as the Alliance Defense Fund (ADF) and the American Center for Law and Justice (ACLJ), that assist them and other plaintiffs in lawsuits against those who work to keep church and state separate or who are perceived as inhibiting freedom of religion—meaning, frequently, the freedom of the Christian far right to impose their religion on others.[1]

The Religious Right is also working to place Bible-believing, evangelical activists in government posts so they can influence legislation and be in a position to advance their careers as politicians and government personnel. For example, Patrick Henry College, which is located in Purcellville Virginia, trains its students for just that purpose. Though this

2. The Consequences of Belief in the Bible

college is small, it is well on its way toward accomplishing its aim by having, as this is being written, several of its graduates hold internships in the White House.[2] In addition, Pat Robertson and the late Jerry Falwell established their respective universities in part to train Bible-believing Christians for placement into government positions with the idea of using those positions to break down the principle of separation of church and state.[3]

In their attempts to prove that that the United States of America was founded as a Bible-based Christian nation, the anti-separationists ignore the plain language of the Constitution and Bill of Rights and resort to specious and irrelevant arguments, which large numbers of believers take as "gospel truth." However, besides the Constitution and Bill of Rights, the actual original intent of the founders is also made quite clear in several other historical documents and letters, including quite pointedly the Treaty with Tripoli, which was passed unanimously by Congress in the year 1797 and signed by President John Adams. It unambiguously states in Article 11 that "... the government of the United States of America is not in any sense founded on the Christian Religion."

Not content with simply doing away with the principle of separation of church and state, some of these groups have the aim of going even further and are working to have their Bible-based religion totally supplant the government. These groups are variously called Dominionists, Christian Nationalists, and Christian Reconstructionists. The Reconstructionist philosophy, for example, is based on the belief that the Church must first set up Christ's kingdom on earth, and then, and only then, will Jesus return. To the Reconstructionists this means that all the nations of the world and every aspect of society must come under biblical law.[4]

The Reconstructionists have made it clear that their goal is to turn the United States of America into a theocracy by removing the Constitution and the Bill of Rights as the source of governmental authority and law. The Reconstructionists would replace those founding documents with the Bible, do away with the principles of democracy, and impose biblical edicts on American society. As one adherent of Reconstructionism stated during a speech he gave at the Christian Coalition's Road to Victory Conference in 1994 and again in the eleventh annual national convention of the Concerned Women for America two weeks later, the laws of society must be based on the position that "... whatever is Christian is legal. Whatever isn't Christian is illegal."[5]

In the Bible-based society that the Christian Reconstructionists envision, the death penalty would be meted out according to biblical mandates. The Reconstructionists list eighteen offenses for which the Bible mandates death, a few of which are cursing or striking one's parent, adultery, unchastity, blasphemy, sacrificing to false gods, desecrating the Sabbath, and propagating false doctrines. As for the last mentioned, one would presume that the Reconstructionists would determine what constitutes false doctrine.

What the Reconstructionists make clear is that all religions that are not based on the Bible are false and that there would be no room for such religions in the society that they are seeking to create. As one Reconstructionist minister has stated, "Nobody has the right to worship on this planet any other God than Jehovah."[6] For that matter, even within the context of Christianity the Reconstructionists have made it clear that their view of the Bible and of Christian doctrine constitutes the only "correct" religion.

If the Reconstructionists or other such groups were to succeed in achieving their goals, the United States of America, and indeed the whole world, would become a repressive theocracy. In such a society, the combination of absolute theological certainty and absolute political power would result in a requirement for absolute religious conformity.[7]

Randall Terry, the founder of Operation Rescue, expresses quite well what he and his fellow Christian nationalists want for America:

> I want you to just let a wave of intolerance wash over you. I want you to let a wave of hatred wash over you. Yes, hate is good. ... Our goal is a Christian nation. We have a biblical duty; we are called by God to conquer this country. We don't want equal time. We don't want pluralism.[8]

There is a parallel to this in the Islamic world. Certain Islamic nations are repressive theocracies in which religious conformity is required and any deviation from that conformity is severely punished. And within the Islamic world, there are extremists who want to convert the whole world to their religion and will use any means to further their goal; September 11 and other acts committed by these extremists are prime examples. Just as the Christian Reconstructionists justify their worldview by making recourse to the Bible, the Islamic extremists base their worldview on the Koran. The two belief systems, in fact, are even closer than one might think, for the Koran follows certain traditions that have their source in the Bible. In fact, Moslems consider themselves to be the inheritors of the biblical God's promise to Ishmael, one of the sons of the patriarch Abraham.[9] The theocratic governments and movements of the Islamic world thus foreshadow in a limited fashion what the future could hold for mankind if the Christian Reconstructionists were to attain their goal of national or world domination.

Although, as a group, the Reconstructionists are presently not well known, they are growing in numbers and are highly influential among the members of the Religious Right. For that matter, many right-wing political figures are sympathetic to the Reconstructionist philosophy and would like to see it become an influence in establishing governmental policy. Because the Reconstructionist philosophy is based upon a total acceptance of the Bible as the word of God, and because the Reconstructionists are determined to bring every aspect of society under biblical law, the question of whether the Bible is, in fact, the word of God gains particular importance. Certainly, it would be prudent to determine, if possible, the answer to that question in order to challenge the Reconstructionists and other like-minded groups of Bible believers in their attempts to impose their beliefs on the institutions of government and society.

The political arena is not the only area in which some Christian groups are trying to foster their Bible-based beliefs upon society. Another example is provided by certain Christian fundamentalists who place a special emphasis on the account of creation in the biblical book of Genesis. The creationists—as these fundamentalists are called—are trying to weaken or to eliminate the teaching of evolution in our nation's schools and to have the biblical account of creation taught in a form that is called "scientific creationism," or, lately, in a modified form called "intelligent design."

Although the creationists call their system of belief "creation science," it is not really science. True science involves looking at a phenomenon or a set of findings relating to the natural world and formulating a theory that will explain its cause or its relationship with other phenomena or findings. The theory then forms the basis for making predictions about what should be observed in further lines of research. If the predictions prove correct, the theory achieves a certain level of confirmation, and the new observations in turn may be incorporated into the theory, further refining it and possibly suggesting even further predictions. However, if the observations are *not* as predicted, then the theory must be either discarded or modified to take the new findings into account. That is the way true science works.

In contrast, the creation "scientists" do not begin by trying to find the answer for the cause of a phenomenon by developing a theory. They already have the answer at the start, an answer that to them is non-theoretical and incontrovertible: the Bible. The approach that the creation "scientists" take is that the evidence must be interpreted in light of the answer they already have—that is, what the Bible says. If the evidence does not support that answer, the creation "scientists" will try to modify the evidence to force it to provide the desired support. If they are unable to do that and the evidence contradicts their Bible-based answer (as it usually does), they will attack the evidence as being unreliable or invalid, or they will simply ignore it. In no case will the creation "scientists" discard or modify—at least not in any significant manner—the answer with which they began. To a creationist, that answer is the absolute truth and must be held at all cost. The methods of the creation "scientists," as well as the absolutist position that they hold, are therefore contrary to the methodology of science.[10]

The efforts of the creationists to impose their views in the nation's schools and on society have many more consequences than would at first appear to be the case. The creationists direct their efforts not only against evolutionary biology, but also against other disciplines that they regard as being inimical to their biblical viewpoint—such as certain aspects of geology, astronomy, physics, and history. In essence, the creationists are trying to "reconstruct" scientific and other disciplines in order to make them fit the biblical mold in the same way that the Christian Reconstructionists are trying to reconstruct society to fit that same mold.

The implications of what may happen if the creationists should succeed in their efforts cannot be minimized or ignored. The creationists have a hard core of determined believers and are assisted by a large segment of the population that is sympathetic to the biblical view of creation. If that view of creation is not correct—and the evidence is overwhelming that it is not—there would be serious ramifications if the creationists prevail.

If they succeed in their efforts, the creationists would tear down the highly detailed, interconnected, and self-consistent view of nature that scientists have constructed over many years of research in multiple disciplines—a view that, in fact, provides the basis for much of our technological and scientific progress. In the place of that scientific view, the creationists would impose an ideological religious view that is masked in scientific-sounding language, but which has no support in, and is contradicted by, the scientific evidence. The result would be a significant reduction in new scientific findings and innovations, for no valid scientific innovations, predictions, or technological advancements can be derived from the Bible-based creation "model."

Even the more "updated" version of creationism, called "intelligent design" or ID—the proponents of which try to avoid the more overtly religious references made by their creationist colleagues—would have a similar effect if it finds its way into the nation's schools and into the population in general. The "IDers" contend that certain aspects of biological systems have what they call "irreducible complexity" and therefore cannot have arisen naturally—hence, they say those systems must have been designed by an intelligence. But what they are really saying, in effect, is that *they* cannot understand how those systems could have arisen naturally; in other words, they are arguing from ignorance, not knowledge. If their views about those systems take hold, scientific advancement would be stifled; after all, if one believes that those systems were "designed" by an "intelligent designer" there would be no sense in making any further investigation. The problem is that there actually are natural explanations that can account for several of the irreducibly complex systems that the proponents of ID say cannot have come about naturally.[11]

Despite the fact that the creationists and their ID counterparts have had many setbacks at the judicial level in their attempts to get state laws passed that would require the teaching of the creation "model" or of intelligent design, they continue to work at the local school-board level to achieve their goals. The creationists and the proponents of ID have had a certain amount of success at that level because, as recent surveys have shown, a great many people are quite unknowledgeable in matters of science and are therefore easily swayed by the arguments put forth by these individuals. Those arguments often sound scientific and convincing, but they are usually characterized by a scant regard for an objective analysis of the evidence and often misrepresent the facts.[12] Since the members of school boards do not necessarily have a better understanding of science than does the general population, and, in fact, may themselves be inclined toward a creationist or ID viewpoint, they have often caused science courses and the books used in those courses to be watered down.

If the creationists and the proponents of intelligent design succeed in getting their views taught in our schools to the extent they seek, our children would be misled concerning the nature of science and what has been learned about our natural world. Again, no real science that would produce tangible and usable results can come from a perspective that is based purely on creationism or ID. The inculcation of students with the tenets of these religious viewpoints would therefore dilute their understanding of the way science works and would impede the scholastic development of those individuals who might seek to follow scientific professions. As a result, our children would inevitably fall behind the children of other nations scholastically, and they would be ill-prepared for living in the technological and competitive world that is coming in their future. Even more significantly, there would be fewer students going into fields that have a relation to evolutionary theory, such as genetics, biology, and geology. This could have a significant deleterious effect on the development of new medical applications and food production, as well as on the search for new sources of petroleum and alternate sources of energy.

Aside from what one may think about groups that are trying to shape the United States into a theocracy or about the creationists and their allies, an uncritical belief that the Bible is the word of God can easily lead one into accepting questionable and even irrational views.

Today, unlike the era when the Bible took form, mankind has enormous technological powers at its disposal and is having a significant and far-reaching effect on the earth and the environment. If we are to survive as a species and are to maintain this world into the future for ourselves and for the other species that inhabit it, we will need to use reason and a true understanding of ourselves and our place in nature in order to make the proper decisions concerning courses of action.

And that is where the danger to mankind lies. If the Bible is not the word of God, it must then be a collection of myths, folklore, superstitious assumptions and speculations, and embellished histories from another age and from an archaic society. As such, it gives us a false view of both ourselves and our place in nature. In that case, using the Bible as an authoritative guide to determine what is valid today in science, government, and social policy, as well as using it to determine how we should look at the world, could result in dire, even devastating, consequences.

For example, because of their belief in the biblical directive, "be fruitful and multiply, and fill the earth," many Christians—fundamentalist and evangelical, as well as Catholic—hold the view that large families are a blessing and that any form of artificial birth control is contrary to the will of God.[13] Indeed, there are those in the Religious Right, including several right-wing politicians, who are trying to limit or ban public access to birth control information, pills, and devices. In fact, as a result of activism by the Religious Right, pharmacists in some states can legally refuse to fill prescriptions for birth control pills or morning-after pills on the grounds that doing so would violate their religious beliefs. In those states, even victims of rape can be denied access to morning-after pills on the basis of the religious whims of pharmacists.

The implications of reduced access to effective means of birth control are enormous. The world is already overpopulated and as the population increases even further it will sooner or later cause enormous suffering. To illustrate the significance of the overpopulation problem, if mankind continues to increase its population at the present annual rate of about 1.5 percent, in 700 years there would be approximately 181 million million (181 trillion, according the numbering system used in the U.S.) people on the earth. That is more than one person for every square yard of land surface on the planet. Obviously, something will happen long before that population level will be reached. Either mankind will learn to stabilize its population at a level that will not overwhelm the environment, or enormous numbers of people will die in wars over scarce resources or by starvation or disease. Certainly, at the present rate of population increase there will be an enormous strain on the earth's natural resources within only a few generations. Even something as basic as fresh water for drink-

2. The Consequences of Belief in the Bible

ing and irrigation is already in short supply in many parts of the world because of population pressure. Moreover, the pressure of current population levels has already resulted in the over-fishing of the world's oceans and has caused the collapse of the marine eco-system in many previously highly productive areas. Increased population will also mean increased pollution with its resultant effects on health and the environment. It is therefore clear that if, as a result the dictates of the Religious Right and of further Christian missionary activity, even more people were to accept a negative view of birth control, population growth could accelerate even faster and bring on even sooner the hopeless overpopulation of the earth and a catastrophic degradation of the environment.

Belief in the Bible can also affect society in other ways. For example, certain preachers, Christian authors, and TV and radio evangelists attempt to interpret contemporary events and to determine the course of future events in light of biblical prophecy. Some of these individuals have gained a large following and have become quite wealthy and influential as a result of the biblically oriented speculations that they have been peddling.[14] Many of these evangelists have considerable influence with certain politicians, and they seek to use that influence to align the government in a way that furthers their own religious agendas and beliefs. What is particularly disturbing about this is that some of these evangelists believe that the biblical Apocalypse is rapidly coming upon us and that they would like to accelerate its coming by aggravating the situation in the Near East and in other trouble spots.[15]

Bible-based speculations are also frequently mixed in with other ideological outlooks, such as the belief that there are secret satanic societies that are out to take over the world. Some believers have even armed themselves and have formed private Bible-based militias to prepare for battle against the coming "takeover." The members of these militias consider themselves to be "agents of God" and have vowed to liberate the nation from a "demonic" government that is, as they see it, dominated by Jews, liberals, and others who do not accept their absolutist views of the Bible.

The environment could also suffer as a result of another aspect of belief in the Bible. Though some believers in the Bible maintain that God has given mankind stewardship over the earth and that we should manage the environment in a responsible manner, other Bible believers are convinced that the Second Coming of Jesus is fast approaching and they assert that the catastrophic events of the Last Days will make pointless any present-day concerns about the environment.

Global warming is a case in point. Some groups of Bible believers dispute the current evidence that human activity has caused global warming and that all of the hoopla over global warming is fostered by liberals who have their own agenda. An example of this is provided by a row between certain Christian Evangelicals who had come to feel that global warming is an issue that should be addressed, and other Christian Evangelicals who dispute that global warming is occurring. In 2004, certain leaders in the National Association of Evangelicals (NAE) approved a policy statement on civic responsibility, which included a section dealing with the environment. Several other Religious Right leaders interpreted the statement as endorsing the global warming thesis and took up arms against it. After several months of discussion, the NAE acquiesced and dropped the stand on global warming.[16]

Another example is provided by the advocates of Christian-based home schooling. This group uses science textbooks designed for home schooling that are primarily based on young-earth creationism and anti-evolution polemics. These text books often also have an anti-global warming outlook, as exemplified by *Science Order and Reality*, which states, "As Christians we must remember that God provided certain 'checks and balances' in creation to prevent many of the global upsets that have been predicted by environmentalists." Another such text book, *Science of the Physical Creation*, states, "All of the scientific evidence gathered indicates that there is no danger of a global warming disaster." The advocates of Christian home schooling are seeking to get the majority of Christian parents to remove their children from public schools, thus causing the shutdown of those schools and dramatically increasing the number of children who are indoctrinated with Bible-based ideology. Home schooling, in fact, has increased significantly and the number of children being home schooled now measures well over two million.[17] This does not bode well for a public understanding of the dangers of global warming or other critical aspects of the natural world.

But the evidence for global warming is substantial and continues to increase. The indications are that global warming is, in fact, accelerating; the Polar Regions are rapidly losing their ice and snow covers, and average temperatures have measurably increased over the past few decades. If mankind were to ignore the reasons for the increase in global warming and do nothing, the results will be disastrous. If the global warming continues unabated and the ice sheets in the Polar Regions and glaciers and snow fields continue to melt, a devastating round of environmental catastrophes would occur. For example, with a complete meltdown of the Antarctic and Greenland ice sheets, the sea level would rise by well over 200 feet, and major coastal cities and a considerable amount of productive farmland would end up under water. The current weather patterns would also change, causing unpredictable climatic changes to the lands that would remain above water. But if these events begin to happen to an extent that would convince even the Religious Right critics that global warming is occurring, many of them would likely tend to disregard as pointless any efforts to do anything about it. In fact, they would probably even welcome the disasters brought on by global warming as being a prelude to the "tribulation" of the apocalyptic "Last Days."

These are only some of many possible and different scenarios that could occur if reason and the findings of science are subverted by Bible-based religious ideology. If politicians holding these views become a power in the government, pertinent environmental considerations could well be disregarded in the legislative process. Indeed, as this is being written, there are already those in government today who hold such outlooks and have tried to counter environmental legislation.

Without a doubt, a great many people gain comfort from their belief in the Bible. Even if it is not the word of God, the Bible provides them with purpose in their lives and support in times of suffering and adversity, and they look to the Bible for answers to the problems that beset them and the society they are a part of. Many critics of belief in the Bible have asked whether the need for those emotional assists is not simply a result of societal or familial inculcation and of a weakness of the spirit in facing reality. On the other hand, many people—even many non-believers in the Bible—have asked what harm is there in such belief if it provides the believers with the means to face the harsh realities of life. However, if the Bible is not the word of God, it cannot provide, with any certainty, the perfect and absolute answers that its believers believe it can. Moreover, for the true believer, belief in the Bible as the ultimate source of truth and knowledge overrides any meaningful consideration of the answers that other, and perhaps more suitable, views and sources of knowledge could provide. But the greater question is whether we can afford such a belief when it distracts us from reality—particularly when that belief itself has the potential of bringing about many of the problems that we would face in our questionable future.

From both the historical and contemporary standpoints, it should be apparent that belief in the Bible is not necessarily non-consequential or benign, and it should be clear that belief in the Bible has a direct and obvious effect on society, regardless of whether the Bible is the word of God or not. Certainly, the efforts of religious groups who are trying to impose their views on society—such as the Reconstructionists, the creationists, and other Bible-believing groups—could result in a loss of religious, intellectual, and social freedom and could perhaps even result in a new series of religious conflicts much like the crusades and inquisitions of the past.

Even more forebodingly, if the Bible is not the word of God and if belief in the Bible is therefore a delusion, such religious impositions could very well undermine the future of mankind.

Coupled with the fact that a great many Bible believers feel compelled to convert the whole world to their belief and are effectively using modern communications technology for that end, all of this should constrain one to be deeply concerned and to seek the answer to the question of whether the Bible can, in fact, be considered the word of God.

With that, let's begin our analysis of the credibility of the Bible from the cosmological perspective.

Part Two

The Modern View of the Cosmos

Chapter 3
The Earth in Space

Recent surveys have shown that a large percentage of present-day Americans have only a vague knowledge about the findings of science or the way science works. In particular, many Americans know little about the findings of science concerning the nature of the cosmos. Shockingly, some people think that the sun goes around the earth and are unaware of the fact that the earth actually orbits the sun in one year or that the stars are other suns. Even many of those who do have some knowledge of the nature of the cosmos have only a limited awareness of all that science has revealed about it. Moreover, some people, because of their religious ideology or world view, reject much of what science has found out about these things.

Of necessity, then, we should define our terms and provide an informative overview of the cosmos. The purpose of this overview will be twofold. First, it will provide a picture that is based on what science has found out about the nature and structure of the cosmos. Second, it will provide a basis for comparison when we analyze the view of the cosmos that is found in the Bible.

To begin, the planet earth is a sphere, or, more correctly, a spheroid because it has a slight bulge around the equator as a result of the centrifugal force caused by its rotation. The earth is approximately 7,900 miles in diameter and rotates on its axis once every 24 hours, which marks the days. It orbits the sun at an average distance of about 93,000,000 miles and with an orbital velocity of about 18 miles per second, or 66,700 miles per hour. At that velocity, it takes the earth one year to complete one revolution around the sun.

The sun, in turn, is approximately 860,000 miles in diameter, which is more than one hundred times the diameter of the earth. It is composed mainly of hydrogen, and the heat and light it gives off are produced by thermonuclear reactions that take place in its core under enormously high temperatures and pressures. In these reactions (which are similar to those that take place in a hydrogen bomb), hydrogen is converted into helium, a process that results in a release of energy.

Our earth is the third planet from the sun. Besides the earth, there are eight other planets (or seven, if one accepts the recent redefinition of what constitutes a planet, which resulted in Pluto's demotion) and various kinds of interplanetary "rubble"—such as planetoids, meteors, asteroids, and comets—that orbit the sun at varying distances. Most of the other planets have their own natural satellites orbiting them, as the earth has the moon. The outermost "planet" (calling it such for the sake of convenience) is Pluto, which orbits the sun at an average distance of about 3.6 billion miles. Pluto's orbit is so large that it takes 248 earth years for Pluto to orbit the sun at a velocity of about 10,000 miles per hour. But Pluto is not the most distant solar system object, for there are large numbers of comets and other objects, some of them similar in size to Pluto, that orbit the sun much further out.

The sun is a star and, likewise, the stars that one sees in the nighttime sky are other suns. With its system of planets, the sun is moving through space along with billions of other stars in a huge, rotating, spiral-armed formation called the Milky Way Galaxy. The Milky Way Galaxy is about 100,000 light years in diameter, which means that light would take 100,000 years to travel from one edge of the galaxy to the opposite edge. The Milky Way Galaxy is so large, in fact, that it takes 250 million years for it to rotate once. There are least 200 billion, and perhaps as many as 400 billion, stars in the Milky Way Galaxy, a large percentage of them densely packed in the central core. The central core of the galaxy cannot be directly seen from the earth because of intervening clouds of interstellar dust and gas, but X-ray and other types of emissions from the core reveal many of its features.

(We should make a note here about how distances are measured in space. As was mentioned in the last paragraph, the Milky Way galaxy is approximately 100,000 light years in diameter. In interstellar space, the distances are so enormous that if they were given in miles the resulting numbers would be so huge they would have little meaning. Instead, astronomical distances are usually given in light years. A light year is simply the distance that light travels in the time of one of our years. Light travels at 186,282 miles per second, or about 5.8 trillion miles per year.)

3. The Earth in Space

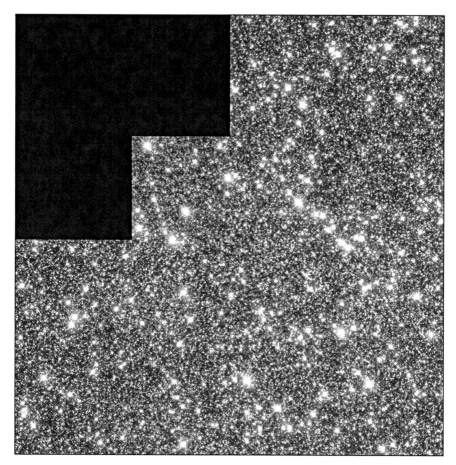

Figure 1. The center of Omega Centauri, a massive globular star cluster. It is located in one of our galaxy's spiral arms and contains several million stars, only a small part of which appear in this photo. Credit: NASA and The Hubble Heritage Team (STScI/AURA).

Our solar system is "out in the sticks," so to speak, because it is located about half way out from the center of the Milky Way Galaxy and between two of its spiral arms.[1] Because the galaxy's core and spiral arms contain most of the galaxy's stars, our immediate stellar neighborhood between those spiral arms has relatively few stars. On a clear, moonless night under very favorable conditions, one might have the impression that there are countless stars in the sky, but, in fact, only about 3,000 individual stars can potentially be seen with the naked eye at any one time on any given night. In total, only about 6,000 individual stars can be seen with the naked eye from the earth in both hemispheres. Most of those stars are also located between the two aforementioned galactic spiral arms and, in astronomical terms, are relatively near our sun. Regrettably, most of us are not able to see many stars at night because of light pollution from the cities and towns in which we live and because of dust and chemical pollution in the atmosphere. As a result of these different types of pollution, we can usually see only a few tens or a few hundreds of stars at night in a typical urban or suburban setting.

The dim band of light called the Milky Way that we might see stretched from horizon to horizon on a clear moonless night is actually one of the neighboring galactic spiral arms. The stars in that spiral arm are so far away we cannot see them individually with the naked eye, but only as the fog of their collective light. However, if we use a telescope, we can see the individual stars in the arm, stars so numerous and close together that those we see with our naked eye in the night sky seem pitifully few in comparison.

Also scattered throughout the galaxy are numerous globular clusters, which are dense groups of stars. Figure 1 is a Hubble Space Telescope composite photograph of a part of the center of one such globular cluster. Called Omega Centauri, this globular cluster is located some 17,000 light-years from the earth in the spiral arm that is nearer to the galactic core from our position. This cluster is about 450 light years in diameter and contains several million densely packed stars. The cluster is so large that only a very small part of it fits within the field of view of Hubble's Wide Field and Planetary Camera, and the Hubble staff had to piece together several separate exposures to make this photo of just a small part of its center (the upper left corner has not been completed in this composite). In fact, the stars are packed so densely in the cluster's center that it is difficult for ground-based telescopes to separate the individual stars, though the high resolution of Hubble's camera is able to do so. Though showing only small part of the cluster's center, the photo contains a view of some 50,000 stars, all packed into a volume only about 13 light-years wide. For comparison, a similarly sized volume around the sun contains only about a half-dozen stars, which goes to shows just how impoverished in stars our local area is. Though the stars in this photo appear to be extraordinarily numerous, they are, in fact, only a very small fraction of the stars in the Milky Way Galaxy.

If, instead of being located between the two galactic spiral arms, the earth were located well within one of them, we would be able to see upwards of hundreds of thousands of stars at night. Even more significantly, if the earth were located within the core of the galaxy, we would be able to see literally hundreds of millions of stars with the naked eye at night. In fact, if the earth were located within the core of the galaxy, there would

Figure 2. M31, the nearest major galaxy to our own. It is over two million light years away and is about 150,000 light years in diameter. The individual small points of light that can be seen on either side of M31 are stars belonging to our own galaxy. The fuzzy elliptical blob to the lower left is a small satellite galaxy of M31. Credit: Robert Gendler / Photo Researchers.

be so many bright stars in the sky there would be no night in the conventional sense of the word. We would even be able to see numerous stars during the day, for several thousand stars would be many times closer and considerably brighter than those stars that we can see in our present night sky.

Our Milky Way Galaxy is by no means the only galaxy. Two small galaxies, called the Large Magellanic Cloud and the Small Magellanic Cloud, are nearby satellites of our own galaxy. Farther away, at a distance of over two million light years, is a large galaxy, even bigger than our own, called M31, which can be seen with the naked eye as a faint patch of light in the constellation of Andromeda. As shown in Figure 2, a telescope reveals this galaxy to be another spiral formation of hundreds of billions of stars. Since M31 is over two million light years away, the light by which we see it left that galaxy over two million years ago. That means we are not seeing it as it exists now, but rather as it existed then. Because M31's galactic disk is seen partially edge-on from our vantage point, its spiral arms are not very distinct in this photo; they are most noticeable in the upper part of the galaxy's image. However, spiral arms are quite distinct in numerous other more distant galaxies. Figure 3 shows just such a galaxy. Called M101, it is about 24 million light years away and is almost twice as big as our Milky Way Galaxy.

Looking through telescopes, we can see multitudes of other galaxies—galaxies by the millions and billions and of various sizes and shapes, including spirals, barred spirals (of which type our galaxy is according to recent findings[2]), globular, elliptical, and irregular. Figure 4 provides a glimpse of only a few of the galaxies in the universe, and these are in a relatively small section of the sky. The farthest known galaxies (actually, galaxies of a type that are called quasars) are about 13 billion light years away, which means that light traveling from them has taken 13 billion years to reach us. Since these objects must have existed 13 billion years ago in order for their light to reach us, that would have to mean that the universe is at least 13 billion years old. According to the scientific evidence, the universe is nearly 14 billion years old, and the earth itself is about 4.5 billion years old, or about one-third the age of the universe.

Since the earth and the other planets in the solar system orbit the sun, which is a star, the question arises as to whether other stars have planets of their own. In fact, there is no reason to believe our sun is unique in that respect among all the stars that exist in the universe, and that conclusion is supported by the recent discovery of planets in orbit around other stars. These planets were not discovered by viewing them directly, but rather by observing specific variations that their gravity causes in the motions of the

Figure 3. Spiral galaxy M101. It is about 24 million light years away and is almost twice as big as our Milky Way Galaxy. Credit: NASA and ESA.

stars that they orbit. This method of finding planets is limited to stars that are within a certain telescope range, but as of this writing, about 200 such planets have been discovered.[3]

That, then, is a basic description of the cosmos according to the current scientific findings. Some people might accept these things, but wonder how scientists came to learn about the distances to the galaxies, and consequently the related ages. Certain religious fundamentalists, on the other

Figure 4. Abell 1689, a Galaxy cluster. Most of the objects in this view are galaxies similar to our own. They are over two billion light-years away. Credit: NASA, N. Benitez (JHU), T. Broadhurst (Racah Institute of Physics/The Hebrew University), H. Ford (JHU), M. Clampin (STScI), G. Hartig (STScI), G. Illingworth (UCO/Lick Observatory), the ACS Science Team and ESA.

hand, reject certain aspects of this view of the cosmos, particularly because of the implication it has for the age of the universe. We should therefore look at the evidence in support of these findings. This evidence will provide a proper understanding of the significance of the modern view of the cosmos when we start our analysis of biblical cosmology.

Chapter 4
The Evidence for the Modern View

Over the course of a year we can see a noticeable change in the brightness of the other planets in the solar system. This is because from our vantage point on the earth there is a significant change in the distances to the other planets as the earth moves in its orbit. Mars, for example, can be seen from a distance of 35 million miles at certain times, and from a distance of over 150 million miles at other times, which causes a significant change in its apparent brightness.

When we look at the stars, however, there is no such noticeable variation in their brightness when the earth travels in its orbit around the sun. The earth's 186-million-mile-diameter orbit must therefore be only a tiny fraction of the distance to even the closest star beyond the solar system, a fraction so small that movement of the earth in its orbit results in insignificant changes in the apparent brightness of any given star.

So how do we know how far the stars are?

There are several ways to measure distances to the stars. One method is by using trigonometry. The earth's orbit around the sun provides a baseline, and astronomers can measure the angle of a star through a telescope, first when the earth is at one position in its orbit, and then when it is in another. By using the angles measured and the known distance of the baseline, astronomers can measure the distance to the star by trigonometric calculation. With this method, astronomers can measure the distance to stars as far away as about 200 light years. Stars that are farther away have too small an angle to measure accurately. There are other methods of measuring the distance to stars that are farther away, but we need not go into all of the details here.

4. The Evidence for the Modern View 39

It turns out that the nearest star outside the solar system is Proxima Centauri, a faint "red dwarf" star that is a little more than four light years away. That means the light from Proxima Centauri takes more than four years to travel the distance to the earth.

But how can we know that the stars are other suns?

There are several lines of evidence that leave little room for them to be anything else. But before we look at that evidence, let's take a look at some of the ways in which stars are categorized.

One way that astronomers categorize stars is by assigning visual magnitude numbers to their degree of brightness as seen from the earth. Under this system, brighter stars have lower numbers and dimmer stars have higher numbers. Generally speaking, magnitude 1 is assigned to stars that are among the brightest that are visible to the naked eye, while magnitude 6 stars are the dimmest stars that can be seen with the naked eye.[1] The brightness of each increase in magnitude is 2.51 times the brightness of the previous magnitude. Therefore, a star of magnitude 3 is 2.51 times brighter than a star of magnitude 4, and a star of magnitude 1 is 100 times brighter than a star of magnitude 6. Stars that are dimmer than the sixth magnitude can be seen only through telescopes. These magnitude numbers refer only to the apparent brightness of the stars—and the planets in our solar system also, since the same system of magnitude is applied to them—as they are seen from the earth. A star that appears bright to our naked eye might actually be smaller and intrinsically less bright, but closer to us, than a much larger and intrinsically brighter star that is much farther away.

For objects brighter than the first magnitude, the numbers go negative. Thus the planet Venus at its brightest has a magnitude of about -4, while the sun has a magnitude of -27. At its magnitude of -27, the sun is about 61 billion times brighter than it would be if it were of the first magnitude.

To continue, the brightness of an object varies inversely according to the square of the distance to that object. For example, if an object of a certain brightness is viewed from a certain distance, and then is viewed from twice that distance, the object will appear to be 2^2 times less bright (that is, one-fourth as bright) as it did at the initial distance. In the case of the sun, in order for it to appear as bright as a first magnitude star, we would have to view it from a distance of 248,653 times its current distance of 93 million miles. That distance would be 2.3×10^{13} miles (that is, 23 followed by 12 zeros), which is approximately four light years.

Now as we noted earlier, the nearest extra-solar star is Proxima Centauri, which is a little more than four light years away. Proxima Centauri is a dim red dwarf star that can be seen only through a telescope. However, near Proxima Centauri and just a little further away is Alpha Centauri, the closest extra-solar star that can be seen with the naked eye. Through a telescope, Alpha Centauri appears as a binary, or a two-star system, consisting of a normal star, called Alpha Centauri A, and a "white dwarf" star, called Alpha Centauri B, in orbit around each other.[2] (This is not unusual. Binary star systems are quite common in the galaxy.) The light produced by Alpha Centauri B is negligible in comparison with that of Alpha Centauri A, which produces most of the light from the system and appears as a single first-magnitude star to the naked eye as seen from the earth.

Now we can see why stars can be regarded as suns similar to our own.

First, in order for Alpha Centauri A to appear as a first-magnitude star at its distance of a little more than four light years, it would have to be actually a little bigger and a little brighter than our sun. That is because our own sun would appear as a first magnitude star at a distance of about four light years.

Second, in orbiting each other, Alpha Centauri A and Alpha Centauri B conform to the law of gravitation, and their orbital dynamics (that is, the orbital speed of the two stars and their distance from one another) are such that the two stars must each have a mass near to that of our sun (specifically, 1.1 solar mass for Alpha Centauri A and 0.89 solar mass for Alpha Centauri B).

Third, spectroscopic studies (a spectroscope is a device that is used to study the characteristics of light emissions by different atoms and molecules) of Alpha Centauri A reveal that it has a chemical composition similar to that of our sun.

These three findings show that Alpha Centauri A has characteristics similar to those of our sun, and show that it is a little larger and a little brighter than our sun. Therefore, it is only reasonable to accept that it is also a sun.

There is yet more evidence that the stars are other suns. This evidence supports the theories of stellar evolution that astronomers have developed, and therefore supports the underlying premise of those theories—which is, of course, that the stars are other suns.

Stars have their beginnings in thin clouds of hydrogen gas that are dispersed throughout vast volumes of interstellar space. A nascent star starts its life cycle when the gas in one of these clouds contracts because of its inherent gravity—perhaps with an initiating push caused by the pressure of the light from nearby stars or by compression resulting from a neighboring supernova explosion (we'll discuss supernovae shortly). As the gas contracts, it becomes denser and its gravitational pull consequently becomes stronger, causing it to contract even more and more until finally it forms a dense mass. At that point, if it has sufficient mass, the resulting pressure and temperature in its center will ignite thermonuclear reactions and it will flare into life as a star.

According to the models of stellar evolution that astronomers have developed, a star's life cycle is dependent upon its initial mass and composition. For example, stars that have a mass within a range between about half that of our sun and a few times more than that of our sun—and, of course, our sun is one of these stars—typically have a life span of several billion years. For a large part of its life span, a star having an initial mass within this range produces energy in its core by a thermonuclear process that converts hydrogen into helium. When its core eventually runs out of hydrogen fuel, the helium in turn provides the fuel for a new series of thermonuclear reactions that result in the synthesis of carbon and oxygen. When this happens, the outer layers of the star cool off and expand and the star becomes a red giant. Eventually, the helium in the core will be used up and the thermonuclear reactions will cease. With the loss of the enormous internal pressure that the thermonuclear reactions had produced, the outer shells of the star collapse and the star gravitationally contracts and becomes a white dwarf star. White dwarf stars are typically about the size of the earth, but are extremely dense, for they still have much of their original mass; a teaspoonful of their material would weigh several tons. (Normal atoms are mostly empty space, but in a white dwarf the gravitational pressure is so great that the atoms are compressed into a much smaller volume, thus significantly increasing the density of the star.) As previously noted, Alpha Centauri B is a white dwarf star.

Stars that have a mass much less than that of our sun have much lower temperatures and pressures within their cores. Consequently, they do not go beyond the hydrogen-burning stage, and they burn their thermonuclear fuel very slowly—so slowly that they have life spans of hundreds of billions of years. These stars are comparatively dim and are called red dwarfs. As was previously noted, Proxima Centauri is an example of this type of star.

Stars that have more than about eight times the mass of our sun have a life span of only a few million years and have a much more spectacular end than the other two types of stars. Stars with such a mass have tremendous temperatures and pressures in their cores, and, as a result, they burn their thermonuclear fuel at a prodigious rate and are extremely hot and bright. Because they burn their thermonuclear fuel at such a high rate, these stars soon run out of hydrogen in their cores. They then start synthesizing heavier elements from the helium that was produced in the initial reactions, and then even heavier elements from those elements.

Because they are more massive and hotter than are the sun-like stars, these stars can carry the thermonuclear synthesis reactions past those that produce carbon and oxygen, and they continue those reactions on up to those that produce iron. At this point, things change drastically, for the thermonuclear reactions that synthesize iron, as well as those that synthesize the elements above iron on the periodic chart, absorb energy rather than release it. Consequently, when the core of such a massive star reaches the stage in which it produces iron, it no longer produces the energy that is required to hold back the outer layers of the star. As a result, those layers collapse under the immense gravitational pull of the star's mass and the star implodes upon itself with a tremendous compression of the star's core.

What happens next depends on just how massive the star is. If the star has a mass between about eight and fifty times that of our sun, the collapsing outer layers will rebound off the core in a tremendous explosion, an explosion that can be so bright it can outshine the combined light of all the other stars in the galaxy in which it appears. Such an exploding star is called a supernova. Moreover, as a byproduct of the tremendous compression of the star's mass during the collapse, elements heavier than iron are synthesized and are then dispersed into space during the subsequent explosion. Not all of the material of the star is necessarily dispersed into space, however; frequently, the core of the star will remain, but in a form quite different from what it was before. During the implosion, the pressures are so great that the electrons and protons in the star's core are forced together and become neutrons. The result is what is known as a neutron star, a dense body only a few miles in diameter but with the mass of a full-sized star. The material that a neutron star is made of is so dense that a teaspoonful of it would weigh as much as a good-sized mountain would weigh on the earth.

On the other hand, if the imploding star has a mass of more than about fifty times the mass of our sun, the density of its core will increase so much that the resulting increase in its gravitational field will cause the star to continue collapsing upon itself. As the star continues its collapse, it becomes even denser and produces an even more intense gravitational field, a gravitational field so powerful that nothing, not even light, can escape. In other words, the star will become what is called a black hole.

Now, the validity of a scientific theory or model is dependent upon its ability to provide verifiable predictions. The models of stellar evolution that astronomers have developed have found a remarkable confirmation as a result of a supernova that was observed in the Large Magellanic Cloud in February of 1987. Prior to the 1987 supernova, all supernovas that had been observed with modern instruments had been in more distant galaxies

4. The Evidence for the Modern View 43

and were too far away for astronomers to make detailed studies of them. (Supernovas have historically occurred much closer and in our own galaxy, but the nearest ones that could be seen with the naked eye, and were not obscured by interstellar dust clouds, occurred hundreds of years ago, long before the development of modern scientific instruments.)

As soon as the 1987 supernova was observed, astronomers all over the Southern Hemisphere (from where the Large Magellanic Cloud is visible) trained their telescopes on it, and instruments that were in orbit around the earth were also aimed at it. The supernova was close enough for scientists to bring several different kinds of instruments to bear and to make observations that were not previously possible to make with other supernovas. These observations confirmed in close detail most of the sequences of events, the time factors, the thermonuclear products, and the types of radiation that were predicted by the models of stellar evolution the astronomers had previously developed. These observations thus verified the general validity of those models.[3]

As a bonus, Supernova 1987A, as it is called, also provided a confirmation of the accuracy of the methods that astronomers use to measure the distances of the stars. Through various methods available to them, astronomers had previously determined that the Large Magellanic Cloud covers a volume of space that is between 150,000 and 190,000 light years away. The star that went supernova in the Large Magellanic Cloud had also been previously observed and was duly noted on star charts; its name was Sanduleak—69 202. It was a very massive star, about 20 times the mass of our sun, and, because it was burning its nuclear fuel at a prodigious rate, was intrinsically about 30,000 times brighter than our sun. Nevertheless, despite the brightness of the star, it could be seen only through powerful telescopes because of its distance.

When the star went supernova and blew up, it produced as much light as 100 billion suns like ours and became visible to the naked eye here on the earth. But even with an intrinsic brightness of 100 billion suns, because of its great distance the supernova appeared to observers on the earth as a star of only medium visual brightness.

Now here is the interesting part. Prior to going supernova, a star will sometimes throw off a large quantity of gaseous material that rapidly moves away from the star in the form of a giant ring or sphere with the star at its center. This particular star had thrown off an encircling ring of material thousands of years before it went supernova, so that the ring had expanded outward from the star quite a distance by the time the star exploded.

When the star went supernova, the light coming to us straight from the explosion reached us on February 23, 1987. However, the light that was reflected off the ring of material encircling the star did not reach us until many weeks later. This was because that light, in going from the explosion to the ring and then to us, had a much longer distance to travel than did the light coming directly to us from the explosion. Because of the time delay between the arrival of the light coming directly to us from the explosion and the arrival of the light that was reflected off the ring, astronomers could accurately determine the radius and therefore the diameter of the ring. Since we know how fast light travels, the time delay could be used to calculate the distance from the supernova to the ring.

It turned out that, when the supernova occurred, the ring was about 1.3 light years in diameter. To determine the distance of the star, the astronomers needed only to make a simple computation using the angular diameter of the ring as it appeared through the telescopes and the now-known actual diameter of the ring. It turned out that the star was almost exactly 169,000 light years away. That measurement verified the distance of the Large Magellanic Cloud that the astronomers had previously arrived at by using other methods, and thus validated the means by which the astronomers measure interstellar and intergalactic distances.[4]

Because the supernova was at a distance of 169,000 light years from the earth, the light from the explosion had been traveling for 169,000 years before it reached the earth. That means the explosion actually occurred 169,000 years ago, which also means we were able to observe an event that took place 169,000 years before we saw it. This, of course, conflicts with and refutes the assertions of most creationists that creation took place only about 10,000 years ago on the basis of what they find in the Bible. One wonders how we could see an event that took place 159,000 years before anything was created. Furthermore, we can observe supernovas in galaxies that are billions of light years away, which means that we can see events that took place billions of years ago.

Supernova 1987A certainly provides substantial evidence for the essential correctness of the current theories of stellar evolution. But those theories, as noted previously, also suggest that objects such as neutron stars should exist as remnants of supernova explosions. And, in fact, over the past forty years several hundred neutron stars have been found, primarily because they rotate extremely fast and because beams of radiation from irregularities on their surface sweep space for enormous distances, sort of like cosmic lighthouses. The effect causes a neutron star to momentarily brighten with clock-like regularity when the beam sweeps past the field of view of telescopes here on the earth; hence, these neutron stars are called pulsars. Pulsars are frequently found within the remnants of supernovas. The Crab Nebula in the constellation of Taurus is an example of such a supernova remnant, and in its center is a pulsar. Incidentally, the supernova explosion that produced that remnant was visible to the naked eye here on the earth in the year 1054.

As we have noted, the current theories of stellar evolution also suggest that objects such as black holes could exist—indeed, even Einstein's theory of relativity predicts that black holes should exist. By their nature, black holes cannot be seen directly. However, the overwhelming gravitational field of a black hole has the potential of being observed because of the effects that it has on the surrounding space. In fact, the gravitational field surrounding a black hole can be so intense that it can swallow up neighboring stars and thus add to the mass of the black hole, resulting in an even more intense gravitational field.

Early in its astronomical observations, the Hubble space telescope uncovered just such a phenomenon. In the core of a giant galaxy called M87, which is about 50 million light years away, is an enormous disk of gaseous material rotating around a massive central object. At a distance of 60 light years from the central object, the disk is rotating at 450 kilometers a second, which is so fast that the disk would fly apart if something was not holding it in place. There is only one kind of object that would have a gravitational pull strong enough to prevent that from happening: a black hole with a mass of about 3 billion times that of our sun.[5] Following that observation, numerous other black holes have been found. In fact, the evidence indicates that most galaxies, including ours, have massive black holes in their centers.

For most people, supernovas, and even neutron stars and black holes, might be taken as mere curiosities. But it is because a star blew up in a supernova explosion several billion years ago that we are here. As a result of the thermonuclear reactions that take place in the cores of massive stars, and also as a result of further thermonuclear reactions that take place in these massive stars when they go supernova, tremendous quantities of elements ranging from helium to uranium are built up from the primordial hydrogen fuel that they contain. During supernova explosions, these elements are spewed into space and, after combining with additional interstellar gas and dust, can eventually condense gravitationally to form new stars, as well as planets.

Our solar system, with its planets, therefore exists because of such a supernova explosion billions of years ago.

The accurate measurement of the distance to the 1987 supernova shows the general accuracy of the means by which astronomers measure the distances to the stars and galaxies. Because astronomers can measure the distances to other galaxies, they are also, in effect, measuring the past. Through telescopes we can see galaxies that are billions of light years away because the light traveling from them has taken billions of years to reach us. This alone indicates that the universe must be billions of years old. In fact, the evidence from various sources indicates that the universe is nearly 14 billion years old.

But what about the evidence for the age of the earth?

There are several lines of evidence that indicate the earth is ancient. Certain geologic processes, for example, lay down layers of strata annually, and those layers may be counted much like tree rings. Such layers are often formed in the bottoms of lakes. During the warm part of the year, algae and other life forms bloom in lake water, and as they go through their life cycles and die off in winter their remains filter on down to leave a thin, dark, carbon-rich layer on the bottom of the lake. In spring, runoff from streams will bring in a thin layer of silt that covers the carbon-rich remains of the previous season with a lighter-colored layer. This process continues year by year, leaving large numbers of alternately colored layers. Eventually the lake will fill in and disappear and the layers will turn into hard stone.

As a result of geological processes occurring over the eons, some such ancient lake beds have uplifted and tilted, which subjected them to crosswise erosion and exposed them for observation. In some areas—e.g., the Green River formation that is exposed in parts of Colorado, Utah, and Wyoming—literally millions of layers can be counted in the uptilted ancient lakebeds. This indicates that the ancient lake, which was quite large, had existed for millions of years.

Other types of formations, such as corals, give a similar picture. Corals are the structural remains of marine invertebrates that live in colonies and grow on the remains of the previous generations. Core borings of coral reefs in the South Pacific have shown that some of these reefs are more than 4000 feet thick. Studies of coral reveal that their growth rates are from about one-quarter inch to one-half inch per year, so at that rate it would have taken more than 100,000 years for these reefs to grow.

One interesting find came about through the study of fossil corals. Because the ocean tides cause mechanical friction and act to slowly brake the earth's rotation, the length of the day slowly increases over geologic time. Currently, the length of the day is increasing by about two milliseconds per century. By extrapolating this change in the length of day backwards in time, scientists have calculated that about 400 million years ago the day would have been about twenty-two hours in length.

That is where the fossil corals enter the picture. A coral adds a microscopic layer to its base structure each day while it is alive. The thickness of these layers in turn varies according to the seasons of the year, so that a yearly pattern is imposed on the daily pattern. Studies of these patterns on some fossil corals that lived about 400 million years ago revealed that there were 400 days in the year at that time. That means the days must have been shorter in order to have more of them per year. In fact, in order to have 400 days per year, the day would have to have been twenty-two hours in length, which is what the scientists estimated from their studies.[6]

Studies of ancient formations called tidal rhythmites support that conclusion. Tidal rhythmites were formed by tidal action that left alternating bands of dark- and light-colored silts and sands on ancient shorelines. Because high spring tides and low neap tides that mark the lunar month cause the bands in the rhythmites to vary in thickness, scientists were able to use the rhythmites to count the number of lunar months per year and, from that, calculate the earth's rate of rotation when the rhythmites were formed. Rhythmites that were formed 900 million years ago indicated that the year had 481 days then, which means the day was eighteen hours long.[7]

Because these studies of ancient fossil corals and tidal rhythmites show that the length of the day at the time of their formation conforms to the calculated expectations, they provide a powerful confirmation for the validity of the processes that scientists use to date ancient rock formations.

But such findings, while providing evidence for an ancient earth, do not give its absolute age. It remains for measurements of radioactive isotopes to do that.

Each elemental atom in the periodic table has a fixed number of protons in the nucleus—the number of protons, in fact, defines the element. However, the elements also come in isotopes, called nuclides, which have varying numbers of neutrons in association with the protons in the nucleus. There are about 1,000 different types of nuclides. Most of these nuclides are found in nature; others have been produced in the laboratory through experiments in nuclear fission, while still others are produced as a matter of course in atomic power plants. Of all of the possible nuclides, many are radioactive and they spontaneously break down, or transmute, to different elements at predictable rates. A given sample of such a nuclide will have a "half-life" of a certain length that depends on the relative stability of the atoms of that nuclide. This means that one-half of a sample of the nuclide will convert to its breakdown product over the time of its half-life. During the next half-life, one half of the remainder of the nuclide sample will break down, and so on. Over several half-lives, most of the original nuclide sample would be gone, and over enough half-lives there would not be a detectable amount of it left.

The half-life breakdown process is fully consistent and predictable. If it were not, atomic power plants and atomic bombs would not work as expected; they would either not work at all or they would spontaneously blow up. Indeed, if the half-life breakdown process were not consistent, the earth would be a very uncomfortable place on which to live, and, in fact, life would likely not be possible.

There are 64 radioactive nuclides that have half-lives greater than 1,000 years. Seventeen of these nuclides have half-lives greater than 50 million years, and 47 have half-lives of between 1,000 and 50 million years.[8]

Of the 47 nuclides having half-lives of between 1,000 and 50 million years, 7 are constantly being formed here on earth because of ongoing processes. One process results from cosmic bombardment of certain kinds of atoms in the atmosphere, which converts them to radioactive nuclides. This process, for example, converts nitrogen into carbon 14. Another process is radioactive decay, which causes certain radioactive nuclides to produce other radioactive nuclides. Since these nuclides are constantly being produced, they must be ignored from consideration in determining the age of the earth. Of the 40 remaining nuclides, any that are found here on the earth would have been present at the earth's formation.

Now suppose the earth were relatively young—say no more than the 10,000 years that most creationists maintain it to be. In such a case, we would expect that traces of all of the 64 radioactive nuclides would be present to some degree in the earth's rocks because not enough half-lives would have passed for all of the nuclides to have disappeared.

What do we in fact find? Well, we can find detectable amounts of the 17 nuclides—all 17—that have half-lives greater than 50 million years, which is not surprising; with such long half-lives, some remnant of the original amount would be left even after several billion years. However, of the 40 other nuclides that would have been present at the earth's formation (ignoring those 7 nuclides that are constantly forming), not a single one can be found in detectable amounts in the earth's rocks, nor in moon rocks or in meteors, for that matter.

There are three hypotheses one could put forth to explain this lack of radioactive nuclides that have relatively short half lives: (1) they were missing by pure chance during the formation of the solar system, (2) the processes that caused the formation of the elements were not capable of producing the short-lived nuclides, (3) the short-lived nuclides were produced along with the extant nuclides, but they were produced so long ago that they have long since decayed away.

There are very good reasons to completely eliminate hypotheses 1 and 2 from any consideration as the cause for the lack of the short-lived nuclides.[9] That leaves hypothesis 3 as the only reasonable cause for their absence.

4. The Evidence for the Modern View

This means that all of these short-lived nuclides must have passed through enough half-lives to cause virtually all of the original material to have converted to breakdown products. The time required for all of these nuclides to have done this is in the order of billions of years. This evidence indicates, then, that the earth is billions of years old.

To get a more exact age for the earth, one can measure the decay products of various series of radioactive isotopes and calculate the amount of time that would have been required to form the products. Some of these measurements require making estimates of the original amount of the parent isotope, but there are ways to directly determine the amount, as exemplified by analysis of rubidium 87.

Rubidium 87 decays directly to strontium 87, with a half life of 47 billion years. In this decay process there are no intermediate products to confuse the measurement. There are also three other isotopes of strontium that are not the products of radioactive decay. Regardless of the original rubidium/strontium ratio in any given mineral sample, the ratios of the four isotopes of strontium were fixed when the mineral was first formed. Thus, in any given sample of mineral with these isotopes, the nonradiogenic isotopes of strontium will remain constant, while the amount of strontium 87 will increase by the exact amount that rubidium 87 decreases.

When a rock forms, it may consist of several different minerals, and each of those minerals may originally have had different rubidium/strontium ratios. Over time, however, the decay of rubidium 87 into strontium 87 will increase the ratio of strontium 87 to the other isotopes of strontium in each of the minerals, and it will change in direct proportion to the rubidium/strontium ratio. If a graph is made that shows the ratio of one of the other isotopes of strontium to that of the strontium 87 in each of the minerals in the rock sample, the values for the ratios in the different minerals will form a straight line, and the intersection of the line with the ordinate of the graph (that is, the Y-coordinate point at the specified X-axis distance) will indicate the ratio of strontium 87 to the other isotope of strontium at the time of the formation of the rock.

This method provides a remarkable verification of the age of the rock. There is no need to estimate the original ratios of rubidium to strontium, because the graph states explicitly what the starting conditions were. If the rock was somehow disturbed by geologic processes during the passage of time, causing strontium or rubidium to be either removed or added, it would be known because the ratios would not form a straight line on the graph. Even different rocks from the same site will give the same result as long as the rocks had not been disturbed.[10]

The results of several different types of radioactive decay measurements show that the oldest rocks that have been found on the earth have an age of about 4.2 billion years, and those rocks certainly formed some time after the earth itself had formed. The same methods have been used to date moon rocks and meteorites. The dates produced by these different types of radioactive decay measurements are remarkably consistent. They indicate that the solar system is about 4.5 billion years old.[11]

Most creationists, of course, reject much of the foregoing. They are committed to a literal interpretation of the Bible and therefore believe that the universe can be no more than about 10,000 years old. They will usually argue that all of the methods that give an age of billions of years for the earth and the universe are unreliable or untestable. For example, they argue that there is no way of knowing if the rates of radioactive decay have always been the same or if the speed of light has always been the same. Of course, they can present no scientific evidence that the rates of radioactive decay or the speed of light have changed. And, in fact, there would have been devastating consequences if the speed of light or the rates of radioactive decay had changed. We'll look at the reasons for this in Chapter 12.

What the creationists try to do is to present their own "scientific" proofs showing that the earth is much younger than 4.5 billion years old and is likely to be only about 10,000 years old. In developing these "proofs," however, the creationists invariably misrepresent or disregard the facts.

For example, a "proof" of a relatively recent creation that creationists have been using since 1973, despite its having been refuted over and over again, is based on the decay of the earth's magnetic field.[12] That decay has been observed from the time scientists began measuring the strength of the earth's magnetic field in the early nineteenth century. By extrapolating this change in the magnetic field exponentially backward in time, creationists say that the magnetic field would have been impossibly high more than about 10,000 years ago. Therefore, they reason, the earth cannot be more than about 10,000 years old.

But that argument is based on the unwarranted assumption that the limited observation of the decay of the earth's magnetic field accurately reflects a condition that has continued ever since the earth formed. In fact, however, studies of paleomagnetism reveal that the earth's magnetic field has not behaved in the way that the creationists allege. (Note that, though the creationists question without compelling evidence whether the rates of radioactive decay and the speed of light have always been constant, they are perfectly content to assume that the rate of decay of the earth's magnetic field has always been constant, even though there is compelling evidence that it has not been. Such is the selective nature of creationists' "evidence.")

Paleomagnetism studies use analyses of the earth's magnetic field that is "frozen" in magnetized particles in ancient rocks or pottery. When a rock is formed by precipitation or crystallization, or by volcanic material solidifying, the magnetic particles in the rock align themselves according to the direction and the strength of the earth's magnetic field as it exists at the time. The same effect occurs during pottery making when the pottery is fired. Studies of the magnetic orientation of the particles in rocks and ancient pottery reveal that 6,500 years ago the magnetic dipole field was about 20 percent weaker than it is now, and that 3,000 years ago it was about 45 percent stronger than it is now. An examination of even older lava flows reveals that the earth's magnetic field has fluctuated in strength for millions of years. These studies have also revealed that the earth's magnetic field periodically reaches a minimum and reverses direction every few hundred thousand years so that the earth's magnetic north pole becomes its magnetic south pole and vice versa.

These findings tie in with the evidence for continental drift. For several decades, a considerable amount of evidence has accumulated showing that the continents are riding on tectonic plates that have slowly drifted over the earth's surface for millions of years. These plates have their origin in areas such as the mid-Atlantic rift, which runs down the whole length of the Atlantic Ocean from the Arctic to the Antarctic regions. The rift is like a giant crack in the earth's crust where molten material is slowly upwelling from the earth's interior. As this material upwells, it solidifies and becomes sea floor, part of it moving in an eastward direction and part in a westward direction. This means that the sea floor on both sides of the rift slowly moves away from the rift as new sea floor forms at the rift.

This same process occurs in several other rifts on the earth's surface and provides the mechanism for moving the plates that carry the continents. At the opposite end of the plates, the sea floor subducts, which causes the existence of the deep ocean trenches, and returns the material to the earth's interior. The effect is much like two conveyor belts placed end-to-end and moving in opposite directions. This process is exceedingly slow, but scientists have measured the results by using highly accurate laser ranging devices and satellites. The plates are typically moving about one-half inch to one inch per year, though a Pacific plate is racing along at about six inches per year, which is one of the reasons the Pacific Rim is so susceptible to earthquakes and volcanism. The movement of these plates is apparently driven by immense cells of slowly circulating fluid material in the earth's hot interior. (In fact, these cells of circulating fluid are what give rise to the earth's magnetic field, and it is likely that disruptions in their circulation are what cause periodic variations in the field.)

One result of this process is that, as the molten material upwells and solidifies at the rifts, magnetic particles within the material align themselves according to the earth's prevailing magnetic field and become frozen in place, thus preserving in the ocean floor a record of the earth's past magnetic field variations over millions of years. Magnetometers lowered from ships and making sweeps of the ocean floor have revealed that, as the distance from the rift increases, this paleomagnetic material alternately reverses its magnetic orientation in regular bands that are parallel to the rift. The bands nearest the rift are the most recently formed, while those farther from the rift are increasingly older in direct proportion to the distance from the rift. The dating of these bands coincides with known reversals of the earth's magnetic field from other studies. These readings also show that the bands in the sea floor on the opposite sides of the rifts are mirror images of each other, just as should be expected considering the sea floor spreading mechanism. Moreover, because of the slowness of the sea floor spreading, the formation of these bands has taken millions of years.[13] Thus, contrary to what the creationists allege, studies of the earth's magnetic field actually provide evidence for an ancient earth.

This evidence for continental drift and the changing magnetic field of the earth has been around for some time. For example, an article in the April 1968 issue of *Scientific American* magazine described the evidence, including the paleomagnetic findings, in support of continental drift.[14] Subsequent studies have reinforced those findings. Nevertheless, the creationists either ignore or belittle all this evidence and continue to assert that the presently observed decay of the earth's magnetic field has continued unabated since the earth's creation.

In effect, this creationist "proof" is like taking two measurements of the tidewater level at the seashore today, one early in the morning and another six hours later, then declaring that, because of the observed rate of change in the sea level, the ocean must have been 1400 feet higher last year than it is today.

The creationists, of course, have other arguments to "prove" that creation occurred relatively recently—for example, arguments based on the thickness of moon dust and the chemical constituents of the oceans. But those arguments have no more value than the magnetic field argument has, and they too have been repeatedly refuted.[15]

Chapter 5
The Big Bang

When Albert Einstein was formulating his general theory of relativity early in the twentieth century, he noted that his equations indicated the universe could not exist in a static condition—the universe had to be either expanding or contracting. This did not seem right to Einstein, because at the time there was no indication that either of those cosmological states actually existed and it appeared that the universe was in a static condition. Therefore, to provide the needed stability, Einstein inserted a factor, which he called the cosmological constant, into his equations.

Eventually, in the 1920s, it was learned that certain nebulae which are visible in telescopes are actually other galaxies. Furthermore, spectroscopic studies of some of these galaxies revealed that their light was shifted toward the red end of the spectrum and that the farther away the galaxy the greater the red shift. This effect was found to be true regardless of the area of the sky in which the galaxies were observed, which in turn indicated that the universe as whole is expanding. This observed expansion of the universe meant that Einstein's original equations were correct. Einstein subsequently stated that his inserting the cosmological constant into the relativity equations was his greatest blunder. (Recent findings have indicated that there actually may be a cosmological constant, though not quite the same that Einstein envisioned. This cosmological constant is apparently causing the expansion of the universe to accelerate.)

But why does the red shift indicate that the universe is expanding?

Normally, a red shift would be considered to be caused by a Doppler effect and would indicate that the source of light is moving away from the observer. It would also mean that, the greater the red shift, the greater the recession velocity. (The classic explanation of the Doppler effect involves listening to the whistle of a passing train. As the train rushes by and moves away, the frequency of the whistle drops to a lower tone. With a receding light source, the light frequencies drop toward the red end of the spectrum. This results in a displacement of the absorption lines of the elements in the light source, as shown in the spectrograph, with respect to the frequencies in which they would normally be found.)

However, the observed galactic red shift is not a Doppler red shift, though the effect is similar and the red shift is often called a Doppler red shift. In fact, the galactic red shift is *not* caused by a movement of the galaxies *through* space; it is a cosmological red shift caused by the expansion of space-time itself. What that means is that galaxies remain more or less within a specific location with respect to the universe as a whole, but that the distances between galaxies are widening because of the expansion of the universe. Thus, the cosmological red shift relates to the difference in the *size* of the universe between the time light is emitted from a galaxy and the time it is observed elsewhere.

For example, when we see a galaxy that has a certain amount of red shift, that means we are looking back in time by a certain amount. If that galaxy is seven billion light years away, say, that means the light was emitted from that galaxy seven billion years ago. But seven billion years ago, the universe was smaller than it is now. During the time that the light has taken to reach us, the universe has expanded, and in doing so it has "stretched" the light emitted by the galaxy, thus lengthening its waves and lowering its frequencies.

A balloon can provide a good example of how the cosmological red shift works. Suppose two dots are inked on a balloon that is blown up half way, and a wavy line is drawn to connect the two dots. If the balloon is subsequently blown up further, the two dots will not move with respect to their positions on the balloon as a whole, but the distance between them will increase as the rubber of the balloon stretches, as will the distance between the crests of the wavy line. Likewise, in the expanding universe the galaxies do not appreciably move with respect to the universe as a whole, but the distances between them are constantly increasing as the fabric of space-time stretches. And, likewise, light waves traveling from one galaxy to another are stretched and thus red shifted.

Though the cosmological red shift has an effect that is similar to that caused by the Doppler effect, there is, in fact, a significant difference. If the cosmological red shift were a true Doppler red shift, it would mean that the galaxies are actually moving through space and away from the "source" of the expansion. It would also mean that the velocities of some galaxies would be approaching the speed of light. However, because the cosmological red shift is *not* a result of a movement of the galaxies

5. The Big Bang

through space, the galaxies are therefore not moving "away" from the source of the expansion, nor are any of them "moving" at near the speed of light—that is, they are not moving at near the speed of light with respect to the space that surrounds them.[1]

There is one other aspect of the cosmological red shift that must be understood: The cosmological red shift is significant only over the large scale of the universe. Galaxies are usually found in gravitationally bound groups. The galaxies within these groups remain relatively close to one another and orbit each other because their mutual gravitational attraction for each other is stronger than the effects of the large-scale expansion of the universe. For example, our Local Group, as it is called, consists of our galaxy, the galaxy M31 in Andromeda, the Large and Small Magellanic Clouds, and a few other nearby galaxies. There are multitudes of such groups of galaxies in the universe. It is these groups of galaxies, rather than individual galaxies, that have their distances from each other increased by the effects of the expansion of the universe.

The consequence of an expanding universe is that in the past the universe must have been much smaller than it is now. Taking this consequence back further still (and according to our current best understanding, which is consistent with a growing number of observations), all the matter and energy presently in the universe must have at one time been concentrated in one small point from which the universe expanded.

It was that point, then, that produced what is called the Big Bang. There are several theories about the source of the point. One suggestion is that the point was what some theoretical physicists call a quantum fluctuation; other suggestions involve various other mechanisms. (Contrary to what the creationists say in denigrating the idea of the Big Bang, these suggestions do not necessarily involve the idea that the Big Bang came from nothing; rather, they envision that the Big Bang was the result of some process that extracted energy from a potentially vast reservoir.) Whatever the source of the point was, theoretical physics describes what must have happened immediately after it came into existence. All of the matter and energy that presently exists in the universe would have been concentrated in a very small volume, and the temperature would have been enormously hot, billions of times hotter than in the core of our sun. Under such conditions, normal matter cannot exist and in the initial microseconds of expansion there would have been only radiant energy and quarks, the basic building blocks of subatomic particles.

As the bubble of the universe expanded, it would have begun to cool down. From about three minutes to about five minutes after the expansion began, according to current theory, the bubble would have cooled down enough for primordial protons and neutrons to coalesce from the quark mixture. A single proton is the nucleus of a hydrogen atom, and during this period of time a certain amount of the hydrogen would have been able to combine in thermonuclear reactions to produce helium and a trace amount of lithium. After five minutes, the bubble would have expanded enough to cool down even more, which would have terminated the thermonuclear reactions. As the universe continued to expand and cool down, the primordial gases eventually formed localized eddies and condensed, through gravitational attraction over time, into galaxies.

It should be emphasized that the expansion of the universe from a point was not an explosion. An explosion implies that material from the source of the explosion is blown out into a larger space. But the material that was formed during the Big Bang was not "blown out" into a larger outside space; instead, the universe carried its own space with it as it expanded from the initial point. In fact, there is no "outside" of the universe, at least not in any sense that can have any practical meaning to us. The term "Big Bang" is therefore an inappropriate appellation. (The term was supposed to have been coined by British astronomer Fred Hoyle, who had an alternative theory concerning the expansion of the universe.)

If the Big Bang had been an actual explosion and the galaxies were moving outward through space, the universe would have a center as defined by the movement of the galaxies away from that center. However, as we previously noted, because the space-time of the universe is expanding, it is carrying the galaxies with it and the individual galaxies remain in the universe at the same approximate coordinates with respect to the rest of the universe. It is sort of like a lump of raisin bread dough that is in the process of rising. As the dough expands in volume, the raisins do not move through the dough; instead, the expanding dough carries them farther apart from one another.

There are some interesting effects that result from this view of the cosmos. The farther we look into the depths of the universe, the farther we are looking back in time, and those galaxies that we perceive as being near the edge of the visible universe are actually near its beginning. Moreover, the universe is finite, but is unbounded and curved in its three physical dimensions, just as the surface of the earth is in its "two" dimensions that curve around the earth's surface. And as the surface of the earth has no center and no edge, the universe likewise has no center or edge. Furthermore, though we have a perception that we are in the center of the universe because the galaxies appear to be moving away from us and because the perceived distant "edge" of the visible universe appears to be equidistant all around us, we would nevertheless have the same perception if we were located anywhere else in the universe. In that sense, every point in the universe is its center, which could lead one to the conclusion that the universe still has the characteristics of the point from which it arose.

The Big Bang theory was derived from the observed expansion of the universe. That expansion, as demonstrated by the cosmological red shift, implies that at one time, nearly 14 billion years ago, the universe began as a point. Aside from the observed cosmological expansion, the evidence that there was indeed a Big Bang comes from the verification of certain predictions that are derived from the Big Bang theory.

One of these predictions derives from the events that would have taken place shortly after the expansion began. From three to five minutes after the quantum fluctuation occurred, as previously noted, thermonuclear reactions would have converted some of the newly formed hydrogen into helium. Theoretical physics predicts that during this time those reactions would have resulted in a mix of approximately 90 percent hydrogen and 10 percent helium by number density. In fact, that is approximately the mix of those two elements that is found in the cosmos. The fraction of a percent of other elements, such as those found in the earth and other planets, is the result of subsequent supernovas that dispersed the thermonuclear ash of the heavier elements that were produced in their cores. The observed cosmological ratio of hydrogen to helium, then, is predicted by and supports the Big Bang theory.

Another prediction that was derived from the Big Bang theory is that there should be a universal remnant of the tremendous heat that existed during the initial moments of the Big Bang. After the Big Bang theory was developed, it was calculated that the tremendous heat would have cooled down as the universe expanded, just as a gas cools down when it expands. Given the amount of cosmological expansion, the calculations indicated that the heat remnant should presently be at a microwave temperature of about three degrees Kelvin above absolute zero. The theory also predicted that this microwave energy should appear to be coming equally from all areas of the sky. At the time these predictions were made, there were no scientific instruments that could measure such a low temperature. Eventually, however, the instruments were developed, and in the 1960s the microwave background radiation, as it is called, was detected and found to be just as the theory predicted. Recent precise temperature measurements from the COBE and other satellites have provided further specific evidence for the Big Bang.[2]

The observed expansion of the universe, the ratio of cosmic elements, and the microwave background radiation, then, all support the Big Bang theory. While it is important to recognize that these agreements between theoretical physics and the observations do not rigorously prove the Big

Bang theory, it is absolutely valid to say that they make the Big Bang both probable and consistent with what is currently known in science. Any alternative theory on the origin of the universe will have to take these findings into account and provide a better explanation for them.

Chapter 6
The Size of the Cosmos

We have seen from the evidence that the universe is immense. But how big is it? Is there any way we can truly comprehend its size?

Not really. About all that can be done is to provide some kind of perspective. Through powerful telescopes, we can observe quasars that are over 13 billion light years away. With a radius of 13.5 billion light years, the visible universe encompasses, from our perspective, a volume of about 1.15×10^{31} (115 followed by 29 zeros) cubic light years. That is 1.13×10^{40} (113 followed by 38 zeros) times larger than the volume of our solar system (if the solar system is taken as having a spherical volume with its circumference defined by the orbit of Pluto) and 9×10^{57} (9 followed by 57 zeros) times greater than the volume of the earth.

But what do those figures really mean? They are quite incomprehensible for us when the greatest distance most of us might deal with is perhaps a cross-continent airline flight.

We might be able to give some meaning to the figures by going the other way. If the visible universe were reduced to the size of the earth, and everything in the universe were reduced proportionally, then on that scale the diameter of our sun would be about one-half that of a normal-sized hydrogen atom (which is the smallest atom and is about 1/250,000,000 of an inch in diameter). On that same scale, the earth would have a diameter that would be more than two hundred times smaller than that of a hydrogen atom.

In actuality, the earth would be considerably smaller because the universe as it exists now is much larger than is our present perception of it. The limits of the universe as we see it now comes from light that began its journey over 13 billion years ago. But the universe has continued expanding over that time, and, if we could see the limits of the universe as they actually exist now, they would appear to be some forty billion light years away. That, of course, would make the earth even more insignificant in the cosmic scheme of things.

Through telescopes, we can see literally billions of galaxies. The total number of galaxies in the universe must be at least several hundred billion. Let's assume there are 100 billion galaxies, which is probably quite conservative. Our galaxy is of average size, and has at least 200 billion stars. Using that figure as the average per galaxy, there must be something in the neighborhood of 2×10^{22} stars in the universe; that is 20,000,000,000,000,000,000,000 stars.

Our sun is a typical star, and there is nothing to indicate that it is unique in any way in comparison with the multitude of other stars in the universe. Therefore, there is no reason to believe that it is unique in having a planetary system. One might therefore wonder how many other planets there might be in the universe.

In order for planets to exist around any star, there must previously have been a nearby supernova that spewed out the heavier elements from its core to provide the material for planet formation. Observations of other galaxies indicate that supernovas occur in a typical spiral galaxy like ours on average about once every 50 years. Since galaxies have existed for billions of years, there must have been a very large number of supernovas that would have spewed a considerable amount of material into interstellar space. In fact, because the early universe was denser than the present-day universe, a larger percentage of the stars that formed then would have been more massive than is now the case, and those stars would have gone supernova in a comparatively short time and spewed their heavy elements into space, thus providing considerable material for planet formation.

It has been proposed that planets have their origins in extensive gravitationally contracted disks of dust and gas surrounding newly formed stars. Over time, the dust and gas in these disks would accrete into larger and larger planetismals. These planetismals would then themselves accrete until they eventually formed full-sized planets.

The Hubble Space Telescope has made some findings that seem to verify this scenario. While even the Hubble telescope is not powerful enough to resolve individual planets orbiting other suns, it has found numerous young stars in the Orion Nebula that are surrounded by dense, flattened disks of dust. Scientists have also detected traces of similar disks in a nebulous region stretching across the constellations Taurus and Aurgis.[1] These disks therefore appear to indicate that planet formation is an ongoing process in our galaxy. Moreover, as we noted in Chapter three, numerous stars have been found to have planets in orbit around them.

6. The Size of the Cosmos

Though the current evidence indicates that stars with planets are relatively common, let us be very conservative and suppose that only one star in one hundred has a planetary system, and that the average number of planets in such systems is five (our own system has at least eight, remember). In that case, our galaxy would have ten billion planets and there would be 10^{21} planets in the universe. That is 1 followed by 21 zeros. Even if we assume that only one star in one thousand has planets, there would be one billion planets in our galaxy and 10^{20} planets in the universe. If, in fact, more stars than in our examples here have planets, the numbers of planets in the universe would be correspondingly greater.

The dimensions of the universe and the numbers of stars and planets it contains are so huge as to surpass our comprehension. But these numbers make one thing clear: On the cosmic scale of things, there is nothing significant or special about our planet earth. This place we call our home planet is lost in the universe with its multitudinous galaxies, stars, and planets, and the duration of our own existence is scarcely a tick in the long-running clock of universal time.

Our sun will end the beneficent part of its life cycle in about five billion years. It will then go through its death throes and eventually turn into a white dwarf that will put out only a tiny fraction of the heat and light it is presently putting out. The earth's passing at that time as a habitable planet will have no significance insofar as the rest of the cosmos is concerned.

Moreover, long after our sun ceases to burn its nuclear fuel and the earth no longer can support life, new stars and their attendant planets will continue to be born from the ashes of future supernovas, repeating the cycle.

This view of the cosmos may be deflating and cold and hard to some people. Mankind seems to have a need to look at itself as having some importance in the cosmic scheme of things and as somehow being above the commonality of nature. That is certainly why some people reject this view of the cosmos and also why they reject biological evolution. These findings of science offend the egos of those who believe themselves to be "a little lower than the angels," and they therefore reject the evidence that gives support to anything that threatens their self-perceived station in life.

However, offended egos are completely irrelevant insofar as the cosmos and the natural world are concerned. The cosmos will go on expanding as it has for billions of years, and life will go on evolving whether or not some members of the human species choose to ignore the evidence for those facets of nature. If some people reject the findings of science concerning the cosmos and the natural world because they are uncomfortable with them, they are simply revealing how shallow and insecure their worldview is.

On the other hand, while not necessarily finding "comfort" (in the religious sense) in what science has learned about nature and the cosmos, those of us who accept and are comfortable with the magnitude of it all can appreciate our place in the cosmos, however small that place may be.

As Carl Sagan said, we are made of star stuff. The atoms in our bodies were produced in the core of a massive star that exploded as a supernova billions of years ago. Some of the material that star ejected during the explosion, along with other interstellar material, subsequently coalesced under its gravity to form the sun, the earth, and the other planets of our solar system, and then—after four billion years of biological evolution—us.

Though our earth and our own existence may be insignificant on the cosmic scale of things, we are a *part* of that magnificent cosmos, and we can look at the grandeur of the universe and be fascinated and exhilarated by what science has found out about it.

Part Three

The Biblical View of the Cosmos

Chapter 7
In the Beginning

There is one overall cosmological view that is found throughout the Bible, from Genesis to Revelation: The earth is the centerpiece of creation and is the focus of God's attention, whereas the rest of the cosmos exists only for the sake of the earth.

Thus, in Genesis, the God of the Bible gave form to the earth on the third day of creation, but he did not create the sun, moon, and stars until the fourth day, and their creation was merely an adjunct to, and was subordinate to, the creation of the earth. Furthermore, as described in the Book of Revelation, after God's salvation plan for mankind is fulfilled and his judgment on the earth and its inhabitants is playing out in the "Last Days," the rest of the cosmos will no longer be necessary: The sun and the moon will be darkened, the stars will fall from the sky, and the heavens will roll up like a scroll.

That thoroughly geocentric view of the cosmos stands in stark contrast to the modern view in which, as we saw in the previous part of this book, the earth is merely an insignificant mote lost in the vastness of space and time—a mote whose existence and eventual passing are of no particular relevance to the rest of the cosmos.

In the context of the biblical worldview, then, why would the God of the Bible create a vast universe such as that which science has uncovered, yet subordinate its very existence to an insignificant speck of dust such as is the earth? And why would the God of the Bible create such a universe—an incomprehensibly immense universe with an unfathomable capacity for a bewildering number of galaxies, suns, and planets—only ultimately to destroy it all when the events of the "Last Days" play out here on the earth?

In answer to those questions, it would not make any sense at all for the God of the Bible to create such a universe. And, in fact, such a universe is totally alien to the biblical worldview; there is absolutely nothing in the Bible to even hint that the biblical God created such a universe. Indeed, in the biblical cosmos there is no need or place for such a universe, for the cosmos that is revealed in the Bible is not the cosmos that science has revealed.

But then, what *does* the Bible say about the cosmos? And what does the Bible say about the sun, moon, and stars? For that matter, what does the Bible say about the earth?

To answer those questions, let us begin by looking at the first two verses of Genesis in the King James Version of the Bible (referred to from now on as the KJV).*

> 1 *In the beginning God created the heaven and the earth.*
> 2 *And the earth was without form, and void; and darkness was upon the face of the deep. And the Spirit of God moved upon the face of the waters.*

The first verse is somewhat problematic. Some hold that it describes the actual creation of the biblical heaven and the earth in their initial form. Others take it as an introduction for what is to follow, since the heaven and the earth will not actually be created and named until later in the chapter. Another interpretation holds that, in the original Hebrew, the first three verses of Genesis consist of a single long sentence describing the state of things when the first act of creation took place. The JPS Tanakh reflects that interpretation. Here is its rendition of those verses:

> 1 *When God began to create heaven and earth—*
> 2 *the earth being unformed and void, with darkness over the surface of the deep and a wind from God sweeping over the water—*
> 3 *God said, "Let there be light"; and there was light.*

The first interpretation can probably be dismissed because the specific acts of creation in the remainder of the chapter are always preceded by the formulaic expression "And God said, . . ." followed by a command. If the

---

* Because it is considered authoritative by most fundamentalists, we will usually use the King James Version of the Bible as the baseline in our study of biblical passages. However, when appropriate and to achieve a degree of consensus, we will also refer to other translations for clarification or alternate interpretations. Note: For titles of Bibles and referenced lexicons, along with their abbreviations, see "Bibles and other Resources used in this Book" following the Introduction.

7. In the Beginning

first verse actually were describing the initial act of creation, it seems it should have stated "In the beginning, God said, let there be the heaven and the earth." That leaves the second and third interpretations, which, though having a somewhat different emphasis, say much the same thing about the initial state of the biblical cosmos.

That initial state of the biblical cosmos is indicated by the second verse. The first part of the KJV rendition of that verse states that "the earth was without form, and void," which apparently was the writer's way of saying that the earth was not yet formed. The original Hebrew words that are translated as "without form" and "void" are *tohuw* and *bohuw*, respectively. Both of these Hebrew words have a common meaning of "emptiness." In addition, *tohuw* has other related meanings, including "formlessness," "confusion," "unreality," "nothingness," "wasteland," and "place of chaos." The writer of Genesis was apparently trying to get across the idea that, before the actual creation of the earth, there was only emptiness and formlessness.

The remainder of verse 2 states: "... *and darkness was upon the face of the deep. And the Spirit of God moved upon the face of the waters.*" It would appear that the formlessness and emptiness that existed at this time consisted of a "deep," or "waters." That is not necessarily inconsistent—after all, water is, by its nature, formless and empty of structure.

So what exactly was the "deep" in the context of the passage?

In referring to the Hebrew lexicons, we find that the Hebrew word that is translated as "the deep" in verse 2 is *tehowm*, which, besides "deep," has "abyss" as one of its meanings. In addition, both BDB and KB also state that *tehowm* has a meaning of "primaeval ocean."

In light of the fact that the creative acts of the biblical God are explicitly stated in the subsequent verses of the chapter, it is noteworthy that verse 2 says nothing about his having created "the deep." The verse, then, provides a description of the initial state of things before God began the creative process—the implication being that the "deep" was a pre-existing cosmic ocean. However, it should be noted that in Proverbs 8:23–24 we find Wisdom, personified, saying, "*I was set up from everlasting, from the beginning, or ever the earth was. When there were no depths, I was brought forth....*" The Hebrew word translated as "depths" there is, again, *tehowm*, which would imply that, at least for the writer of Proverbs, even the primaeval deep had a beginning at some point in time; nevertheless, that passage still implies that "the deep" existed before God began the creative process. (Babylonian mythology, which appears to have influenced ancient Hebrew thought in several ways, also posited the idea that a primal cosmic ocean existed prior to the creation of the earth and the heaven. See Appendix D for more information about this.)

Going on, we find that Genesis 1:3 describes the first act of creation:

And God said, Let there be light: and there was light.

The biblical God's first creative act is—fittingly—the creation of light. The Hebrew word translated as "light" in this passage is *'owr*, which has several meanings, including specifically "light"; but also included in the meanings of *'owr* are the light of lamps, the light of day, and the light of the heavenly luminaries that are the sun, moon, and stars.

Note that the passage specifically states that God said "let there be light." Again, in the first two verses of the chapter, God made no such command concerning the deep, which is further evidence that the deep was pre-existing.

Now, we can continue with verses 4 and 5, which state:

> 4 *And God saw the light, that it was good: and God divided the light from the darkness.*
> 5 *And God called the light Day, and the darkness he called Night And the evening and the morning were the first day.*

Here, God divided the light from the darkness (*choshek* in the original Hebrew), which resulted in the first day (*yowm* in the original Hebrew). Note that verse 5 specifically states that the "light" the biblical God created was "Day"—meaning the light of daytime—and that the darkness and light in combination, as evening and morning, were called a "day." For the ancient Hebrews, the coming of the evening darkness marked the end of one day and the beginning of the next day. Thus, the division of the first "day" (and, for that matter, each of the remaining days of the creation week) into "the evening" (or Night) and the "morning" (or Day) would indicate that a normal day (i.e., the equivalent of a 24-hour day) is meant. We, of course, use the word "day" in a similar way—that is, as meaning the period of daylight during a 24-hour period, as well as the 24-hour period itself.

But there is a peculiarity about this biblical division of light from the darkness to create the first day. Today, we know that it is the light of the sun that causes daytime, and that it is the rotation of the spherical earth (that is, an earth with a definite form) that "divides" nighttime from daytime for any given location on the earth. However, at this point in the biblical creation process, not only has the sun not yet been created, but the earth is also still "without form" during this initial division of day and night. Nevertheless, the passage specifically states that the "light" that the biblical God created *was* "Day," meaning the light of daytime.

If we think about it from the standpoint of our present-day knowledge of the cosmos, Genesis 1:3–5 makes little sense. However, the passage most likely did make sense to the ancient Hebrews who lived when it was written, which would mean that they had a different view of the cosmos than we have. The only way for us to make sense of the passage, then, is by temporarily putting aside our present-day knowledge of the cosmos and trying to understand the passage from the perspective of the ancient Hebrews. That means we must begin by reading the plain words of the passage and taking them at face value. In doing that, the most logical conclusion we can arrive at is that the ancient Hebrews considered the light of day—specifically, the light of the daytime sky stretching from horizon to horizon—to be separate from, and not a result of, the light that is produced by the sun. That is, they considered the light of the daytime sky to be an entirely separate creation in and of itself, and that its existence is not dependent upon the sun.

After all, the sky turns bright some time before the rising of the sun, and the sky remains bright for some time after the setting of the sun. Therefore—as the ancient Hebrews would have perceived it—the light of the daytime sky is not derived from the sun. (Of course, the ancient Hebrews did not know about the atmospheric refraction of the sun's light when the sun is just below the horizon, which we know is what causes the sky to become bright before the sun rises and to remain bright for some time after it sets.)

In having God create the light of day before he created the sun, the writer of Genesis was simply taking a clue about the order of the creation process by following the observable celestial order of things as nighttime passes into daytime. Thus, the darkness of night would represent the original cosmological darkness before the creative process began. The lightening of the sky in the morning before the appearance of the sun would represent the creation of light as the first act of creation. Finally, the appearance of the sun at sunrise would represent the creation of the sun after the creation of the light. Thus, according to the perspective of the ancient Hebrews, the light of the daytime sky would be an entity unto itself, and, as an entity unto itself, the light of day could be created before the creation of the sun, and even before the creation of the earth. This also explains the why the ancient Hebrews reckoned the day as beginning at night, and a day is described as "the evening and the morning."

That conclusion helps us to understand the 16th verse of Genesis 1, where we read that God created the "greater light" (i.e., the sun) to "rule the day," just as a king rules over his kingdom. And, just as the arrival of an earthly king is preceded by those of his subjects who announce his coming, so the rising of the sun is preceded by and is announced by the light of day—that is, the predawn lightening of the sky. Also, just as the king's retinue trails the king when he departs, so does a retinue of daylight continue to lighten the sky after the setting of the sun.

Genesis 1:3–5 therefore makes perfect sense if it is understood in terms of our conclusion about the ancient Hebrew concept of the cosmos. Within that concept, the biblical God could create the light of "Day" before he created the sun because the light of day was not considered to be derived from the sun. Indeed, because Genesis indicates that the biblical God created the light of day before he created the sun, the light of day must have been considered to be an entity unto itself.

In fact, that conclusion is verified by other passages in the Bible. For example, it appears the ancient Hebrews thought that, subsequent to its creation, the light of day became attached to the firmament of heaven when it was created, thus giving the firmament its own source of light independent of the sun. As it states in Daniel 12:3:

> *And they that be wise shall shine as the brightness of the firmament....*

But an even more significant indication of the independent nature of the light of the daytime sky in the biblical cosmos is found in Ecclesiastes 12:2:

> *While the sun, or the light, or the moon, or the stars, be not darkened, nor the clouds return after the rain.*

The Hebrew word translated as "light" in that passage is *'owr*, the same word that occurs in the original Hebrew of Genesis 1:3–5 and is there called the light of day. As the sun, moon, and stars are observable in the sky, the light referred to in the Ecclesiastes passage must certainly also be observable in the sky, and the only light that it could be is the light of the daytime sky. The light referred to in the passage is not indicated as coming from the sun, but rather as being another source of light besides that of, and having an equal standing with, the heavenly luminaries. Indeed, if the sun had been considered to be the source of the light of day, there would have been no need to include the light of day in the passage, for the darkening of the sun would have also darkened the light of day. This passage would thus appear to support our conclusion that the light of the daytime sky was considered to be separate from that of the sun and therefore to be an entity unto itself.

Another passage that provides a similar view is Psalm 74:16:

> *The day is thine, the night also is thine: thou hast prepared the light and the sun.*

The Hebrew word translated as "light" in that passage is *ma'owr*, which is derived from *'owr* and also means "light," "a source of light," or "luminary." The Hebrew word translated as "prepared" is *kuwn*, which means "to prepare" or "to establish." Again, the "light" in that passage is not indicated as being "of" the sun, but rather is indicated as being another source of light besides the sun, as it is in the Ecclesiastes passage. In fact, the passage indicates that the light and the sun were each prepared separately. (It should be noted that some versions of the Bible translate *ma'owr* as "moon" in the passage. However, the lexicons do not give "moon" as a meaning for *ma'owr*. Moreover, that verse is the only instance in which these versions of the Bible have *ma'owr* translated as "moon." The moon *is* called the lesser light [*ma'owr*] in Genesis 1:16, but its use there is in a descriptive term rather than appellative, and the sun is likewise called the greater light [*ma'owr*] in that verse. Is the sun then the greater "moon"? If the moon were intended as being the counterpart to the sun in Psalm 74:16, the specific Hebrew word for moon, *yareach*, surely would have been used instead of *ma'owr*, since the specific Hebrew word for "sun," *shemesh*, is used for *that* luminary in the passage. By way of illustration, in Ecclesiastes 12:2, which we looked at previously, the specific Hebrew words for both the sun and the moon are used: "... *the sun* [shemesh], *or the light* ['owr], *or the moon* [yareach]....")

The conclusion that we can therefore derive from Genesis 1:3–5 and the other passages quoted above is that the writers of the Bible did not consider the light of the daytime sky to be derived from the sun, but rather that it was an entity unto itself and remained so even after the sun was created. Moreover, as we shall see, that conclusion about the independent nature of the light of the daytime sky fits into the biblical cosmos quite well.

Of course, from our modern understanding of the cosmos, we know that the light of the daytime sky *does* have its source in the sun, and we can conclude that the different view the Bible presents is based on a cosmos that is different from the modern view of the cosmos.

There is another point that might be noted concerning Genesis 1:3–5. Some biblical apologists who have tried to reconcile the Bible with the findings of science have argued that the light that came forth as a result of God's decree in that passage can be equated with the light of the Big Bang, and that the Bible is therefore scientifically accurate. However, given the language of the passage, it is quite clear that the God-decreed light was meant to be taken as ordinary daylight—specifically, what would be the light of the earth's daytime sky—and that it was divided from the darkness to make the evening and the morning, or night and day, for the earth. The word for "day," *yowm*, is, in fact, used in both occurrences of the word in the original Hebrew of verse 5: "*And God called the light Day, and the darkness he called Night. And the evening and the morning were the first day.*" From the modern perspective, the light of the earth's daytime sky obviously cannot be applied to the whole universe or to the Big Bang. On the other hand, it could be applied to the cosmos of the Bible, a geocentric cosmos that, as we shall see, is quite small and self-contained when compared to the universal cosmos that modern science has uncovered. Moreover, from the modern frame of reference, the Big Bang occurred billions of years before the formation of the earth. Therefore, the light that Genesis describes as coming into existence at God's decree simply cannot be equated with the light of the Big Bang. It should also be noted that, though some Bible believers try to equate the light mentioned in Genesis 1:3 with the light of the Big Bang, many other Bible believers reject the idea that there was a Big Bang or that creation occurred billions of years ago.

That brings up an often-asked question: according to the Bible, how long ago did the creation occur?

The traditional fundamentalist answer to that question comes from adding up the years associated with the generational "begats" that are listed in the Bible. If the summation of those years is used to develop a timeline backwards from certain historical clues that are also given in the Bible, the result is a creation date, according to the Bible, that took place a little over 6,000 years ago. This adding of the begats is fairly straightforward, and a large number of those who believe that the Bible is literally true also believe that the earth is only about 6,000 years old because of the biblical timeline.

However, some "scientific" creationists come up with different dates. They realize that, from the perspective of human history, 6,000 years is a little too short to justify. These creationists will therefore say that the earth is a little older, but no more than about 10,000 years old. These creationists are called young-earth creationists (or YECs). Other creationists interpret the length of the days of the creation week more loosely and accept the scientific evidence indicating that the earth is much older, though they still generally believe in the steps of creation that are outlined in the Bible and try to interpret them to make them fit the findings of science. These creationists are called old-earth creationists (or OECs).

Some Bible believers justify believing that the biblical creation occurred considerably more than about 6,000 years ago because of the reference in 2 Peter 3:8 that "one day is with the Lord as a thousand years, and a thousand years as one day." These individuals interpret that as meaning the creation days could have spanned an indeterminate period of time—perhaps equivalent to several thousand years, or even millions of years, each. However, the purpose of that verse is simply to put forth the idea that God is timeless. The fact of the matter is that the Hebrew word translated as "Day" in Genesis 1 is *yowm*, which is exactly the same word that is repeatedly used elsewhere in the Old Testament for a normal day—either the period from sunrise to sunset (i.e., daytime), or from sunset to sunset (i.e., a normal 24-four hour day).

Some biblical apologists argue that *yowm* can be used in the sense of a much longer and vague period of time, even a period of time that can encompass several years. That is true, but, whenever *yowm* is used in such a way, the context would make it clear that something other than a normal 24-hour day is meant. That the writer of Genesis 1 in fact meant the days of the creation week to be taken as normal 24-hour days is shown contextually by his reference to "the evening and the morning" to mark those days. That term, "the evening and the morning," specifically defines a normal 24-hour day. That the sun was not created until the fourth day has no relevance, because, as we noted, it appears that the ancient Hebrews considered the light of the daytime sky to be separate from that of the sun. In that scenario, a full day consisted of both the light of daytime and the dark of nighttime, so the sun was not necessary to mark a full day.

Moreover, the belief that the biblical days of creation could have encompassed thousands of years, perhaps even being equivalent to eons, has several serious problems. If the length of each of the days in the creation week were the equivalent of an inordinate length of time, that would mean that the earth would have had days and nights that were also of correspondingly long duration. After all, each creation "Day" consisted of nighttime and daytime, i.e., the biblical "evening" and "morning." There would be no justification in ignoring this facet of the biblical days of creation, for the first chapter of Genesis is quite specific in stating that each day of creation consisted of an "evening" and a "morning." If, from the biblical perspective, that were not the case, it would have been pointless for the author of Genesis 1 to have divided each of the biblical days of creation into an evening and a morning.

Therefore, if each biblical "Day" were an eon that was, for example, 1,000,000 years in length, then the period of daytime would have been 500,000 years and the period of nighttime would also have been 500,000 years. This would have a significant implication for the results of the remaining acts of creation. To illustrate why this would be, let's take a look at the specific acts of the creation during the creation week, starting with the creation of the firmament, which is the next act of creation after the creation of light. These verses encompass Genesis 1:1 through 2:3. For the sake of brevity, only the essentials are shown here:

> 1:6 *And God said, Let there be a firmament in the midst of the waters, and let it divide the waters from the waters.*

> 1:8 *And God called the firmament Heaven. And the evening and the morning were the second day.*
> 1:9 *And God said, Let the waters under the heaven be gathered together unto one place, and let the dry* land *appear: and it was so.*
> 1:10 *And God called the dry* land *Earth.* . . .
> 1:11 *And God said, Let the earth bring forth grass, the herb yielding seed,* and *the fruit tree yielding fruit after his kind, whose seed* is *in itself, upon the earth: and it was so.*
> 1:13 *And the evening and the morning were the third day.*
> 1:14 *And God said, Let there be lights in the firmament of the heaven to divide the day from the night.* . . .
> 1:16 *And God made two great lights; the greater light to rule the day, and the lesser light to rule the night:* he made *the stars also.*
> 1:19 *And the evening and the morning were the fourth day.*
> 1:20 *And God said, Let the waters bring forth abundantly the moving creature that hath life, and fowl* that *may fly above the earth in the open firmament of heaven.*
> 1:23 *And the evening and the morning were the fifth day.*
> 1:24 *And God said, Let the earth bring forth the living creature after his kind, cattle, and creeping thing, and beast of the earth after his kind: and it was so.*
> 1:26 *And God said, Let us make man in our image, after our likeness.* . . .
> 1:27 *So God created man in his* own *image.* . . . ; *male and female created he them.*
> 1:31 . . . *And the evening and the morning were the sixth day.*
> 2:2 *And on the seventh day God ended his work which he had made; and he rested on the seventh day from all his work which he had made.*

Since, according to Genesis 1:11–13, God created all of the different kinds of plants on the third day, one might ask how those plants would have survived the cold darkness of the nights of the creation week if those nights lasted 500,000 years. And, according to Genesis 1:15–19, God created the sun and the other "lights" of heaven on the fourth day. One might therefore ask how could the plants have survived the scorching that would have resulted from a daytime that consisted of the equivalent of 500,000 years of continual sunlight for each of the remaining days of the week. Moreover, according to Genesis 1:20–31, the sea creatures and birds were created on the fifth day, and the animals, "creeping things" (insects, etc.), and man (male and female) were created on the sixth day. Finally, according to Genesis 2:1–2, God rested on the seventh day. The living things that were created on fifth and sixth days would therefore also have had to deal with days and nights that were each the equivalent of 500,000 years until the end of the creation week. The same situation would occur if the days of the creation week are taken to be any other excessively long period of time.

The only reasonable conclusion that can be drawn from this is that, because the days in the creation week each consisted of an evening and morning, any attempt to hold that those days were anything other than normal 24-hour days raises some serious problems and simply does not make sense.

In any case, if one asserts that the days of creation were something other than normal days of twenty-four hours, what does one do with verses such as Exodus 20:11, in which the Sabbath day is established? That verse states, *"For in six days the Lord made heaven and earth, the sea, and all that in them is, and rested the seventh day"* (KJV). If the days of creation were not normal 24-hour days, then the institution of the Sabbath in that verse has no basis in its reference to the days of creation.

The conclusion can only be that anyone who interprets the biblical days of creation as being considerably longer than normal 24-hour days is doing so in light of modern cosmological knowledge and not in light of what the writers of the Bible believed about the nature of the cosmos and of the creation process. Even aside from that, attempts at "interpreting" passages in the Bible to make them conform to the present-day understanding of the cosmos does violence to the original meaning of those passages. The authors of the Bible knew what they meant when they wrote their narratives, and when someone, today, interprets the plain language of those narratives in order to make them conform to modern knowledge, it turns the Bible, in effect, into something that is different from what its original authors had in mind.

This means that, on the basis of what we find in the Bible, we must conclude that creation, according to the Bible, occurred only about 6,000 years ago. The scientific evidence, of course, provides absolutely no support for such a recent creation. As we noted in the previous part of this book, the scientific evidence indicates that the universe is more than 13 billion years old and that the earth is about 4.5 billion years old.

To complicate matters, it should be noted that Genesis 2:4–22 presents a different view and order of creation than does Genesis 1:1 through 2:3. Here are the relevant verses from Genesis 2:

> 4 *These are the generations of the heavens and of the earth when they were created, in the day that the LORD God made the earth and the heavens,*

> 5 *And every plant of the field before it was in the earth, and every herb of the field before it grew. ...*
> 7 *And the LORD God formed man of the dust of the ground, and breathed into his nostrils the breath of life; and man became a living soul.*
> 8 *And the LORD God planted a garden eastward in Eden; and there he put the man whom he had formed.*
> 9 *And out of the ground made the LORD God to grow every tree that is pleasant to the sight, and good for food. ...*
> 15 *And the LORD God took the man, and put him into the Garden of Eden to dress it and to keep it.*
> 19 *And out of the ground the LORD God formed every beast of the field, and every fowl of the air; and brought them unto Adam to see what he would call them: and whatsoever Adam called every living creature, that was the name thereof.*
> 21 *And the LORD God caused a deep sleep to fall upon Adam, and he slept: and he took one of his ribs, and closed up the flesh instead thereof;*
> 22 *And the rib, which the LORD God had taken from man, made he a woman, and brought her unto the man.*

According to these verses in Chapter 2 of Genesis, God created the heavens and earth in one day. He then, presumably on the same day, formed the man (the man only) as the next act of creation *before*—and in contradiction to the order of creation in Genesis 1—he created the plants and animals. God then formed the first woman *after* he created the plants and animals.

Actually, those contradictions highlight an essential fact about the Bible: Its authors frequently worked with earlier stories and documents that they had available to them. In the case of the discrepant accounts of creation in Genesis 1:1 through 2:3 and Genesis 2:4–22, there is compelling evidence that those accounts are based on two different creation stories that were derived from two different source documents. Narratives from those two source documents are found not only in chapters 1 and 2 of Genesis, but also, along with narratives from certain other source documents, in several other chapters of Genesis and in other books of the Old Testament as well. It was the incorporation of these different sources into the biblical text that resulted in many of the inconsistencies and contradictions that can be found in the Bible. An analysis of the story of the Flood, for example, reveals quite clearly that it was constructed from two different sources, for the author of the book of Genesis interspersed the two sources in the narrative about the Flood even though they contradict each other in several aspects.

We'll take this up again in more detail in Appendix A. The point to be made at this time is that this evidence shows that the first two chapters of Genesis present what are actually two separate and different creation stories that have their origins in two separate traditions. This, of course, should give one pause to wonder about the credibility of either story.

Chapter 8

The Firmament of Heaven

After the God of the Bible created the light and divided it from the darkness, he performed the next creative act, which is described in Genesis 1:6–8:

> 6 *And God said, Let there be a firmament in the midst of the waters, and let it divide the waters from the waters.*
> 7 *And God made the firmament, and divided the waters which were under the firmament from the waters which were above the firmament; and it was so.*
> 8 *And God called the firmament Heaven. And the evening and the morning were the second day.*

The "firmament" that is referred to in that passage from the KJV is central to the cosmological viewpoint of the Bible. In the passage, the biblical God calls the firmament "heaven," and today, because of that passage, the word "firmament" has an acquired meaning that relates it to the sky.

In trying to harmonize the Bible with the findings of science, some biblical apologists say that the firmament can be identified with the upper atmosphere, whereas others identify it with extraterrestrial space. Both of those positions, however, have real problems. Most of those who say that the firmament is the upper atmosphere also say that in the beginning there was a canopy of water just above the upper atmosphere, the idea for which they derive from verse 1:7, which speaks of "the waters which were above the firmament." This canopy of water, they say, supplied the water that caused the biblical Flood. However, Genesis 1:14–17 states that the sun, moon, and stars were set *in* the firmament. If the firmament is supposed to be the upper atmosphere, and the canopy of water was above the firma-

8. The Firmament of Heaven

ment, the sun, moon, and stars would have had to be under the canopy of water and in the earth's atmosphere. Leaving open for the moment the question of how it fits in with the biblical view of the cosmos, that construct obviously cannot be applied to the modern view of the cosmos.

On the other hand, if the firmament is taken as being the extraterrestrial space of the modern view of the cosmos, the waters above the firmament must be beyond the farthest stars, more than 13 billion light years away. That being the case, how could those waters have supplied the water for the biblical Flood?

In attempting to fit the firmament of the biblical cosmos into the modern view of the cosmos, then, these Bible believers actually make it clear that it does not fit and that the structure of the biblical cosmos is totally different from the structure of the actual cosmos—the cosmos that we know of.

So what is the biblical firmament supposed to be? And what does the word "firmament" really mean?

For starters, the word "firmament" is the English form of the Latin word *firmamentum*, which is the word used in that passage in the Latin Vulgate Bible. Latin dictionaries define *firmamentum* as "a prop or a support," and it means just what it seems to mean—i.e., something that is "firm" or "solid."

The passage has God saying, "*Let there be a firmament in the midst of the waters, and let it divide the waters from the waters.*" The passage goes on to state: "*And God made the firmament, and divided the waters which were under the firmament from the waters which were above the firmament.*" Note that, in contrast to the passage describing the creation of the light, this passage indicates that the firmament did not simply come into existence by a fiat decree from God. The passage states that God "made" the firmament. So what exactly did the biblical God make when he "made" the firmament?

To determine the answer to that question, we must first look at how the firmament fits into the context of the passage. To begin, the passage states that the firmament was made to be in the "midst" of the waters. According to *Strong's Concordance*, the Hebrew word translated as "midst" is *tavek*, which has meanings that include "midst," "middle," and "between (of things arranged by twos)." KB states that *tavek* means "in the middle of." The picture that Genesis 1:6–8 presents, then, is that the waters of the "deep"—i.e., the pre-existing cosmic ocean that was mentioned in Genesis 1:2—are now divided by the firmament of heaven, so that the firmament of heaven has water "above" it and water "under" it. In other words, just as the passage states, the firmament of heaven is between the waters above it and the waters beneath it.

The next two verses, which in part describe the events of the third day, now come to bear.

> 9 *And God said, Let the waters under the heaven be gathered together unto one place, and let the dry land appear: and it was so.*
> 10 *And God called the dry land Earth; and the gathering together of the waters called he Seas: and God saw that it was good.*

Genesis 1:6–10 therefore indicates that there is a three-tier stratification of the cosmos at this point: There are the waters above the firmament of heaven, the firmament of heaven itself, and the dry land and the waters, called the seas, beneath the firmament of heaven. This makes it clear that the biblical firmament of heaven *is* something firm or solid. By supporting, or holding back, the waters "above" it, the solid firmament provides a space under it so that the dry land can appear. The context of this passage therefore supports the meaning of the Latin-derived "firmament" as "a prop or support."

Doubtlessly, most Bible believers will refuse to accept the conclusion that the Bible indicates that the biblical firmament of heaven is a solid structure above the earth, so we should try to confirm that conclusion by further research.

Before we continue, however, it is important to bear in mind that we have not yet established what the Bible says about the shape of the earth. As we shall see later on, the shape of the biblical earth has a bearing on the matter, because, in the cosmos described in the Bible, the shapes of the earth and of the "firmament of heaven" have a specific relationship to each other.

We have seen that the English word "firmament" is derived from the Latin word *firmamentum* and signifies something firm or solid that is used as a support. But the original language of the book of Genesis was not Latin: It was Hebrew. So what is the original Hebrew word that was translated as "firmament"?

The Hebrew word that is translated as "firmament" is *raqiya`*, which *Strong's Concordance* defines as "an *expanse,* i.e., the *firmament* or (apparently) visible arch of the sky." That definition, however, is not really all that helpful. We have already examined the word "firmament" and are trying to get more information about the original Hebrew word. Specifically, we need to find out what is the nature of the biblical *raqiya`*, or firmament.

As noted, *Strong's Concordance* also defines *raqiya`* as the "visible arch of the sky" (editorializing the definition by adding in the parenthetical "apparently"), but that part of the definition does not add much more information.

8. The Firmament of Heaven

Note also that *Strong's Concordance* defines *raqiya`* as meaning "expanse," and, in fact, several versions of the Bible use that word instead of "firmament" in the passage. For example, the NIV translates verse 6 as follows:

And God said, 'Let there be an expanse between the waters to separate water from water.'

But does the word "expanse" convey an accurate or a meaningful interpretation of what the writer of Genesis meant?

For starters, the word "expanse" by itself is vague in meaning. An expanse has to be *of* something. We can talk about an expanse of desert, or an expanse of ocean, but what is the "expanse between the waters"? An "expanse" of *what* between the waters? The word "expanse" by itself says nothing about what its makeup is, and saying that it is between the waters provides no further information about its inherent nature other than it would have to be something that would be able to hold the waters apart. At least the word "firmament" indicates something firm or solid, which would explain how it could "separate the waters from the waters."

So what is the nature of the "firmament," or *raqiya`*, of Genesis 1:6–8? Certainly, there is nothing in the passage to indicate that it is to be taken as the vast vacuum of extraterrestrial space, or even of the atmosphere as some Bible believers aver. Again, the context of the passage indicates that the *raqiya`* is something that is used as a solid support to hold back the waters above.

Going back to the entry for *raqiya`* in *Strong's Concordance*, we find that it contains a note stating that the word is derived from *raqa`*. The entry for that Hebrew word indicates it is a verb having several meanings, including "to *pound* the earth; . . . to *expand* (by hammering); . . . to *overlay* (with thin sheets of metal)." The entry also states that *raqa`* is translated elsewhere in the Bible as "beat, make broad, spread abroad (. . . into plates), stamp, stretch." These definitions and usages indicate that *raqa`* is a verb that has a meaning of spreading out some medium by a physical action, such as beating or hammering.

An example of the usage of this verb is found in Exodus 39:3: "*And they did beat* [raqa`] *the gold into thin plates.*" Another example is found in Jeremiah 10:9: "*Silver spread* [raqa`] *into plates is brought from Tarshish.*" In that passage from Jeremiah, *raqa`* is translated as "spread"; however, for silver to be "spread" into plates it must be beaten out, and the NRSV translates the verse accordingly: "*Beaten* [raqa`] *silver is brought from Tarshish. . . .*" Another example is in Numbers 16:39 (here again from the NRSV): "*So Eleazar the priest took the bronze censers that* had been presented by those who were burned; and they were hammered [raqa`] *out as a covering for the altar.*"

These verses, as well as several others, show that the verb *raqa`* is used in the sense of beating or hammering out metal into thin plates or sheets. Therefore, the sense of *raqiya`*, as a noun derived from *raqa`*, could be taken as referring to a material that has been spread out, or "expanded," by being hammered out or beaten out into a thin plate or sheet—that is, hammered out into an "expanse" of the material. This interpretation of *raqiya`* would seem to fit the idea that it, and hence the "firmament," is more properly characterized as something that is firm or solid. However, to see whether we have accurately interpreted the meaning of *raqiya`*, we must look to other lexicons to see if the sense we have derived from the verb form of the word is valid.

In BDB we find the following definition of *raqiya`*: "extended surface, (solid) expanse (as if *beaten out...*) ...; 1. (flat) *expanse* (as if of ice...), as base, support...." The entry also equates the Hebrew word with the Greek word *stereoma* (given in Greek letters in the entry) and with the Latin word *firmamentum*. The entry goes on to state: "Hence... 2. the vault of heaven, or 'firmament,' regarded by Hebrews as solid, and supporting 'waters' above it."

The entry in the BDB lexicon specifically states that the meaning of *raqiya`* is an "extended surface" and a "(solid) expanse (as if beaten out)." That makes it quite clear that we have correctly derived the meaning of the word and that it does mean a solid material that has been beaten out or spread out into an "expanse" to form the heaven above the biblical earth. The second definition in the entry further confirms our understanding that the biblical firmament is a solid structure in the form of a vault—like an upside-down bowl—that supports, or holds back, the waters above it.

KB confirms that interpretation and defines *raqiya`* as follows: "the beaten metal plate, or bow; firmament, the firm vault of heaven...." This lexicon also equates the firmament with the Latin *firmamentum* and with the Greek *stereoma*. An etymological dictionary of the Hebrew language provides the following definition of *raqiya`*: "1 extended surface, expanse. 2 firmament, sky. [From *raqa`*; lit. 'something beaten out'.]"[1]

These references thus confirm that *raqiya`* has the literal meaning of "something beaten out," such as a piece of metal that has been beaten out into a sheet or plate. If one continues hammering a metal plate on one side repeatedly, the plate will gradually turn into a bowl. That, in fact, was how metalworkers in some early cultures made metal bowls and is how some metalworkers still make them. Interestingly, in ancient Phoenician, a Semitic language closely related to Hebrew, a word having the same root as *raqiya`* means "a hammered-out bowl."[2] A large bowl turned upside

8. The Firmament of Heaven

down becomes a vault, which would conform to the BDB definition of *raqiya`*.

To continue building our case, we should now turn to the Septuagint. (As was previously noted in Chapter 1, the Septuagint is a translation of the Hebrew Scriptures into Greek that some Hebrew scholars began during the third century B.C.E. in Alexandria, Egypt.) In the Septuagint, the Greek word that is translated from the Hebrew word *raqiya`* in the Genesis passage is *stereoma*, the same word that was noted in the above-given entries from the Hebrew-English lexicons in their exposition of *raqiya`*. By turning to Friberg, we learn that the word *stereoma* means a "strictly *solid body* or *part*, as what has been made solid or firm." BDAG defines it as "in var[ious] senses of someth[ing] solid." Thayer provides the meaning as "that which has been made firm," "the arch of the sky, which in early times was thought to be solid," and "that which furnishes a foundation; on which a thing rests firmly, support." These meanings correspond closely to those for the Latin word "firmamentum," which means a "prop" or "support." The Septuagint therefore shows that the third century B.C.E. Hebrew scholars who translated the book of Genesis into Greek considered the biblical *raqiya`*, or "firmament," to be a solid or firm structure above the earth—a firm structure that provides a support to hold back the waters above.

The real question, however, is whether that understanding of *raqiya`* as meaning a solid heaven above the earth can be contextually confirmed elsewhere in the Bible.

As a matter of fact it can! For starters, let's go back to Genesis 1:7, which states: *"And God made the firmament. . . ."* Again, note that the passage says that God "made" the firmament. The Hebrew word translated as the verb "made" is `asah*, the meanings of which include "to do," "to fashion," "to make," and "to work."

Psalm 102:25 goes further and indicates that the God of the Bible made the heavens with his hands:

> *Of old hast thou laid the foundation of the earth; and the heavens are the work of thy hands.*

The Hebrew word translated as "the work" in that passage is *ma`aseh*, the meanings of which include "deed," "work," and "thing made."

Another passage indicating that God made heaven with his hands is Isaiah 45:12, in which God is speaking:

> *I have made the earth, and created man upon it: I, even my hands, have stretched out the heavens, and all their host I have commanded.*

There, God says he stretched out the heavens with his hands. The Hebrew word translated as "stretched out" is *natah*, which also has the meaning of to "spread out." The NJB, in fact, uses "spread out" in its translation of the verse. The above two verses therefore substantiate the idea that God had physically beaten out the firmament of heaven, which, of course, would cause the beaten material of heaven to "spread out" and to become an "expanse."

But even more significantly, in Job 37:18 we read the following:

> *Hast thou with him spread out the sky, which is strong,*
> *and as a molten looking glass?*

The Hebrew word that is translated here as the verbal "spread out" is *raqa`*. That is the same word that we have already noted means "to beat out" or "to spread out" (as the silver in the Jeremiah verse was spread by being beaten out into thin plates), and from which the Hebrew word *raqiya`*—translated as "firmament"—is derived.

In addition, the Hebrew word that is translated as "strong" in this verse is *chazaq*, which, as well as meaning "strong," has alternate meanings of "hard" and "firm." This verse therefore presents the concept that the sky is made of a strong, firm, or hard material that was beaten out. Certainly, if the firmament of heaven is to hold back the waters above the firmament, it must be both hard and strong. That being the case, the Hebrew word *chazaq* is quite fitting in describing the nature of the firmament.

Note that the passage says the sky is "as a molten looking glass." The English of the King James translation obscures the sense of the passage somewhat. The Hebrew word that is translated as "looking glass" is *re'iy*, the meaning of which is "mirror." By the time of King James, mirrors were made of glass and were called, as the translation indicates, "looking glasses." In biblical times, however, mirrors were not made of glass; they were made of highly polished metal plates, usually bronze. For modern ears, the KJV therefore gives a somewhat misleading and confusing translation of the Hebrew word for mirror.

The translation of this verse in the KJV also causes confusion because of a shift in the meaning of the word "molten" since the days of King James. To us today, if something is molten, it has been melted and is in a fluid or flowable state. But in the time of King James, the word "molten" also meant "cast," or "that which has been cast from melted metal." The Hebrew word that is translated as "molten" in Job 37:18 is *yatsaq*, which, in fact, means "cast." An example of its use as such can be found in Exodus 25:12: *"And thou shalt cast four rings of gold for it...."* In 2 Chronicles 4:3 the same Hebrew word is again translated as "cast": *"And under it was the similitude of oxen.... Two rows of oxen were cast, when it was cast."*

Job 37:18 thus indicates that the sky is made of a strong, solid material and is similar to a mirror made of cast metal that has been beaten out into a thin plate or bowl and polished.

So Job 37:18 further confirms our understanding of the meaning of the original Hebrew word for "firmament" and gives action and substance to that understanding. The NEB translation of that verse verifies that understanding, for it follows the meaning of the key Hebrews words more closely and is clearer than is the version in the KJV:

> *Can you beat out the vault of the skies, as he does,*
> *hard as a mirror of cast metal?*

So there we have it. The sky is a vault and is hard. Note that, in the NEB translation of the verse, the Hebrew verb *raqa`*, the meaning of which is "to beat out," is translated as "beat out" rather than as "spread out" as it is in the KJV.

Other versions of the Bible also confirm that Job 37:18 describes heaven as being a solid structure. For example, here is the NIV translation of the passage:

> *. . . can you join him in spreading out the skies,*
> *hard as a mirror of cast bronze?*

The NKJV makes a similar statement:

> *With Him, have you spread out the skies,*
> *Strong as a cast metal mirror?*

The NRSV states:

> *Can you, like him, spread out the skies,*
> *hard as a molten mirror?*

And the JPS Tanakh states:

> *Can you help him stretch out the heavens,*
> *Firm as a mirror of cast metal?*

So the relevant passages in the Bible indicate that God "made" the "firmament" by beating out a piece of hard material, such as a plate of metal, into the shape of a giant bowl. He then placed this bowl upside down as an expansive vault, or a support, to hold back the waters above the vault. Thus, the original Hebrew words tell the story and show that the sense of the "firmament" as being a firm, or solid, structure that is used as a support is correct. Along with the other passages that we previously looked at, Job 37:18 thus makes it clear that the God of the Bible "made" the *raqiya`*, or the firmament of heaven, just as it states in Genesis 1:7: *"And God made* [`asah] *the firmament* [raqiya`]."

There are still other verses in the Bible which support the idea that heaven is a solid structure above the earth, though again we need to dig out the meaning by looking at the original Hebrew words. In Proverbs 8:27–28 of the KJV, for example, we find the speaker, who (as we previously noted) is supposed to be Wisdom personified, saying:

> 27 *When he prepared the heavens, I was there: when he set a compass upon the face of the depth:*
> 28 *When he established the clouds above: when he strengthened the fountains of the deep.*

Note that the word "established" appears in verse 28: *"When he established the clouds above...."* But therein lies an inconsistency: The Hebrew word that is translated as "established" in that verse is *'amats*, and that verse is the *only* instance in the entire KJV in which *'amats* is translated as "established." Nor is it translated as "establish" anywhere in the KJV. In the KJV Old Testament, the English words "established" and "establish" together appear more than 100 times, and the Hebrew word that is most frequently translated as "established" or "establish" in the KJV is *kuwn*.[3] That Hebrew word in fact occurs in verse 27, above, where it is translated as "prepared," which is also a proper, though less frequent, translation for *kuwn*.

The question therefore arises as to whether "established" is a suitable translation for *'amats* in verse 28. Significantly, according to BDB, *'amats* has a specific meaning of "be stout, strong. . . ." BDB further states about *'amats*: "1. make firm, strengthen [followed by direct object]. . . ." In *Holladay*, *'amats* is translated as "c) make strong Pr 8:$_{28}$ (clouds)." According to KB, the meaning of *'amats* is "be hard" and "to **make firm** Pr 8$_{28}$ (clouds)" (the emphasis is in the original).

Thus, *'amats*, the Hebrew word that is translated as "established" in Proverbs 8:28 of the KJV, has a more proper meaning of "make strong," "strengthen," "make firm," or "harden." More pointedly, as the above reference in BDB notes, when *'amats* is followed by a direct object, it specifically means "make firm, strengthen." That applies to verse 28, in which "clouds" is the direct object of the word in question. The passage can be more properly translated, then, as "When he made firm the clouds above."

But that is not all. The Hebrew words that are most commonly translated as "cloud" or "clouds" in the KJV Old Testament are `*anan* and `*ab* (or their plural forms), which together are translated as such more than 100 times. However, in Proverbs 8:28 the word translated as "clouds" is *shachaqim*, which has an alternate meaning of "skies" or "heavens," as exemplified in Psalm 77:17 (which also shows the original Hebrew word for "clouds"): *"The clouds [`ab] poured out water: the skies [shachaqim] sent out a sound: thine arrows also went abroad."* Moreover, in every place where the words "sky" or "skies" appear in the KJV, the original Hebrew word is *shachaq* (or its plural form, *shachaqim*), as, for example, in Job 37:18, which we looked at previously: *"Hast thou with him spread out the sky, which is strong, and as a molten looking glass?"*—or, as in the NEB translation: *"Can you beat out the vault of the skies, as he does, hard as a mirror of cast metal?"*

The question is, then, in Proverbs 8:28 should *shachaqim* have been more properly translated as "skies" rather than as "clouds"? In looking elsewhere in the KJV, we find that *shachaqim* is translated as "clouds" only ten times, and in most of those occurrences it could as well have been, or could have been better, translated as "skies" or "heavens." For example, we find in Psalms 57:10: *"For thy mercy is great unto the heavens, and thy truth unto the clouds."* Would not "skies" be better than "clouds" in that verse? One might ask whether God's "truth" disappears on a cloudless day. The passage is also an example of biblical poetic reiteration in which the second part of a verse reiterates the first part. In fact, the ASV, the NEB, the NIV, and the NAB all translate *shachaqim* as "skies" in that passage, and the JPS Tanakh translates it as "sky." The NIV, for example, renders the verse as *"For great is your love, reaching to the heavens; your faithfulness reaches to the skies."*

Psalm 18:11, like Psalm 77:17, which was cited above, also makes a clear distinction between the words in question: *"He made darkness his secret place; his pavilion round about him were dark waters and thick clouds [`ab] of the skies [shachaqim]."* We would therefore be justified in translating *shachaqim* as "skies" in Proverbs 8:28.

Now, if, instead of "prepared," we use the more usual translation of "established" for *kuwn* in verse 27, and if we revise verse 28 according to our analysis of the original Hebrew of that verse, Proverbs 8:27–28 could be restated as follows:

> 27 *When he established the heavens. . . .*
> 28 *When he made firm the skies above. . . .*

That wording is, in fact, exactly the same as that given for Proverbs 8:27–28 in the ASV, the ESV, the NRSV, the NASV, and the NAB. The JPS Tanakh wording is similar:

> 27 *I was there when He set the heavens into place; . . . .*
> 28 *When He made the heavens above firm, ...*

Proverbs 8:27–28 thus supports the concept of a solid or firm biblical heaven and hearkens back to Genesis 1:7–8, in which, as we saw, God made the "firmament" (*raqiya`*, meaning "something beaten out") of heaven as a solid or "firm" structure:

> 7 *And God made the firmament, and divided the waters which were under the firmament from the waters which were above the firmament.*
> 8 *And God called the firmament Heaven.*

In addition, when we previously analyzed Job 37:18, we saw that the NEB more properly translates that verse as

> *Can you beat out the vault of the skies, as he does, hard as a mirror of cast metal?*

Thus, three different passages in the Bible use three different words that have related meanings to portray the nature of the sky, and all three passages indicate that the biblical sky, or heaven, is strong, hard, or firm. Certainly the sky, or heaven, would need to have all of those attributes in order to hold back the waters above.

The idea that the "heaven" above the earth is made of a solid material is not isolated to the books of Genesis, Job, and Proverbs. A further confirmation that *raqiya`*, or "firmament," has the meaning of something solid is found in a vision that is described in the Book of Ezekiel. In Ezekiel 1:22–26, there is a description of a "likeness of four living creatures" with something over their heads that is like the firmament of heaven with the throne of God placed on top of it.

> 22 *And the likeness of the firmament upon the heads of the living creatures was as the colour of the terrible crystal, stretched forth over their heads above....*
> 26 *And above the firmament that was over their heads was the likeness of a throne....*

The same verses in the NEB are perhaps a little clearer:

> 22 *Above the heads of the living creatures was, as it were, a vault glittering like a sheet of ice, awe-inspiring, stretched over their heads above them....*
> 26 *Above the vault over their heads there appeared, as it were, a sapphire in the shape of a throne....*

The NJB is even more specific about the nature of the *raqiya`* in its translation:

8. The Firmament of Heaven

> 22 *Over the heads of the living creatures was what looked like a solid surface glittering like crystal, spread out over their heads, above them, ...*
> 26 *Beyond the solid surface above their heads, there was what seemed like a sapphire, in the form of a throne.*

In this passage, the Hebrew word *raqiya`* is translated as "firmament," "vault," and "solid surface" in the respective versions. That is the same original Hebrew word translated as "firmament" in Genesis 1:6–8. Here, in the book of Ezekiel, the Hebrew word *raqiya`* is used to describe a likeness of the sky, a solid sky upon which sets the throne of God. Note that the translation in the KJV likens the firmament to crystal, while the NEB translation likens it to ice. The Hebrew word, *qerach*, that is translated as "crystal" and "ice" respectively in the two versions can be translated either way, and both translations imply something solid, as in the NJB.

This likening of the firmament to crystal or ice indicates that the author of the Book of Ezekiel had a concept of the heavens that was similar to one of the cosmological views of the ancient Greeks, in which the sun, moon, planets, and stars were set in concentric crystalline spheres with the earth at the center.

Given the Hebrew view of the cosmos, it would make a certain amount of sense to regard the sky above as being composed of crystal. If one stands on the seashore and looks to the horizon, one can see the blue ocean and the blue sky meeting there. A scientifically naive person might reason that, since the ocean is blue and is water, the sky must therefore also be water because it is also blue, and, after all, water does fall from the sky as rain. But if the sky is water, what is holding the water up? Why, if the roof of the sky were a giant, transparent upside-down bowl of crystal, it would hold back the water above it, and the water could be seen through it!

Another apparent reference to the crystalline nature of the vault of heaven can be found in Exodus 24:10, which states (here, the RSV):

> *and they saw the God of Israel. Under his feet there was something like a pavement of sapphire stone, like the very heaven for clearness.*

Since, in the biblical view, the throne of God is placed upon the vault of heaven (as the passage from Ezekiel indicates), the upper surface of the vault of heaven would be a floor under the feet of God. Thus, the implication is that the pavement of sapphire, a blue gemstone, is the blue vault of heaven.

Revelation 4:1–6 makes a similar statement:

> 1 *After this I looked, and, behold, a door* was *opened in heaven.* ...
> 2 ... *and behold, there was a throne set in heaven.* ...
> 6 *and before the throne, as it were a sea of glass like a crystal.* ...

The "sea of glass like a crystal" that was before the throne would be analogous to the "pavement of sapphire" of the Exodus passage or the firmament of crystal of Ezekiel 1:22. The passage from Revelation therefore also suggests that the vault of the sky is like glass or crystal and is the floor of heaven.

In contrast to some of the other verses that we have looked at in the Bible which imply that the firmament of heaven is a beaten metal plate, these verses indicate that the firmament of heaven is "like" crystal or "like" a sapphire, i.e., a blue gem stone. But these differing views about the nature of the firmament should not be surprising. There was no requirement that all of the ancient Hebrews and early Christians over the centuries of their existence must have had exactly the same concepts about the nature of the cosmos. As it is implied in Job 37:18, some of the writers of the Bible believed the firmament to be metallic in nature, whereas others, as exemplified in the above passages, believed it to be like crystal, glass, or a gemstone. In any case, all of these passages indicate that the biblical vault of heaven is made of a solid material.

There are a couple of other points concerning the biblical heaven that should be made clear. In the Bible, heaven is frequently referred to in the plural—i.e., "heavens." In fact, the original Hebrew word for heaven is *shamayim*, which is plural. There is not a complete consensus on the exact meaning of *shamayim*. Some, as KB notes, have thought that it means "place of the waters" (from *mayim*, "waters," referring to the waters above the heavens) though others do not agree. If it does mean "place of the waters," that would explain the plural nature of the word and would indicate that it does not necessarily mean that there are several heavens. KB, in fact, also has a note that it probably does not mean a number of different heavens but is an expression for the superlative. If the word does imply more than one heaven; it could perhaps refer to the two separate heavens of the day and night skies, or it could refer to several successive heavens, one above the other. It could also mean all of the separate structures above the surface of the earth: the air where the birds fly, the firmament, and the waters above the firmament.

For that matter, several cultures of the time believed that there were multiple heavens. For example, the ancient Babylonians believed there were seven successive heavens, one above the other, and may have provided the ancient Hebrews with the source of the idea. Also, in the cosmos of the ancient Greeks, as already noted, seven concentric crystalline spheres surrounded the earth.

8. The Firmament of Heaven

In those cultures that believed in multiple heavens, the third heaven was usually considered the most glorious. Thus, in 2 Corinthians 12:2, Paul wrote: *"I knew a man in Christ above fourteen years ago, ... such an one caught up to the third heaven."* The idea of a more glorious heaven is also implied in Deuteronomy 10:14: *"Behold, the heaven and the heaven of heavens is the Lord's thy God, the earth also, with all that therein is"* (though the word heaven appears in the singular twice in English translation there, the plural form appears in both instances in the original Hebrew). The "heaven of heavens" is one that has a higher standing than the other heavens. In the biblical cosmos, the open space under the firmament where the birds fly and where the clouds are could be considered the first heaven, the firmament could be considered the second heaven, and God's dwelling place above the firmament could be considered the third heaven.

There are numerous other passages in the Bible relating to the nature of the firmament of heaven. For example, Genesis 1:20 states:

> *And God said, Let the waters bring forth abundantly the moving creature that hath life, and fowl that may fly above the earth in the open firmament of heaven.*

However, that wording of the verse from the KJV (and the wording in several other versions of the Bible, as well) does not accurately reflect the wording of the original Hebrew. The original Hebrew word translated as "open" is *paniym*, which actually means "face" and also "surface." For that matter, none of the lexicons referred to here give "open" as a meaning for *paniym*. (The KJV, as well as some other versions of the Bible, translates *paniym* as "open" in numerous other verses, but the English translations in which it is so translated are idiomatic, such as this one found in Ezekiel 32:4: "*... I will cast thee forth upon the open field. ...*" In this case, the more correct translation from the Hebrew would be "I will cast thee forth upon the face of the field," or "I will cast thee forth upon the surface of the field.") Moreover, the word "open" has a connotation of being more insubstantial than does "face" or "surface," so the use of "open" in Genesis 1:20 of the KJV obscures the sense of the original. With that understanding, we find that the NKJV gives a more accurate translation of the verse:

> *Then God said, "Let the waters abound with an abundance of living creatures, and let birds fly above the earth across the face of the firmament of the heavens."*

In speaking of the birds flying "across the face of the firmament," this passage is, in effect, emphasizing that the firmament is a distinct physical structure above the earth and that it has a surface. Otherwise, why would the verse not simply say "... and let birds fly across the skies above the earth"? This verse therefore fits in quite well with the understanding that the firmament is a solid vault above the earth.

Isaiah 34:4 is another verse which indicates that the firmament of heaven is a solid structure. In describing the "Day of the Lord," this verse states:

> *And all the host of heaven shall be dissolved, and the heavens shall be rolled together as a scroll: and all their host shall fall down, as the leaf falleth off from the vine, and as a falling fig from the fig tree.*

The "host of heaven" is a reference to the stars of the sky. The NIV accordingly provides this translation of the passage:

> *All the stars of the heavens will be dissolved and the sky rolled up like a scroll; all the starry host will fall like withered leaves from the vine, like shriveled figs from the fig tree.*

Revelation 6:13–14 reiterates that concept (here the KJV):

> *13 And the stars of heaven fell unto the earth, even as a fig tree casteth her untimely figs, when she is shaken of a mighty wind. 14 And the heaven departed as a scroll when it is rolled together; and every mountain and island were moved out of their places.*

These verses put forth the idea that the sky is a thin, solid material—it can be rolled up like a scroll. It is significant that, in biblical times, scrolls were occasionally made of metal plates that were beaten so thin they could be rolled up; one of the Dead Sea scrolls was made in such a manner. Therefore, it is apparent that the writers of the Isaiah and Revelation passages believed that heaven is a solid vault—a vault made from a metal plate that God had beaten thin—above the earth. They were therefore using, in their own minds, an apt illustration for their view of what would happen to that solid vault on the Day of Judgment.

That brings up another point. The idea that the heavens can be rolled up like a scroll certainly negates the idea that the firmament of heaven consists of the vast vacuum of interstellar space, as some biblical apologists have averred it to be. How can the vacuum of interstellar space be "rolled up"? For something to be rolled up implies that it must be composed of a thin, solid material.

These passages indicating that the heaven can be "rolled up" certainly posit a view of the cosmos that is different from our own. That the biblical stars, as the "host of heaven," can fall to earth also shows that the biblical view of the cosmos cannot be equated with the modern view of the cosmos. Therefore, the biblical firmament of heaven, in which the "lights" that are the stars are set, certainly cannot be understood as being the same as the modern view with its vast vacuum of interstellar space and its massive stars that are actually other suns.

In Psalm 144:5, we find another example of the idea that the heaven is a solid or firm structure: *"Bow thy heavens, O LORD, and come down: touch the mountains, and they shall smoke."* Likewise, in 2 Samuel 22:10, we find: *"He bowed the heavens also, and came down; and darkness was under his feet."* The Hebrew word translated as "bow" and "bowed" in these verses is *natah*, which has a large number of meanings, including "bend" or "incline," as well as "bow." BDB defines *natah* as "bend down" and *"bend down* heavens." The NAB translates the passage as *"He inclined the heavens and came down...."* Even more pointedly, the JPS Tanakh translates the Samuel passage as *"He bent the sky and came down...."* The picture here is that God brought himself down to earth by bowing or bending heaven downward, thus having it carry him down. The idea that heaven or the sky can be made to bow or bend down implies that it is made of a material that can bend or warp—something, for example, like a thin metallic vault.

There is another biblical description that should be mentioned concerning the firmament. The firmament, or heaven, has "windows" in it that God opens to let some of the waters "above" the firmament fall as rain. These windows are mentioned in Genesis 7:11 and 8:2 (here from the NRS):

> 7:11 *In the six hundredth year of Noah's life, in the second month, on the seventeenth day of the month, on that day all the fountains of the great deep burst forth, and the windows of the heavens were opened.*
> 8:2 *the fountains of the deep and the windows of the heavens were closed, the rain from the heavens was restrained,*

The Hebrew word translated as "windows" is *arubbah*, which has the meanings "lattice," "window," and "sluice." The windows of heaven are also mentioned in Isaiah 24:18 (again, from the NRS):

> *Whoever flees at the sound of the terror shall fall into the pit; and whoever climbs out of the pit shall be caught in the snare. For the windows of heaven are opened, and the foundations of the earth tremble.*

And also Malachi 3:10 (here, again the NRS).

> *Bring the full tithe into the storehouse, so that there may be food in my house, and thus put me to the test, says the LORD of hosts; see if I will not open the windows of heaven for you and pour down for you an overflowing blessing.*

The firmament of heaven also has doors. Psalms 78:23–24 provides an example (here, the KJV):

> 23 *Though he had commanded the clouds from above, and opened the doors of heaven,*
> 24 *And had rained down manna upon them to eat, and had given them of the corn of heaven.*

The mentions of windows and doors in the firmament of heaven imply that the firmament is a solid material. If the firmament were the vacuum of space, or even the upper atmosphere, it would be rather hard to picture its having "windows" or "doors" in it.

The firmament of heaven not only has windows and doors, it also has "chambers." Psalms 104:13, for example, states: *"He watereth the hills from his chambers...."* The Hebrew word translated as "chambers" is `aliyah, which means "roof room" or "roof chamber." Since the firmament of heaven could be considered the "roof" of the world, its rooms would be "roof rooms."

Job 9:9 states that God *"... maketh the Bear, Orion, and the Pleiades, And the chambers of the south."* The Bear, Orion, and the Pleiades refer to celestial constellations. The original Hebrew word translated there as "chambers" in that verse is *cheder*, which means, in fact, "chamber," and refers to heavenly chambers that are used as storehouses for rain, snow, or wind.

Job 37:9 states: *"Out of the south cometh the whirlwind: and cold out of the north."* The Hebrew word translated as "south" in that passage is *cheder*, which, as noted above, actually has a meaning of "chamber." In this case, the KJV translators apparently carried over the "chambers of the south" from Job 9:9 and shortened it to "the south" in Job 37:9. The NRSV gives a more accurate translation of the verse than does the KJV: *"From its chamber comes the whirlwind, and cold from the scattering winds."*

In Jeremiah 10:13, we find:

> *When he uttereth his voice, there is a multitude of waters in the heavens, and he causeth the vapours to ascend from the ends of the earth; he maketh lightnings with rain, and bringeth forth the wind out of his treasures.*

8. The Firmament of Heaven

The Hebrew word that is translated "treasures" is *'owtsar*, which means "storehouses," and several versions of the Bible use that word in the passage, as for example the RSV: "....*He makes lightnings for the rain, and he brings forth the wind from his storehouses.*" The idea expressed in the passage, then, is that the storehouses of heaven hold the wind until it is to be used. Job 38 shows this even more clearly. In that chapter, God scolds Job by sternly questioning him about his knowledge, or lack thereof, concerning various aspects of creation and the cosmos. With each question, God compares Job's lowly human abilities with his own divine creative and commanding powers. Relating to the subject at hand, for example, God asks in verses 22 and 23:

> 22 *Hast thou entered into the treasures of the snow? or hast thou seen the treasures of the hail,*
> 23 *Which I have reserved against the time of trouble, Against the day of battle and war?*

Again, *'owtsar*, the Hebrew word translated as "treasures," means "storehouses," and, again, several translations of the Bible use that word. The NRSV, for example, states: "*Have you entered the storehouses of the snow, or have you seen the storehouses of the hail....*" Some might say this is merely poetic language. But this is supposed to be God speaking, and, poetic language or not, one might ask whether God is then deceiving Job concerning the existence of the storehouses. Certainly, the writer of the passage had no intent of showing God as being a deceiver, and God's stern rebuke of Job in this chapter of the Bible calls for hard facts, not unrealistic fictions. Furthermore, God is specifically saying that he has reserved the contents of the storehouses against the time of trouble. That indicates the storehouses for the snow and hail are to be taken as real structures. This chapter of Job may have been written in the form of poetry, but what it expresses must have been taken as reality at the time it was written—a view of reality that is completely different from the modern view.

Joel 2:10 provides another description in the Bible that supports the concept of a solid heaven:

> *The earth shall quake before them; the heavens shall tremble: the sun and the moon shall be dark, and the stars shall withdraw their shining:*

For the heavens to "tremble," they would need to be made of a solid material.

Another indication that the firmament of heaven is made of a physical material can be found in Isaiah 64:1:

> *Oh that thou wouldest rend the heavens, that thou wouldest come down, that the mountains might flow down at thy presence.*

The Hebrew word that is translated as "rend" in that passage is *qara`*, which has meanings that include "rend," "tear," and "to split asunder." The idea is that heaven can be split open to let God come down. The same idea is found in the New Testament. When they described the baptism of Jesus by John the Baptist, the writers of the first three gospels also showed that they believed heaven to be a solid vault over the earth. In Luke 3:21–22, we find that when Jesus was baptized "the heaven was opened" and the "Holy Ghost" descended in a shape like a dove. Again, the idea that heaven could be "opened"—perhaps meaning the opening of one of the previously mentioned doors of heaven—to allow the "Holy Ghost" to come down implies a belief in a solid heaven, or sky. If the writers of the gospels had the same understanding about the nature of the sky as we do, why would they say "the heaven was opened"?

At this point we might note that the pseudepigraphical Book of Enoch, which was written within two centuries before the Christian era, presents additional, and significant, evidence that the ancient Hebrews believed heaven to be a solid structure above the earth. (See Appendix C.)

Now that we have a better understanding of *raqiya`*, the Hebrew word that is translated as "firmament" in the King James Version of the Bible, we have a clearer picture of the creation of the heavens as given in Genesis 1:6–8: The "firmament" of heaven is a solid vault or dome.

The NEB translation of those verses in Genesis supports that understanding:

> 6 *God said, 'let there be a vault between the waters, to separate water from water.'*
> 7 *So God made the vault, and separated the water under the vault from the water above it, and so it was;*
> 8 *and God called the vault heaven. Evening came, and morning came, a second day.*

Some other translations of the Bible also use the words "vault" or "dome" in place of the word "firmament" in the creation account. For example, the NRSV states:

> 6 *And God said, "Let there be a dome in the midst of the waters, and let it separate the waters from the waters."*
> 7 *So God made the dome and separated the waters that were under the dome from the waters that were above the dome. And it was so.*
> 8 *God called the dome Sky. And there was evening and there was morning, the second day.*

8. The Firmament of Heaven

The NJB likewise states:

> 6 *God said, "Let there be a vault through the middle of the waters to divide the waters in two." And so it was.*
> 7 *God made the vault, and it divided the waters under the vault from the waters above the vault.*
> 8 *God called the vault "heaven". Evening came and morning came: the second day.*

In his noteworthy recent translation of the first five books of the Bible, Everett Fox renders verses 6 and 7 of Genesis 1 as follows:

> *6. God said: Let there be a dome amid the waters, and let it separate waters from waters.*
> *7. God made the dome and separated the waters that were below the dome from the waters that were above the dome. It was so.*

A footnote for "dome" in the verses states: "dome: Heb. *raki'a*, literally a beaten sheet of metal."[4]

We should, however, be clear on one thing: Although the Hebrew word *raqiya`*, translated as "firmament" in the KJV, and "vault" or "dome" in the above quotations from other versions of the Bible, does not *specifically* have the meaning of "vault" or "dome," the inherent sense of *raqiya`* is that it is a solid structure, and the form of that structure as the "arch of the sky" would most naturally be that of a "vault" or "dome." To the modern mind, the words "vault" or "dome" provide a better understanding of what the ancient Hebrews considered the nature of the sky to be than does the word "firmament." One needs only to think of a dome on a building, such as the U.S. Capitol building. By standing inside the building, under the vaulted dome, one would get a good idea of what the ancient Hebrews considered the sky to be—a vaulted dome holding back or supporting the waters above the dome. In that sense, translating *raqiya`* as "vault" or "dome" is justified.

To be sure, in an attempt to make the Bible appear to conform to, or at least not contradict, our present understanding of the nature of the sky, some other modern translations of the Bible subvert the real meaning of the Hebrew word *raqiya`*. The New Living Translation, for example, translates verse 6 as "*And God said, 'Let there be space between the waters.'*" The Life Application Bible translates that verse as "*And God said 'Let the vapors separate to form the sky above the ocean below.'*" Such inaccurate translations can only confuse those who are trying to understand what the writers of the Bible actually meant. Worse, they may deceive some people into believing that the biblical cosmos is compatible with the findings of modern science, when in fact it is not.

Chapter 9

The Circle of the Earth

We have so far determined that, according to the Bible, the firmament of heaven is made of a solid material that God had beaten out in the form of a vault to separate the waters above the firmament from those below the firmament. We also noted what the book of Genesis has to say about the formation of the earth under the firmament of heaven on the third day. Here again are the relevant verses in Genesis 1:

> 9 *And God said, Let the waters under the heaven be gathered together unto one place, and let the dry land appear: and it was so.*
> 10 *And God called the dry land Earth; and the gathering together of the waters called he Seas: and God saw that it was good....*
> 13 *And the evening and the morning were the third day.*

The above passage states that the waters were gathered together into one place under the heaven to form the seas so that the dry land, called "Earth," could appear. But what is the shape of the biblical earth?

Isaiah 40:22 provides the answer:

> It is *he that sitteth upon the circle of the earth, and the inhabitants thereof are as grasshoppers; that stretcheth out the heavens as a curtain, and spreadeth them out as a tent to dwell in.*

In this verse, the Hebrew word that is translated as "sitteth" is *yashab*, the meanings of which include "to dwell" and "to sit." The Hebrew word that is translated as "upon" is *al*, the meanings of which include "upon" and "above." The passage could therefore be translated as indicating that God "sits [or dwells] above the circle of the earth." In fact, several ver-

9. The Circle of the Earth

sions of the Bible use the term "above" rather than "upon" in this verse, as for example, the NKJV:

> It is *He who sits above the circle of the earth, And its inhabitants* are *like grasshoppers, Who stretches out the heavens like a curtain, And spreads them out like a tent to dwell in.*

Because this passage mentions "the circle of the earth," Bible believers often refer to it as proof the Bible teaches that the earth is round. They further say that this passage therefore proves that the Bible must be God's word because at the time it was written no one but God could have known that the earth is round.

But that argument is totally without merit.

In the first place, as will be shown in Chapter 13, the ancient Greeks knew that the earth is a sphere, and they arrived at that knowledge through reason and scientific methodology. Therefore, Isaiah 40:22 can hardly be taken as proof that the Bible must be a revelation from God because of its allusion to a "round" earth.

In the second place, there is more than one kind of roundness, and neither that verse nor *any* other verse in the Bible indicates that the earth is a sphere. In the original Hebrew of that verse, the word that is translated as "circle" is *chuwg*, which, in fact, has a primary meaning of "circle"; moreover, none of the lexicons referenced in this book give "sphere" as a meaning for *chuwg*. As the meaning of *chuwg* is significant in our analysis of the biblical cosmos, and as the word occurs in the original Hebrew of several of the passages we will examine, we should take a close look at the definitions for it that are given in the lexicons. Doing so, we find that there are two separate definitions of *chuwg*, depending on its usage as a verb or as a noun. Here are the definitions from the lexicons.

Strong's Concordance provides the following interpretation of *chuwg* as a noun: "circle." It provides the following as a verb: "describe a circle."

KB has the following interpretation of *chuwg* as a noun: "**circle** ... the earth conceived as a disc **Is 40**$_{22,}$ ... the horizon on the sea **Pr 8**$_{27}$; ... the vault of heaven **Jb 22**$_{14}$." It provides the following as a verb: "to describe a circle."

BDB gives the following interpretation of *chuwg* as a noun: "vault; — only of vault of the heavens." It gives the following as a verb: "draw round, make a circle."

From these definitions, it is clear that *chuwg* has a primary meaning of "circle." Note that KB and BDB also equate the usage of *chuwg* as a noun with the "vault of heaven." There is a good reason for this, and we shall see why further on.

But most significantly, these definitions make it clear that the term "circle of the earth," rather than referring to a spherical earth, actually indicates that the biblical earth has the shape of a circle or—in the physical aspect of a circle—a flat disk. In any case, it is not really all that unusual that the author of Isaiah 40:22 would refer to the "circle of the earth" and yet not conceive of the earth as being a sphere. After all, if one stands on the top of a moderately high mountain and turns around while looking to the distance, the horizon will appear to be a great circle at the apparent limits of the earth—a great circle where the rim of the perceived flat disk of the earth appears to meet the vault of the sky.

There are, in fact, numerous passages in the Bible which support the conclusion that the biblical earth is flat. That conclusion is hardly new, of course, and has often been expressed by non-believers and even some believers in the Bible. Unsurprisingly, however, most of those who believe that the Bible is the word of God take umbrage at any suggestion that the Bible indicates the earth is flat. Their position is that, if the Bible is the word of God, it must be believed in all things. Therefore, if there are irrefutable facts, such as the sphericity of the earth, that appear to contradict what is found in the Bible, the Bible must be "interpreted" to conform to those facts—and indeed it must be shown that those facts were to be found in the Bible even before they became generally known to mankind.

The creationist author, Henry M. Morris, for example, had this to say:

> The word "compass" in Proverbs 8:27 and the word "circle" in Isaiah 40:22 are both translations of the same Hebrew *chuwg*, an excellent rendering of which is "circle." It could well be used also for "sphere," since there seems to have been no other ancient Hebrew word with this explicit meaning (a sphere is simply the figure formed by a circle turning about its diameter).[1]

Morris was, of course, doggedly trying to salvage something from those passages in order to make it appear that the Bible is saying that which, in fact, it does not. His statement that *chuwg* could also be used for "sphere" because there seems to have been no other ancient Hebrew word having that meaning is by his own whim and simply reflects his own need for it to have that meaning. In addition, his statement that a sphere is simply a figure formed by a circle turning about its diameter is obfuscatory in relation to the referenced verses and is meant to draw attention away from the plain language of those verses.

Even if *chuwg* could be interpreted to mean "sphere" as well as "circle," it would not necessarily follow that the biblical earth has the shape of a sphere. To determine which interpretation would be correct, one would have to look elsewhere in the Bible for passages that would provide the needed evidence pointing one way or the other. But, again, *chuwg* does *not* have a meaning of "sphere"; it has the meaning of "circle." And again, there are a great many passages in the Bible that indicate the earth is flat and has the shape of a disk, and there is not one single passage in the Bible indicating that the earth has the shape of a sphere.

On top of that, Morris is wrong when he says that ancient Hebrew did not have a word with the meaning of "sphere." In Isaiah 22:18 we find the following, here from the KJV: "*He* [God] *will surely violently turn and toss thee like a ball into a large country; there shalt thou die. . . .*" A ball certainly is a sphere. The Hebrew word translated as "ball" in the passage is *duwr*, the meanings of which include, depending on the context, either "ball" or "circle," unlike *chuwg*, which specifically means "circle." The NIV translation of the passage is more understandable and shows the contextual meaning better: "*He will roll you up tightly like a ball and throw you into a large country. . . .*" The ASV, the NAB, and the NASV have wording that is similar to that of the NIV. In the context of being rolled up tightly, "ball" would be the proper interpretation of *duwr* in the passage.

The Hebrew scholars who translated their scriptures into the Greek Septuagint prior to the Christian era provide further support for the conclusion that the biblical earth is a flat disk. For their translation of Isaiah 40:22, they rendered the Hebrew *chuwg* into the Greek *gyro*, which has the meaning of "circle" or "ring," instead of using *sphaira*, which is the Greek word that specifically means "sphere" or "ball." Since these Hebrew scholars could possibly have been familiar with the suggestion by the Greek scientists that the earth is a sphere (they did their translation in Alexandria, Egypt, at a time when it was heavily influenced by Greek culture), it is significant that they committed themselves to maintaining in their translation their own understanding about the shape of the earth as they derived it from their Hebrew Scriptures. Even if they were not aware of the Greek idea that the earth is a sphere, they could still have used the Greek word for "sphere" in their translation if they had understood that the biblical earth has that shape. The fact that they did not do so is therefore quite telling.

If the biblical earth has the shape of a sphere, why then does the Bible not plainly make statements to that effect? Why does Isaiah 40:22 not say, for example, that God "sits on [or "above"] the ball [*duwr*] of the earth"? That passage does not say any such thing—indeed, no passage anywhere in the Bible says any such thing—because the ancient Hebrews did not conceive of the earth as a sphere. Rather, they believed that the earth has the shape of a flat disk and they incorporated that belief throughout their scriptures.

As noted above, Henry Morris tried to show that Proverbs 8:27 also refers to a spherical earth. We looked at that verse in the last chapter when we were examining what the Bible says about the nature of the firmament, but let's take another look at it to see what it says about the shape of the earth. The KJV wording of that verse states:

> *When he prepared the heavens, I was there: when he set a compass upon the face of the depth.*

Again, the Hebrew word translated as "compass" in this verse is *chuwg*, the same word translated as "circle" in Isaiah 40:22. The Hebrew word translated as "set" is *chaqaq*, which has several meanings, including "to decree," "to set," "to inscribe," "to trace," or "to mark out."

Before we continue, in light of what Proverbs 8:27 states concerning God's setting a "compass" on the face of the depth (or "deep" in some other versions of the Bible), we might note that Isaiah 44:13 provides an interesting piece of information relating to the meaning of *chuwg*:

> *The carpenter stretcheth out his rule; he marketh it out with a line; he fitteth it with planes, and he marketh it out with the compass. . . .*

The word translated as "compass" there is *mechuwgah*, which is derived from *chuwg* and means "circle instrument" (BDB), or "an instrument for marking a circle" (*Strong's Concordance*). Of course, the word "compass" in modern usage also refers to an instrument that is used for marking out a circle.

To understand how Proverbs 8:27 indicates that the biblical earth has the shape of a circle or disk, we need to put it into a more complete context by looking at verses 28 and 29 as well. However, when we examined Proverbs 8:27–28 in the last chapter, we saw that some other versions of the Bible give a more accurate translation of the original Hebrew of those verses than the KJV does. Verse 28, remember, actually indicates that the sky was "made firm," "hard," or "strong," which provides further evidence that the biblical "firmament" is meant to be understood as a solid vault or dome over the earth. In light of that, let's take a look at the NRSV reading of Proverbs 8:26–29 in order to get a more complete understanding of the context:

> 26 *While as yet he had not made the earth, nor the fields, nor the highest part of the dust of the world.*
> 27 *When he established the heavens, I was there, when he drew a circle on the face of the deep,*
> 28 *when he made firm the skies above, when he established the fountains of the deep,*

> 29 *when he assigned to the sea its limit, so that the waters might not transgress his command, when he marked out the foundations of the earth.*

Note that *chuwg* has been translated as "circle," instead of "compass," in this version of the passage, and also that *chaqaq* has been translated as "drew."

As was noted in the last chapter, the speaker in Proverbs 8 is supposed to be Wisdom personified. In Proverbs 8:26–29, above, "Wisdom" is providing part of a description of the creation, which, except for the mention of the circle that God drew on the face of the deep, closely follows Genesis 1:6–10, as the following makes clear.

First, Proverbs 8:26 obviously refers to the time before the creation.

Second, note that verse 27 mentions "the face of the deep." The Hebrew word that is translated as "the deep" there, and also in verse 28, is *tehowm*, the same word translated in Genesis 1:2 as "the deep"—referring to the abyss of the primaeval ocean. This passage thus agrees with that passage of Genesis, which indicates that the creation process began in the midst of a pre-existing great cosmic ocean when "*. . . darkness was upon the face of the deep.*"

Third, as we saw in the last chapter, the first part of verse 28—*when he made firm the skies above*—refers to the creation of the firmament of heaven as it is stated in Genesis 1:6–8: "*And God said, Let there be a firmament in the midst of the waters. . . . And God called the firmament Heaven.*"

Fourth, the latter part of verse 28 and the first part of verse 29 state: "*. . . when he established the fountains of the deep, when he assigned to the sea its limit, so that the waters might not transgress his command, when he marked out the foundations of the earth.*" This part of the passage reiterates Genesis 1:9–10, in which God gathered the waters under the heaven into one place to be the sea and had the dry land of the earth appear. In both Proverbs 8:29 and Genesis 1:10, the Hebrew word translated as "sea" is *yam*, as opposed to *tehowm*, which refers to the pre-existing "deep" in which the heaven and earth was created. In the biblical context, "the sea," *yam*, primarily refers to the seas of the earth, such as the Mediterranean.

Now that we have the context of Proverbs 8:26–29, we can have a better understanding of verse 27. With that understanding, when we now combine verse 27 with verses 28 and 29 we have the following sequence. When the biblical God "established the heavens," he first "drew a circle on the face of the deep." Next, he "made firm the skies above." And finally he "assigned to the sea its limit" and "marked out the foundations of the earth." That sequence of actions indicates that the biblical God drew the circle on the face of the deep in order to provide the starting point for the creation process. Since the biblical God's next act was to make "firm the skies above," the circle must have been used as the base for the "firm," or solid, vault of the skies (i.e., the firmament).

Now, if a vault has a circular base, that base encompasses an area that is, in effect, a disk. The implication of Proverbs 8:27, therefore, is that when God subsequently formed the earth to fit under the vault of heaven, he made it in the shape of a disk. Thus, the circle that God drew on the face of the deep to be the base for the vault of heaven also demarcated the limits of the disk of the earth when it was subsequently created. In other words, the base of the vault of heaven and the rim of the disk of the earth were joined, or bound, together in a common circle—the circle that God initially drew on the face of the deep.

We can now understand that, in the biblical cosmos, the "circle of the earth," which is mentioned in Isaiah 40:22, and the circle (or the base) of the vault of the heaven, which is mentioned in Proverbs 8:27–28, are essentially the same circle. As we have already seen in the description of the vision that is narrated in Ezekiel 1:22–27, the biblical God sits on his throne on the top of the vault of heaven. And because the "circle" or the base of the vault of heaven is the same as, or is joined with, the "circle" of the earth, the God of the Bible is literally—as the verse in the KJV of Isaiah 40:22 indicates—sitting on (or above) the "circle of the earth." The conclusion, then, is that—since the circular base of the vault of heaven encompasses it—the "circle of the earth" is a flat disk. Furthermore, the common circle that binds together the disk of the biblical earth and the base of the vault of heaven appears as the far horizon when viewed from a mountaintop.

These conclusions are corroborated by what the KB lexicon states concerning the meaning of *chuwg*: ". . . **circle** . . . the earth conceived as a disc Is 40_{22}, . . . the horizon on the sea Pr 8_{27}; . . . the vault of heaven Jb 22_{14}. . . ." In that entry, however, the lexicon does not account for the fact that, as we found above, Proverbs 8:27 is actually describing part of the creation, and that the "deep" (i.e., *tehowm*, the primaeval ocean, in the original Hebrew) is what is being referred to in the passage and not simply the "sea" (*yam*). As Proverbs 8:27 actually indicates, God drew the "circle" on "the face of the deep"—which the lexicon entry calls, imprecisely, "the sea"—*before* he created the firmament of heaven. For that matter, as we noted, God drew the circle on the deep *before* he gathered the waters under the firmament of heaven into one place to form the "seas" (*yamim*, i.e., the gathered-together "seas" named as such in Genesis 1:10) and had the dry land of the earth appear. Since the "sea" was not formed until verse 29, the lexicon is erroneously applying the word "sea" to verse 27.

Thus, contrary to what the lexicon implies concerning Proverbs 8:27, the God of the Bible inscribed the circle on the face of the deep before he formed the sea and the earth, and before the earth's horizon came into being. After the firmament of heaven was created and the dry land was formed, the "circle" of the far horizon could appear to be—in part, anyway—on the land as well as on the sea. In fact, since Genesis 1:9 states that the waters under the heaven were gathered together into "one place" so the dry land could appear, it was probably considered that the dry land enclosed the gathered-together waters, which God called the seas. Thus, if one sailed far enough on the sea, one would come to the "circle of the earth"—that is, a ring of dry land—that was located along the rim of the disk of the earth where it meets up with the vault of the firmament of heaven. This ring, or circle, of dry land, along with the habitable land area, would—in the biblical cosmos—have enclosed the gathered-together waters beneath the firmament of heaven.

That brings us back to the three separate meanings for *chuwg*, as "circle," in the verses referenced in the KB lexicon. It should be clear by now that all three meanings—the circle of the earth (Isaiah 40:22), the far horizon (Proverbs 8:27), and (the base of) the vault of heaven (Job 22:14)—in fact refer to the same circle.

But that is not all. Job 26:10 is another passage in the Bible that mentions the circle that God drew on the face of the primaeval cosmic ocean, and it provides some additional relevant information. Here is the KJV rendering:

> *He hath compassed the waters with bounds, until the day and night come to an end.*

Again, the original Hebrew word that is translated as "compassed" is *chuwg*, which, as we have seen so many times, has the meaning of "circle." The KJV rendering of Job 26:10, however, does not completely reflect the wording of the original Hebrew. That incompleteness disguises the fact that this verse, insofar as the circle is concerned, is actually very similar to Proverbs 8:27: "*When he established the heavens, I was there, when he drew a circle on the face of the deep.*" The ESV rendering of Job 26:10 shows better the similarity (the ASV, NRSV, JPS 1917, NIV, NKJV, and RSV also have wording that is similar to that of the ESV):

> *He has inscribed a circle on the face of the waters at the boundary between light and darkness.*

An analysis of the original Hebrew wording shows how the ESV rendering of the verse is more complete than is that of the KJV, and that it provides a clearer picture of what was in the mind of the author. The Hebrew word that is translated as "inscribed" in the ESV version of Job 26:10 is *choq*, which is a form of the word *chaqaq*—translated as "drew" in Proverbs 8:27—and has the meanings of "something prescribed," "ordained," and "decree." The Hebrew word translated as "face" (which the KJV omits) is *paniym*, which, in fact, means "face." The Hebrew word translated as "waters" is *mayim*, which is frequently translated as "the seas" in the Bible, but which is also frequently applied to the waters of the cosmic ocean. The Hebrew word translated as "boundary" is *taklith*, which has the meanings of "boundary," "end," and "limit." The Hebrew word that is translated as "day" in the KJV and "light" in the ESV is *'owr*, which, as we previously saw, means "light." And the Hebrew word translated as "night" in the KJV and "darkness" in the ESV is *choshek*, which, as we also saw, means "darkness."

Job 26:10, like Proverbs 8:26–29, is also referring to a part of the creation story and adds further information. In Genesis 1:3–4, remember, God's first act of creation was to bring forth light, which he subsequently "divided" from the darkness:

> *And God said, Let there be light. . . . And . . . God divided the light from the darkness.*

It would appear from Job 26:10 that God divided the light from the darkness by inscribing a circle on the "face of the waters" to be a boundary between them. This has a counterpart in Proverbs 8:27 where God is also specifically described as inscribing a circle, the same circle, on the face of the waters. The waters would therefore refer to the waters of the primaeval ocean of Genesis 1:2: "*. . . and the spirit of God moved upon the face of the waters.*"

Taken together, Job 26:10 and Proverbs 8:27–29 therefore retell with a slightly different slant the creation described in Genesis 1:1–10. These passages also appear to indicate that some of the writers of the Bible placed a special significance on the circle that is manifested by the far horizon—in their minds, a cosmic circle. It is perhaps not surprising that they did so, for, in the biblical cosmos, the far horizon defines the limits of both the earth and the sky, and it encompasses everything that the ancient Hebrews could perceive. The far horizon also brings the first light in the morning and retains for a time the last vestige of daylight before the arrival of the darkness of night. Thus the writers of these passages, in describing their view of creation, had their God begin the creation process by inscribing, or setting, the cosmic circle on the waters of the deep—that is, the primaeval ocean.

9. The Circle of the Earth

With that, we can now more fully understand the place that this circle had in the biblical creation process. To sum up, Job 26:10 and Proverbs 8:27–29, taken together, indicate that the inscribed circle had a three-fold purpose during the creation process. On the first day of creation, the circle provided the means by which God divided the light from the darkness. On the second day of creation, it provided the base for the firmament—i.e., the vault—of heaven when it was created. And, on the third day of creation, it marked out the extent of the foundations of the earth and limited the area of the seas under the firmament of heaven. Thus, the circle that the biblical God initially inscribed on the face of the deep ultimately tied together the main cosmological objects of creation and became the circle of the far horizon.

(It should be noted that some biblical apologists say that the circle referred to in Job 26:10 is the circle that demarcates the night and day sides of the spherical earth—that is, the earth we know of today. However, that circle invariably crosses both land and sea; at no time does it encircle only the sea as did the circle mentioned in Job 26:10. It should therefore be clear that the circle mentioned in Job 26:10 would have been drawn when only the abyss of the deep existed and while the earth was still "without form" before its creation. The circle of Job 26:10 would thus be the same circle that is referred to in Proverbs 8:27, the circle that provided the base for the vault of heaven, and the same circle that defined the limits of the earth when it was subsequently created.)

Job 22:14, as indicated in the entry in the KB lexicon that was noted above, also mentions a circle that is relevant to the discussion at hand. Here is the passage in the KJV:

> *Thick clouds are a covering to him, that he seeth not; and he walketh in the circuit of heaven.*

The passage states: "*. . . and he walketh in the circuit of heaven.*" The original Hebrew word that has been translated as "circuit" is, again, *chuwg*, the same word translated as "circle" in Isaiah 40:22, Proverbs 8:27, and Job 26:10. The "circuit of heaven" would therefore be literally translated as the "circle of heaven," as it is in the NKJV:

> *Thick clouds cover Him, so that He cannot see, And He walks above the circle of heaven.*

The passage could be taken as meaning that God walks around the circular rim of the base of the vault of heaven. The NJB version of the passage, in fact, translates *chuwg* as the "rim"—meaning the base—of the vault of the heavens:

> *The clouds, to him, are an impenetrable veil, as he goes his way on the rim of the heavens.*

However, other versions of the Bible translate *chuwg* in this passage as "vault," as, for example, in the ASV:

> *Thick clouds are a covering to him, so that he seeth not; And he walketh on the vault of heaven.*

The ESV, the NAB, and the NASV also translate *chuwg* as "vault" in this passage, and the NRSV translates it as "dome." These translations would indicate that the ancient Hebrews expanded the meaning of *chuwg* to apply by association, not only to the circular base of the vault of heaven, but also to the vault of heaven itself. More specifically, it would be reasonable to conclude that the ancient Hebrews simply used the term "circle of heaven" to mean all of that which was encompassed by the heavenly part of the "circle" (i.e., the far horizon), which, of course, would be the whole heaven. Similarly, they would have used the term "circle of the earth" to mean all of that which was encompassed by the earthly part of the "circle" (again, the far horizon), which of course would be the whole earth in its form as a disk. Still, the words "vault" and "dome" provide us with a suitable understanding of what would have to be the necessary nature of the "circle of heaven" in the biblical cosmos.

We have seen that there are passages in the Bible which indicate that God physically made the firmament of heaven. Interestingly, there are also similar passages which indicate that God physically made the earth, and these passages also provide information on the shape of the earth in the biblical cosmos. In Isaiah 44:24, for example, we find:

> *Thus saith the LORD, . . . I am the LORD that maketh all things; that stretcheth forth the heavens alone; that spreadeth abroad the earth by myself;*

In that passage, God says he "maketh all things," and he specifically refers to the heavens and the earth. The original Hebrew word that is translated as "spreadeth" is *raqa`*, which as we have seen, means "to beat out," "to *pound* the earth; . . . to *expand* (by hammering)." That same Hebrew word was used, as we saw, in Job 37:18 to describe God's having beat out the vault of heaven like a mirror of cast metal. The NJB rendition of Isaiah 44:24 reflects that aspect of the original Hebrew word with respect to the earth:

> *Thus says Yahweh, . . . I, Yahweh, have made all things, I alone spread out the heavens. When I hammered the earth into shape, who was with me?*

The earth's being hammered into shape, or being "spreadeth abroad," indicates that God beat out the earth flat.

9. The Circle of the Earth

Statements in the Bible that refer to the extent of the surface of the earth lend further support to what we have so far determined about the shape of the biblical earth. When we, today, want to make a statement that applies the object of discussion to the whole surface of the earth, we will say "all around the earth," "all around the world," or "all around the globe." When we make such statements, we are acknowledging the fact that the earth has the shape of a sphere. When we look at the Bible, however, we can find no such expressions. Instead, we find corresponding expressions that conform to the belief that the earth is flat. Deuteronomy 28:64 is one such passage, for it mentions the ends of the earth—meaning the rim of the disk of the earth:

> *And the LORD shall scatter thee among all people, from one end of the earth even unto the other; and there thou shalt serve other gods. . . .*

Also, in Deuteronomy 13:7 we find

> *Namely, of the gods of the people which are round about you, nigh unto thee, or far off from thee, from the one end of the earth even unto the other end of the earth.*

Another example can be found in Jeremiah 25:32–33:

> *32 Thus saith the LORD of hosts, Behold, evil shall go forth from nation to nation, and a great whirlwind shall be raised up from the coasts of the earth.*
> *33 And the slain of the LORD shall be at that day from one end of the earth even unto the other end of the earth: they shall not be lamented, neither gathered, nor buried; they shall be dung upon the ground.*

The Hebrew word that is translated as "end" in both of these passages is *qatseh*, which has a meaning of "end" or "extremity." The Hebrew word translated as "coasts" in Jeremiah 25:32 is *yerekah*, which has the meanings of "flank," "side," and "extreme parts." The NJB renders that verse as "*. . . a mighty tempest is rising from the far ends of the earth.*" Certainly, none of these passages can be applied to a spherical earth.

Some biblical apologists have said that the term "the ends of the earth" refers only to the edges or shoreline of the continent that the peoples of the Bible lived on. Others have said that the term "the ends of the earth" is merely a figure of speech. These views, however, do not find support in biblical usage. For example, we find the following in Isaiah 40:28:

> *Hast thou not known? hast thou not heard, that the everlasting God, the LORD, the Creator of the ends of the earth, fainteth not, neither is weary?. . .*

Is that passage positing that the God of the Bible created a mere figure of speech? Or is it instead declaring that the God of the Bible created something real? Is it saying that God created only the continent on which the peoples of the Bible lived? Or is it saying that God created the whole earth? The Hebrew word that is translated as "ends" in this passage is *qatsah*, which is the feminine form of *qatseh*. Concerning *qatsah*, *Strong's Concordance* makes this statement: "This word refers primarily to concrete objects." It should be clear from this passage that the term "the ends of the earth" was used to mean the whole earth within its farthest limits. That term is therefore contextually similar to the "circle of the earth," which likewise was used to mean the whole earth within its farthest limits. Thus Isaiah 40:28 is declaring that God created the whole earth all the way to its ends—i.e., the rim of the disk of the earth—and not just a part of the earth.

Isaiah 45:22 provides another example of the usage of the term "ends of the earth" as meaning the whole earth:

> *Look unto me, and be ye saved, all the ends of the earth: for I am God, and there is none else.*

There the biblical God is commanding the people of the whole earth, as encompassed by its ends, to look to him and be saved.

Psalm 98:3 also uses the same language:

> *He hath remembered his mercy and his truth toward the house of Israel: all the ends of the earth have seen the salvation of our God.*

There, "all the ends of the earth" would necessarily mean that the whole earth—a flat earth—has seen the salvation of God.

1 Samuel 2:10 provides yet another example:

> *The adversaries of the LORD shall be broken to pieces; out of heaven shall he thunder upon them: the LORD shall judge the ends of the earth. . . .*

There the biblical God is saying that he shall judge the whole earth as encompassed by its ends.

And Psalm 67:7:

> *God shall bless us; and all the ends of the earth shall fear him.*

There, all the earth, as encompassed by its ends, shall fear the God of the Bible.

9. The Circle of the Earth

In Psalm 2:8, we find God saying (here, the NIV):

Ask of me, and I will make the nations your inheritance, the ends of the earth your possession.

There, God is telling David, the anointed king of Israel, that he has but to ask and he would give him all of the earth, all the way to its ends, to rule over.

In the New Testament, the same language is found—here, Acts 13:47:

For so hath the Lord commanded us, saying, I have set thee to be a light of the Gentiles, that thou shouldest be for salvation unto the ends of the earth.

It is clear from that passage that salvation shall be to the whole earth as encompassed by its ends.

If, instead of "the ends of the earth," the term "the whole earth" were used in these passages it would at least have left open the question of the shape of the biblical earth. The term "the whole earth" is, in fact, used in several locations elsewhere in the Bible, so it is curious that the verses above, which *do* provide an indication of the actual shape of the biblical earth, use a term that puts forth the idea that that shape is flat. Because the term "ends of the earth" is used in place of "the whole earth" in these and in numerous other passages, it should be clear that the biblical earth does have ends and is therefore flat.

The Hebrew word that is translated as "earth" in all of the above passages is *'erets*, which does have many meanings, including "land," "earth" (dirt), "earth" (as opposed to heaven), "world," "country," "territory," and "piece of ground." In specific contexts it can mean the whole area, including the seas, under the biblical heaven or firmament. We have already seen a very clear example of this in Isaiah 40:22, which states: "*It is he that sitteth upon the circle of the earth.*" In the context of that passage, the whole area of the earth, including its seas, extends all the way to where the "circle" of the earth and the "circle" of the base of the vault of heaven are bound together. What the meanings of the biblical *'erets* do *not* include is that of an earth conceived of as a spherical planet. Such a concept would have been totally alien to the writers of the Bible.

In Job 28:23–24, we find that "*God . . . looketh to the ends of the earth, and seeth under the whole heaven.*" Again, the Hebrew word that is translated as "ends" in this passage is *qatsah*. The passage is in the form of biblical poetic reiteration. An expression in the form of poetic reiteration consists of parallel parts that mean essentially the same thing. In this case, God looks to the ends of the earth and sees under the whole heaven. If

God sees everything under the whole heaven all the way to the ends of the earth, that would have to include the seas of the earth. Therefore the ends of the earth would have to enclose and include the seas. It certainly does not mean that God can see only to the shoreline of the land the biblical peoples lived on.

Job 37:2–3 likewise mentions the ends of the earth:

> 2 *Hear, oh, hear the noise of his voice, And the sound that goeth out of his mouth.*
> 3 *He directeth it under the whole heaven, and his lightning unto the ends of the earth.*

The Hebrew word that is translated as "ends" here is *kanaph*, another word that has a meaning of "end" or "extremity." Does that passage mean that the God's "lightning" goes only to the end of the land the peoples of the Bible lived on and stops at the sea? Or does it mean that it extends to the horizon at the "ends" of the whole earth, including the sea?

In Psalm 135:7 we find:

> *He causeth the vapours to ascend from the ends of the earth; he maketh lightnings for the rain; he bringeth the wind out of his treasuries.*

The Hebrew word translated as "vapours" is *nasiy'im*, which as well as "vapors" also means "clouds," and the Hebrew word translated as "treasuries" is *otsar*, which also means "storehouse." The NIV accordingly renders the verse as follows:

> *He makes clouds rise from the ends of the earth; he sends lightning with the rain and brings out the wind from his storehouses.*

Jeremiah 10:13 makes a similar statement. Here again is the NIV rendering:

> *When he thunders, the waters in the heavens roar; he makes clouds rise from the ends of the earth. He sends lightning with the rain and brings out the wind from his storehouses.*

In Chapter 9 we noted other references to the "storehouses" in the firmament of heaven in which the snow and hail were stored for their eventual use. These verses add the storehouses of the winds to the mix. In these verses we also see added references to the "ends of the earth" from which the clouds rise. In declaring that God makes the clouds rise from the ends of the earth, these verses give substance to the reality of the ends of the earth in the biblical cosmos. This also shows that the biblical cosmos is different from the actual cosmos, for the clouds do not come from "ends of the earth" in the actual cosmos. The clouds are actually formed as a result of evaporation of water from lakes, streams, oceans, plants, animals, and the ground; in the upper atmosphere, the evaporated water eventually condenses to form clouds. The ancient Hebrews, of course, knew nothing of this process. To them, the wind-blown clouds would appear to come from the far horizon—the ends of the earth—where their God kept them until he caused them to rise and move over the earth.

In Psalm 48:10 we find:

According to thy name, O God, so is thy praise unto the ends of the earth: thy right hand is full of righteousness.

Would the praise of God stop at the land's edge by the sea? Or would it encompass the whole of the earth all the way to its ends, as defined by its rim, and including the seas?

In Romans 10:18 we find the following:

. . . Yes verily, their sound went into all the earth, and their words unto the ends of the world.

The Greek words translated as "earth" and "world" in this passage are *ge* and *oikoumene*, respectively. Both Greek words have a range of meanings all the way from a limited area of land to the whole earth or whole world. The Greek word translated as "ends" is *peras*, which has a meaning of "extremity" or "end." Again, the context of the passage would need to refer to the whole earth all the way to its ends.

As previously noted, in Job 38 the biblical God rebukes Job by sternly questioning him about his knowledge, or lack thereof, concerning various aspects of the creation and the cosmos. With each question he compares Job's lowly human abilities with his own divine creative and directive powers. In Job 38:12–13, for example, God asks (here from the KJV):

12 Hast thou commanded the morning since thy days; and caused the dayspring to know his place;
13 That it might take hold of the ends of the earth, that the wicked might be shaken out of it?

The NAB is a little clearer in showing the meaning of "since thy days":

Have you ever in your lifetime commanded the morning and shown the dawn its place

The original Hebrew word that translated as "dayspring" in verse 12 of the KJV passage is *shachar*, which means "dawn" (specifically the first light of the morning), which is the word used in the NAB version. Several other versions of the Bible also use that word in this passage.

All of the versions of the Bible referenced in this book attribute the shaking of the earth to the dawn (or the equivalent term). However, the words "that it might" in verse 13 of the KJV, or the equivalent words in the other versions of the Bible, do not appear in the original Hebrew of the passage, but are interpolated. The verse, as it is translated, also contradicts the whole sense of Job 38, which is that it is God himself who does everything described therein. For that matter, as we shall see, there are several passages elsewhere in the Bible indicating that it is God who shakes the earth. Therefore, the original intent of verse 13 would have been for it to be a continuation of the "Hast thou" of verse 12, as in the following revision:

> 12 Hast thou commanded the morning since thy days; and
> caused the dayspring to know his place;
> 13 And [hast thou] taken hold of the ends of the earth, that
> the wicked might be shaken out of it?

Thus, in asking Job if he has done these things, God is indicating that he himself does them. That revised reading of the passage certainly appears to more correctly retain the sense of what the intent of the chapter is about. The alternative to that revision, of course, is that one would have to accept that, in the biblical cosmos, the dawn actually can take hold of the ends of the earth and shake it.

Nevertheless, regardless of whether verse 13 indicates that it is God who takes hold of the ends of the earth to shake it, or that it is the dawn, the passage indicates that the biblical earth is flat, for a spherical earth does not have "ends" to take hold of.

These are but a small sampling of the scores of references to the "ends of the earth" that can be found in the Bible. If the biblical earth has the shape of a sphere as Bible believers aver, why are there not at least some passages in which the terms "all around the earth" or "from the other side of the earth" can be found?

From those and many other passages, it is clear that the writers of the Bible plainly believed that the "ends of the earth" had a physical reality. The implication of that belief is that a flat, disk-shaped earth is a part of the biblical view of the cosmos.

Since, in the biblical cosmos, heaven is a solid vault that has its circular base placed on the rim of, or on the "ends" of, the disk of the earth, the base of the vault of heaven also prescribes the farthest limits of the surface of the earth. Moreover, since the rim of the disk of the earth is called the "end of," or "the ends of," the earth, the circular base of the vault of

9. The Circle of the Earth

heaven could likewise be called the "end of," or "the ends of," heaven. Accepting this, we should therefore not be surprised to find biblical passages reflecting that concept. And, in fact, we do.

In Psalm 19:1–6, for example, we find:

> 1 *The heavens declare the glory of God; and the firmament sheweth his handywork.* . . .
> 4 *Their line is gone out through all the earth, and their words to the end of the world. In them he set a tabernacle for the sun,*
> 5 *Which is as a bridegroom coming out of his chamber, and rejoiceth as a strong man to run a race.*
> 6 *His going forth is from the end of the heaven, and his circuit unto the ends of it: and there is nothing hid from the heat thereof.*

The first verse is in the form of a biblical parallelism, for the second part of the verse has an expression that is similar to what is described in the first part. The Hebrew word translated as "firmament" in the verse is, again, *raqiya`*. The Hebrew word translated as "line" in verse four is *qav*, which has the meaning of a "measuring cord" or "measuring line." Verse four is therefore indicating that the heavens measure, or encompass, the whole surface of the earth. That makes sense if the heavens are viewed as a vault, the base of which is setting on the rim of the earth's disk, and the line is stretched from one side of the base to the other. Verse four continues by indicating that the declarations of the glory of God by the heavens (from the first verse) go out to the end (*qatseh*) of the world. Verse four completes by stating that God set in the heavens a tabernacle for the sun. As verses five and six describe the beginning of the sun's daily journey, the "tabernacle" would have to be at the earth's eastern horizon. Verse six states that the "going forth," or rising, of the sun is "from the end [*qatseh*] of the heaven," which would be the eastern "end," or the base, of the vault of heaven where it rests on the rim of the disk of the earth. The "circuit" of the sun then brings it to the western "ends" (*qatsah*) of the vault of heaven.

In view of this passage, the term "end of the heaven" therefore explicitly means the end of the vault of the firmament of heaven—meaning its base—where it rests upon the rim of the disk of the earth and forms the horizon. The conclusion is that the terms "end of the heaven" and "end of the earth," as they relate to the biblical cosmos, mean essentially the same thing and are represented by the far horizon. Given that, we should not be surprised to find those terms used interchangeably in the Bible. And, in fact, we do find such usage. One example is found in Deuteronomy 30:3–4. Moses had told the Israelites that, if they ever were to become unfaithful, God would curse them and drive them to another land; but if they subsequently turned back to God

> 3 *That then the LORD thy God will turn thy captivity, and have compassion upon thee, and will return and gather thee from all the nations, whither the LORD thy God hath scattered thee.*
> 4 *If any of thine be driven out into the outmost parts of heaven, from thence will the LORD thy God gather thee, and from thence will he fetch thee.*

Here, Moses is saying that, if the Israelites were driven to the "outmost parts of heaven," God would gather them from that place. The Hebrew word that is translated as "outmost" is qatseh, which, as we saw previously, has the meaning of "end" or "extremity" (and is the same original Hebrew word used in the term "end of the earth"). The meaning of the passage is clear: It is saying that the Israelites would be gathered from the farthest extremities of the earth, even from as far as where the ends of (i.e., the base) of the vault of heaven rest on the rim of the disk of the earth. Verse 4 would therefore be properly translated as "If any of thine be driven out to the ends of heaven, from thence will the LORD thy God gather thee. . . ." The Darby version of the Bible in fact reflects that interpretation:

> *Though there were of you driven out unto the end of the heavens, from thence will Jehovah thy God gather thee, and from thence will he fetch thee;*

The NJB also reflects that interpretation:

> *Should you have been banished to the very sky's end, Yahweh your God will gather you again even from there, will come there to reclaim you.*

Following the KJV wording, some versions of the Bible use either "the outmost parts of heaven" or "uttermost parts of heaven" instead of "the ends of heaven" in the passage. In contrast, some other versions of the Bible—e.g., the NASV, the NRSV, and the JPS Tanakh—translate the term in question as "the ends of the earth" or "the ends of the world," or something similar, rather than "ends of heaven." Here, for example, is the NASV rendering:

> *If your outcasts are at the ends of the earth, from there the LORD your God will gather you, and from there He will bring you back.*

However, the original Hebrew word that these versions translate as "earth" (or "world" in some versions) is *shamayim*, which, in fact, means "heavens." The translators of these versions of the Bible apparently thought that the reader would be confused by the usage of the term "ends of heaven" with respect to the limits of the surface of the earth, so they reworded the passage in order to prevent such confusion. Nevertheless, in making the substitution, these translations show that the terms "ends of the earth" and "ends of heaven" are interchangeable in meaning.

Another reference to the "end of heaven" is found in Nehemiah 1:9:

But if ye turn to me, and keep my commandments, and do them; though there were of you cast out unto the uttermost part of heaven, yet will I gather them from hence, and will bring them unto the place that I have chosen to set my name there.

The Hebrew word that is translated as "uttermost" in this passage is, again, *qatseh*, meaning "end," and that is how the NJB interprets this passage: "*. . . even though those who have been banished are at the very sky's end.*" Again, the passage refers to where the base of the vault of heaven meets the rim of the disk of the earth, which would be the farthest extent of the earth. As with Deuteronomy 30:4, several versions of the Bible reflect the KJV wording by using either "the outmost parts" or "uttermost parts" instead of "the ends of" in the passage. And again, some other versions of the Bible translate the term in question as "the ends of the earth" or "the ends of the world," or something similar, rather than "ends of heaven." However, the original Hebrew word that these versions translate as "earth" or "world" in Nehemiah 1:9 is *shamayim*—meaning "heavens"—just as it is in Deuteronomy 30:4.

In Deuteronomy 4:32, Moses is talking to the people of Israel and provides a particularly good description showing that the biblical earth is flat:

For ask now of the days that are past, which were before thee, since the day that God created man upon the earth, and *ask from the one side of heaven unto the other, whether there hath been* any such thing *as this great thing* is, *or hath been heard like it?*

Here, the extent of the earth is from one "side" (*qatseh* again) of heaven to the other side of heaven—that is, from one side of the vault of heaven where it rests on one side of the earth's disk to the other side of the vault of heaven where it rests on the other side of the earth's disk. In this case, several versions of the Bible do use "end of heaven" rather than "side of heaven" in this passage. Here, for example, is the NRSV:

> *For ask now about former ages, long before your own, ever since the day that God created human beings on the earth; ask from one end of heaven to the other: has anything so great as this ever happened or has its like ever been heard of?*

So in this passage Moses is saying that the people of Israel should ask the question over the whole surface of the earth; that is, from one end of the vault of heaven where it rests on one end of the flat earth all the way to the other end of the vault of heaven where it rests on the other end of the earth.

Yet another example is found in Isaiah 13:4–5, in which the term "end of heaven" describes the extent of the surface of the earth from which God would gather the nations to destroy Babylon:

> *4 The noise of a multitude in the mountains, like as of a great people; a tumultuous noise of the kingdoms of nations gathered together: the LORD of hosts mustereth the host of the battle.*
> *5 They come from a far country, from the end of heaven, even the LORD, and the weapons of his indignation, to destroy the whole land.*

There, the host of the battle comes from a country so far away that it is situated by the rim of the earth upon which the base—i.e., the end—of the vault of heaven rests. If the biblical view of the earth is that it is a sphere, why does this passage not say that God would gather the host from the other side of the earth, or from all the way around the earth?

All of these passages are meaningless in the context of the modern view of the cosmos, but they make perfect sense in the biblical view, in which the base of the vault of heaven is resting on the rim of the disk of the earth. These passages certainly cannot be applied to a spherical earth.

There are still other passages in the Bible that use the term "ends of heaven." Jeremiah 49:36 in the KJV, for example, states:

> *And upon Elam will I bring the four winds from the four quarters of heaven, and will scatter them toward all those winds; and there shall be no nation whither the outcasts of Elam shall not come.*

The Hebrew word translated as "quarters" in that passage is *qatsah*, which, again, actually has the meanings of "end" or "extremity." And, in fact, the lexicons do not give "quarter" as a meaning for *qatsah*. Therefore, the proper translation should be the "four ends of heaven," rather than the "four quarters of heaven." The "four ends of heaven" would, of course, refer to the base of the vault of heaven at the four cardinal directions. The NASV, the NAB, and the Darby, in fact, use the term "four ends of heaven" in their translations of the passage. Here is the NASV rendering:

And I shall bring upon Elam the four winds From the four ends of heaven, And shall scatter them to all these winds; And there will be no nation To which the outcasts of Elam will not go.

But perhaps the most significant use of the term "end of heaven," as meaning the farthest limit to the surface of the earth, is found in Matthew 24:31, where we find Jesus saying the following:

And he shall send his angels with a great sound of a trumpet, and they shall gather together his elect from the four winds, from one end of heaven to the other.

This passage refers to the "elect"—i.e., those righteous people on the earth who would be gathered together in the Last Days. Here, Jesus is saying that the elect are to be gathered "from the four winds, from one end of heaven to the other." The Greek word translated as "end" in this passage is *akron*, which, according to *Strong's Concordance* has the meaning of "the *extremity*." BDAG provides the meanings as "*Extreme limit, end.*"

Again, that passage makes no sense in relation to the modern view of the cosmos, but it makes perfect sense in relation to the biblical view. Matthew 24:31 thus indicates that the elect would be gathered from the whole breadth of the flat, disk-shaped earth and from the four cardinal directions (which, again, is what the term "four winds" means in the biblical context): from the east, even from the farthest point of the east where the rim of the disk-shaped earth meets the base, or the end, of the vault of heaven, and likewise from the west, the north, and the south.

Consequently, that passage also indicates that the biblical Jesus, who is supposed to be God incarnate, believed that the earth is flat and that heaven is a solid vault resting on the rim of the flat earth.

There are yet other passages in the Bible showing that the rim of the disk of the biblical earth joins the base of the vault of heaven. In Isaiah 11:12, for example, we find:

He ... will assemble the outcasts of Israel, and gather together the dispersed of Judah from the four corners of the earth.

The original Hebrew word for "corner" in that passage is *kanaph*, which, as we previously saw, means "edge" and "extremity," but it also has the meaning of "corner." Again, the four corners of the earth would be at the edge or the rim of the flat, disk-shaped earth where it meets the solid vault of heaven at the four cardinal directions.

Revelation 7:1 also mentions the four corners of the earth:

> *And . . . I saw four angels standing on the four corners of the earth, holding up the four winds of the earth, that the wind should not blow on the earth, nor on the sea, nor on any tree.*

The Greek word that is translated as "on" in the phrase "on the four corners of the earth" is *epi*, which means "upon," "on," "at," or "by." The Greek word translated as "corners" is *gonias*, which indeed has a meaning of "corners." Given what we have learned about the relationship between the vault of heaven and the disk of the earth, the phrase would be more properly translated as "*at* the four corners of the earth," or "*by* the four corners of the earth," rather than "*on* the four corners of the earth." The ASV, ESV, NAB, NIV, NJB, and NRSV, in fact, all use "at" rather than "on" in their translations of the passage. For example, the NRSV renders the passage as, "*After this I saw four angels standing at the four corners of the earth, holding back the four winds of the earth. . . .*" A spherical earth does not have "four corners" at which to stand, so this passage can therefore make sense only for the biblical view of the cosmos. According to Revelation 7:1, the angels would be standing by the rim of the earth's disk where it meets the base of the vault of heaven at the four cardinal directions. (Incidentally, as shown in Appendix C, the Book of Enoch sheds light on why the angels of Revelation 7:1 were standing at the four corners of the earth to hold up the winds.)

Revelation 20:7–8 in the KJV states that Satan would "*deceive the nations which are in the four quarters of the earth.*" The Greek word that is translated as "quarters" in this passage is *gonias*, the same word translated as "corners" in Revelation 7:1. The lexicons, in fact, do *not* give "quarters" as a meaning for *gonias*, and all of the previously mentioned versions of the Bible that used "at" in Revelation 7:1 also use "corners" instead of "quarters" in this passage. The sense of the passage is that Satan would deceive the all of the nations of the world, even those at the extreme limits of the disk of the earth at the four cardinal directions.

We can see from the foregoing passages that the phrases "the four ends of heaven" and "the four corners of heaven" refer to the base of the vault of heaven at the four cardinal directions. The phrases "the four ends of the earth" and "the four corners of the earth" would likewise refer to the rim of the disk of the earth at the four cardinal directions. Since, in the biblical cosmos, the base of the vault of heaven rests on the rim of the disk of the earth, the "four ends of heaven" would match up with and be synonymous with the "four ends of the earth." These passages therefore provide further evidence for our understanding of the nature of the biblical cosmos.

9. The Circle of the Earth

From the textual evidence, it should be clear that, in the biblical cosmos, the solid vault of heaven rests on the rim of the disk of the earth. Amos 9:6, in fact, supports that construct. First, let's take a look at the KJV rendering of the passage:

> *It is he that buildeth his stories in the heaven, and hath founded his troop in the earth; he that calleth for the waters of the sea, and poureth them out upon the face of the earth: The LORD is his name.*

The Hebrew word that is translated as "stories" in the passage is *ma`alah*, which has several meanings, including "stairs," and "stories of heaven" (meaning floors of heaven). The original Hebrew word translated as "troop" in this passage is *'aguddah*, which has an underlying meaning of "binding together." *Strong's Concordance* defines *'aguddah* as "a *band, bundle, knot,* or *arch.*" BDB provides the following meanings for *'aguddah*: "1. ... *bands, thongs.* ... 2. ... *bunch of hyssop.* ... 3. *band of men.* ... 4. *vault* of the heavens (as fitted together, constructed...) Am9^6." KB provides this definition: "... 4. Vaults of the heavens Am 9_6." Note that the last two mentioned lexicons provide "vault of the heavens" as a definition for *'aguddah*, and they specifically refer to Amos 9:6 as an example of the usage. Moreover, as we saw in the last chapter, *Strong's Concordance* defines *raqiya`*, the original Hebrew word for "firmament," as "an *expanse,* i.e., the firmament or (apparently) visible arch of the sky." Thus, the "arch" in the definition of *'aguddah* in *Strong's Concordance* would hearken back to *Strong's* "visible arch of the sky" in its definition of *raqiya`* and would equate with the "vault of heaven" in the other definitions.

So the question is whether *'aguddah* of Amos 9:6 is more correctly translated as "troop" or as "arch" or "vault." Significantly, in Job 19:12 there is another reference to the "troops" of God: *"His troops come together, raise up their way against me, and encamp round about my tabernacle."* However, the Hebrew word translated as "troops" (appropriately, according to the context) in that passage is not *'aguddah*, but rather *geduwd*. That Hebrew word specifically does mean "troop" or "band of men" and is used with those meanings in more than thirty other locations in the Bible. One might therefore ask why it was not used in Amos 9:6 if that passage does, in fact, refer to the "troops" of God. Moreover, translating *'aguddah* as "troop" in Amos 9:6 leaves the purpose of the troop unanswered. What is the troop to be used for? In Job 19:12, quoted above, the purpose of the troop is made clear; it is not clear in Amos 9:6.

As this verse from Amos is cosmological in nature, the "vault (or arch) of heaven" is certainly closer in meaning to the original intent of the passage than is the "troop" of the KJV. This understanding would also be supported by the original Hebrew word translated as "founded" in the passage. That word is *yacad*, which has the meanings of "to found, fix, establish, lay foundation." That word, for example, is used in Psalm 24:2: *"For he hath founded [yacad] it [the earth] upon the seas, and established it upon the floods."* In addition, the Hebrew word that is translated as "in" in Amos 9:6 is *al*, which has the meanings of "upon," "above," and "over." The sense of the passage, then, is that it relates to establishing or laying the foundation of a structure—that is, the vault of heaven—upon the earth. For that matter, by using "vault of heaven" in the second part of the verse, it would pair up with, in the form of a conventional biblical reiterative parallelism, the "stories" of heaven in the first part to complete the description of the heavenly structure. Thus the cosmological parallelism would have a purpose in the passage, whereas, as pointed out above, the "troop of God" does not.

Several versions of the Bible agree with the celestial translation of "vault" for *'aguddah* in Amos 9:6 and also clearly show the proper parallelism of the passage. For example, here is the NRSV translation of the passage:

> *who builds his upper chambers in the heavens, and founds his vault upon the earth. . . .*

The Jerusalem Bible provides a similar translation:

> *He has built his high dwelling place in the heavens and supported his vault on the earth. . . .*

Also the NASV:

> *The One who builds His upper chambers in the heavens And has founded His vaulted dome over the earth. . . .*

The JPS Tanakh has:

> *Who built His chambers in heaven And founded His vault on the earth, . . .*

Other versions of the Bible provide similar wording (e.g., the ASV, ESV, NAB, NIV, NJB, and RSV). These translations indicate that the vault of heaven is founded or supported upon the earth. In this physical arrangement, of course, the rim of the disk-shaped earth would provide the foundation, or the support, for the rim, or base, of the vault of heaven. (This view of the biblical cosmos is also supported by the Book of Enoch, which specifically describes the placement of the solid heaven on the rim of the earth. See Appendix C.)

9. The Circle of the Earth

If there still is any question as to whether the biblical "circle of the earth" refers to an earth that has the shape of a flat disk or to an earth that has the shape of a sphere, it should be resolved by several other relevant passages that can be found in the Bible.

In Matthew 4:8, for example, we find:

Again, the devil taketh him [Jesus] *up into an exceeding high mountain, and sheweth him all the kingdoms of the world, and the glory of them;*

The cosmology of this passage presupposes a flat earth, for on such an earth one could see all the kingdoms of the world from a mountain if it were high enough. On a spherical earth one can see only a relatively limited area from a mountain, no matter how high it is.

Similarly, in Daniel 4:10–11, Nebuchadnezzar asks Daniel to interpret what he saw in a dream:

10 *Thus* were *the visions of mine head in my bed; I saw, and behold, a tree in the midst of the earth, and the height thereof was great.*
11 *The tree grew, and was strong, and the height thereof reached unto heaven, and the sight thereof to the end of all the earth.*

First, this passage states that the tree was "in the midst of the earth." The NJB translates it as "the middle of the world." The NAB has "the center of the world," and the NRS has "the center of the earth." The passage makes no sense if the biblical earth is a sphere, for the surface of a spherical earth has no "middle" or "center." However, the surface of a disk-shaped earth *would* have a "middle"—i.e., its center.

Second, the passage states that the tree could be seen to the "end of all the earth." The surface of a spherical earth has no "end," but the surface of a disk-shaped earth would have an "end"—i.e., the rim of the disk. Thus, if it were high enough, Nebuchadnezzar's tree could be seen from the whole surface of a flat, disk-shaped earth. The passage does explicitly state that the tree could be seen from "all" the earth—specifically, all the way to its end. On the other hand, if the tree were growing on a spherical earth, it could be seen from only a relatively limited area of the earth's surface, no matter how high it grew. The passage therefore indicates that the tree grew in the middle, or center, of a flat, disk-shaped earth and could be seen all the way to the rim of the earth.

Since, in the biblical narrative, Nebuchadnezzar's vision is taken to be prophetic, it must also be taken as having been induced by the biblical God. And, in fact, previously in the book of Daniel, Daniel had been taken before Nebuchadnezzar to interpret an earlier dream that the Babylonian ruler had. Daniel prefaced his interpretation of the dream with the following (in Daniel 2:28):

> *But there is a God in heaven that revealeth secrets, and maketh known to the king Nebuchadnezzar what shall be in the latter days. Thy dream, and the visions of thy head upon thy bed, are these;*

There, Daniel specifically states that God revealed these things to Nebuchadnezzar, which means that, according to the biblical God, the earth is flat. Therefore, the vision that Nebuchadnezzar had about the tree, with all of its flat-earth ramifications, supports what we have already discerned about the biblical cosmos.

In Psalm 103:12 we find the following:

> *As far as the east is from the west, so far hath he removed our transgressions from us.*

This passage can have meaning only for a flat earth, for on such an earth the easternmost point *would* be the maximum distance from the westernmost point. On the other hand, from any point on a spherical earth (except for the North and South poles, which have no east or west), east and west eventually meet on the opposite side of the earth.

All of these passages, and many others, therefore answer the question of whether the Bible indicates that the earth is flat or whether it indicates that the earth is a sphere. If the biblical "circle of the earth" refers to a spherical earth as so many Bible believers aver, why then are there no passages in the Bible that conform to the idea of a spherical earth? Why is it that, of all of the numerous passages in the Bible that provide any insight into the shape of the biblical earth, every single one of them indicates that the earth is flat? The only conclusion that one can reasonably come to is that the biblical "circle of the earth" refers to a flat, disk-shaped earth.

We have so far determined that the biblical cosmos consists of a solid vault of heaven that is placed rim-to-rim upon a flat, disk-shaped earth. All of the material that we have gleaned from the Bible leading up to that conclusion also makes clear another conclusion: the biblical heaven is physically near and above the earth. As it states in Genesis, the God of the Bible made the firmament and "called the firmament Heaven." The God of the Bible then said:

> *Let the waters under the heaven be gathered together unto one place, and let the dry land appear.*

Isaiah 40:22, which we took a look at near the beginning of this chapter provides another example:

> It is *he that sitteth upon the circle of the earth, and the inhabitants thereof are as grasshoppers; that stretcheth out the heavens as a curtain, and spreadeth them out as a tent to dwell in.*

In that verse, God is described as sitting or dwelling above the circle of the earth, and the inhabitants of the earth appear no bigger than insects to him.

As Isaiah 40:22 indicates, since the biblical heaven is just above the earth, God can look down on the earth and observe the doings of men. For example, we find the following in Psalm 14:2:

> *The LORD looked down from heaven upon the children of men, to see if there were any that did understand,* and *seek God.*

Also in Psalm 33:14:

> *From the place of his habitation he looketh upon all the inhabitants of the earth.*

Because the ancient Hebrews believed that heaven was a place not so far away above the earth, they thought it could be possible to physically reach it if they only had the means. For example, in Genesis 28:12 we find this about Jacob:

> *And he dreamed, and behold a ladder set up on the earth, and the top of it reached to heaven: and behold the angels of God ascending and descending on it.*

Another such passage is 2 Kings 2:11:

> *And it came to pass, as they still went on, and talked, that, behold, there appeared a chariot of fire, and horses of fire, and parted them both asunder; and Elijah went up by a whirlwind into heaven.*

Of course, the biblical God would not like it if mere mortals tried to reach heaven uninvited. In Amos 9:2, we therefore read, "... *though they climb up to heaven, thence I will bring them down.*" And, in Genesis 11:1–9, we read about the inhabitants of the plain of Shinar who constructed a city and a tower that they intended to build with "its top in the heavens." The God of the Bible, of course, did not like his territory being encroached upon, so he dispersed the people of the plain of Shinar and confused their language. Because heaven *is* close to the earth in the biblical cosmos, the biblical God felt compelled to act. If the biblical heaven were not physically close to the earth, the biblical God could simply have let the inhabitants of the plain of Shinar labor in vain and learn that they could never reach heaven by their own efforts. One also wonders why the God of the Bible took steps to prevent the inhabitants of the plain of Shinar from engaging in their futile labors to reach heaven, but did not prevent our own modern-day flights into space—particularly since, as it states in Psalm 115:16, *"The heaven, even the heavens, are the LORD's: but the earth hath he given to the children of men."*

All of this makes it clear that, contrary to what most modern Bible believers believe, the heaven of God's abode as it is described in the Bible is not some ethereal place beyond human ken and outside the physical world that we perceive. Rather, the biblical heaven is a physical part of the biblical cosmos, just as the earth is.

Before we end this chapter, we should look at one more passage in the Bible that relates to what we have discussed. In Job 38:4–5 we find God asking Job the following:

> 4 *Where wast thou when I laid the foundations of the earth? Declare, if thou hast understanding.*
> 5 *Who hath laid the measures thereof, if thou knowest? or who hath stretched the line upon it?*

The Hebrew word translated as "line" in that passage is *qav*, which, as we saw in relation to Psalm 19:1–6, has the meaning of a "measuring cord" or "measuring line." The implication of this passage is that no one but the God of the Bible can measure the extent of the earth. (Note also that the passage asks who has stretched the line *upon* the earth, rather than *around* it.) However, as will be seen in Chapter 13, the ancient Greek astronomer Eratosthenes did a pretty good job of measuring the circumference of the earth—that is, the circumference of the *sphere* of the earth—well before the end of the Old Testament era. Thus, it appears that the ancient Greeks knew more about the size and the shape of the earth than did the God of the Bible.

Chapter 10
The Pillars of the Earth

The cosmos that we have constructed so far from the words of the Bible is now beginning to look more complete. In the biblical cosmos the earth has the shape of a flat disk, and a solid vault—the firmament of heaven—is placed rim to rim on the disk of the earth. We have also established that the biblical earth and firmament of heaven are positioned in the midst of a cosmic ocean.

As strange as this concept of the cosmos may seem to us today, in ancient times it was hardly unique. In fact, this concept of the cosmos was not limited to the Bible or to the ancient Hebrews, for the Babylonians viewed the cosmos in a similar way. Of particular significance, the Babylonians, like the ancient Hebrews, believed that the flat earth and the solid vault of heaven were formed within a preexisting cosmic ocean.

Such a similarity in the cosmological views of the ancient Hebrews and the Babylonians can hardly be a coincidence, as the two peoples were frequently in contact with each other and shared many concepts and myths. Indeed, much of the Old Testament reached its final form during and after the Babylonian exile of the Israelites, and several books of the Bible bear recognizable vestiges of Babylonian influence. (See Appendix D for some information relating to this subject.)

As we have already seen, the Bible contains several references to the cosmic ocean that was a part of the cosmology of the two cultures. Genesis 1:6–7, remember, specifically states that, when it was made, the firmament of heaven separated the waters of the cosmic ocean, so there were waters above the firmament and waters below the firmament:

> 1 *In the beginning God created the heaven and the earth.*
> 2 *And the earth was without form, and void; and darkness*

> *was upon the face of the deep. And the Spirit of God moved upon the face of the waters. . . .*
> 6 *And God said, Let there be a firmament in the midst of the waters, and let it divide the waters from the waters.*
> 7 *And God made the firmament, and divided the waters which were under the firmament from the waters which were above the firmament; and it was so.*

The "waters" mentioned in this passage are those of the cosmic ocean—which is called "the deep" (again, *tehowm* in the original Hebrew) in verse 2—in the midst of which the God of the Bible created heaven and earth. Besides the above passage, there are several other biblical references to the waters of the cosmic ocean above the firmament of heaven. Psalm 148:4 provides an example:

> *Praise him, ye heavens of heavens, and ye waters that be above the heavens.*

Note specifically that the waters are "above" the heavens. This should make it clear that the biblical cosmos is quite different from the modern view of the cosmos.

Psalm 104:1–3 provides yet another reference to the waters of the cosmic ocean. In the context of the passage, these waters would also have to be those above the firmament of heaven.

> *. . . O Lord, . . . who stretchest out the heavens like a curtain: who layeth the beams of his chambers in the waters. . .*

In this passage, the Hebrew word translated as "beams" is *qarah*, which refers to structural beams or timbers for a building. The Hebrew word translated as "chambers" is *`aliyah* and specifically means "roof room" or "roof chamber." The passage therefore indicates that the God of the Bible built his heavenly chambers in the cosmic waters above the "roof" of the world (i.e., the firmament of heaven).

We had noted previously that there are several mentions of the "windows of heaven" in the Bible, and that the God of the Bible opened these windows in the firmament to let some of the waters above the firmament fall as rain to the earth. An example is found in Genesis 7:11–12, which specifically states that the God of the Bible opened the windows of heaven to bring the waters of the Flood:

> 11 *In . . . the same day were all the fountains of the great deep broken up, and the windows of heaven were opened.*
> 12 *And the rain was upon the earth forty days and forty nights.*

The Hebrew word translated as "windows" there is *'arubbah*, which has the meanings of "window," "sluice," and "floodgate." Thus, in the above passage from Genesis 7, when the windows, or floodgates, of heaven were opened, the waters of the cosmic ocean above the firmament of heaven came through the openings to bring the rains of the Flood.

As well as being above the firmament of heaven, the waters of the cosmic ocean were also believed to extend below, or under, the disk of the earth. In verse 11, the Hebrew word that is translated as "fountains" also has the meaning of "springs" or "wells." To the ancient Hebrews, the fact that there are springs and wells in the earth from which water came was explained by the existence of the great reservoir of water—the "great deep"—beneath the earth. Indeed, *tehowm*, the Hebrew word for "the great deep," is further defined by *Strong's Concordance* as "the *main* sea or the subterranean *water-supply*"—i.e., the waters beneath the earth.

The NRSV version of Genesis 7:11 provides a translation that shows these things more clearly than does the KJV:

> *... on that day all the fountains of the great deep burst forth, and the windows of the heavens were opened.*

Thus, for the Flood, the waters of the abyss, or the great deep, beneath the earth "burst forth" through the flat disk of the earth as huge springs or fountains to inundate the land.

Deuteronomy 8:7 is another verse indicating that the deep under the earth provides the source of water for springs (here, the NJB):

> *But Yahweh your God is bringing you into a fine country, a land of streams and springs, of waters that well up from the deep in valleys and hills,*

The Hebrew word that is translated as "the deep" in that verse is, again, *tehowm*. Other versions of the Bible do not accurately provide the sense of the original, as most of them translate *tehowm* as "springs" instead "the deep." However, the lexicons do not provide "spring" as a definition of *tehowm*. Rather, as the NJB indicates, it is "the deep" under the earth that is the *source* of the springs. Some versions of the Bible, such as the KJV, translate *tehowm* as "depths" in the passage: "*... a land of brooks of water, of fountains and depths that spring out of valleys and hills.*" That is somewhat close to the original, but it still does not convey the full sense of the passage in the original Hebrew.

But, of course, there is actually no "great deep" under the earth that provides the source of the water that comes from springs and wells. Rather, rainwater permeates through porous strata in the earth and collects in favorable places where it seeps out as springs or where it can be collected in wells.

The overall picture of the biblical cosmos that we have, then, consists of several levels or tiers. These tiers include the waters above the firmament of heaven, the firmament of heaven itself, the open area under the firmament of heaven, the disk-shaped earth, and the waters under the disk of the earth. Exodus 20:4 mentions three of those tiers:

> *Thou shalt not make unto thee any graven image, or any likeness of any thing that is in heaven above, or that is in the earth beneath, or that is in the water under the earth.*

Deuteronomy 5:8 makes a similar statement:

> *Thou shalt not make thee any graven image, or any likeness of any thing that is in heaven above, or that is in the earth beneath, or that is in the waters beneath the earth*

Deuteronomy 33:13 also refers to these three tiers of the biblical cosmos and also indicates that the waters of the deep are underneath the earth. This is the NRSV wording:

> *And of Joseph he said: Blessed by the LORD be his land, with the choice gifts of heaven above, and of the deep that lies beneath;*

There, the land is the reference point, with the heaven above the land and the deep beneath the land. The Hebrew word that is translated as the "deep" is, again, *tehowm*, referring to the waters of the abyss under the earth. Thus, the passage is meant to show that the blessing of God encompasses the whole of creation, from heaven above the earth, to the earth itself, and to the watery abyss under the earth.*

The existence of the three tiers is also implied in the New Testament, as in Revelation 5:13, which hearkens back to Exodus 20:4 and Deuteronomy 5:8, mentioned above:

> *And every creature which is in heaven, and on the earth, and under the earth. . . .*

---

* Ecclesiasticus 16:18 of the Apocrypha also indicates that the heavens, the earth, and the deep were considered to be the three main constituents of creation:
> *Behold, the heaven, and the heaven of heavens, the deep, and the earth, and all that therein is, shall be moved when he shall visit.*

In the Greek of the Septuagint, the word translated as the "deep" is *abussos*, meaning "abyss," which, as we saw, is also a term applied to the Hebrew *tehowm*.

These passages therefore illustrate the three-tier structure of the biblical cosmos. (See Appendix C for a quite graphic description of the three tiers of the biblical cosmos in the Book of Enoch.)

Concerning the cosmic waters, we find the following in Ecclesiastes 1:7:

> *All the rivers run into the sea; yet the sea is not full; unto the place from whence the rivers come, thither they return again."*

There, the rivers of the earth, after receiving their waters as rain from the windows of heaven and as streams from the springs of the earth, return the waters to the sea, which the writer of the passage apparently viewed as being connected to the cosmic ocean. The cosmic ocean is therefore the "place from whence the rivers come" and the place "thither they return again" so the sea does not overfill onto the land.

Some biblical apologists say that this passage from Ecclesiastes describes the hydrological cycle that present-day science describes—i.e., water evaporates from the surface of oceans and lakes, condenses as rain, and flows in rivers back to the oceans and lakes. However, the passage specifically indicates that the writer believed that the waters go to some "place" other than the visible seas. But there is no such "place" in the hydrological cycle that present-day science describes; therefore, the writer of this passage in Ecclesiastes had a different cosmological view than that described by present-day science. That is, he believed that the waters of the rivers return to the cosmic ocean.

Moreover, another passage, Isaiah 55:10, specifically contradicts the present-day view of the hydrological cycle:

> *For as the rain cometh down and the snow from heaven, and returneth not thither, but watereth the earth, and maketh it bring forth and bud, and giveth seed to the sower and bread to the eater;*

This passage specifically indicates that the fall of rain and snow is one-way and that the water of the rain and snow does not return to heaven. But according to the present-day view of the hydrological cycle, the water does return—that is, it evaporates from the oceans and surface of the earth, and is also given off by living things, and goes into the atmosphere where it condenses and falls as rain, which wets the land and flows back into the sea where it again evaporates and repeats the process over and over again. Isaiah 55:10 therefore not only contradicts the present-day view of the hydrological cycle, it also contradicts Ecclesiastes 1:7, which indicates that the waters derived from the rain and snow *do* return to the place they come from. Nevertheless, though it contradicts Ecclesiastes 1:7, Isaiah 55:10 also implies the existence of the cosmic ocean. Since, according to Isaiah 55:10, the water does not return to where it came from, the implication is that the author of that book thought the water comes from a supply that is inexhaustible: that is, again, the cosmic ocean above the firmament of heaven.

The fact that the Bible states there are waters beneath the earth poses a problem and raises a question. Psalm 24:1–2 illustrates the problem:

> 1 *The earth is the Lord's, and the fulness thereof; the world, and they that dwell therein.*
> 2 *For he hath founded it upon the seas, and established it upon the floods.*

The words "seas" and "floods" in this passage refer to the waters of the great deep, or the part of the cosmic ocean under the earth, and "upon" which the earth is founded. The Hebrew word translated as "founded" in verse 2 is *yacad*, which, as we have seen, has a meaning of "fix" as well as "lay foundation." The Hebrew word translated as "established" is *kuwn*, which, as we previously saw, has a meaning of "to be firm, be stable, be established." The Hebrew word that is translated as "upon" in both occurrences is *al*, one of the meanings of which is "upon." BDB provides the following as a definition of *al*: "Upon, of the substratum *upon* which an object in any way rests, or *on* which an action is performed." Thus, according to Psalm 24:1–2, the biblical earth is positioned "upon" the waters of the great deep. The NEB translation of that passage also makes it clear that the waters of the great deep are beneath the earth: "*The earth is the Lord's. ... For it was he who founded it upon the seas and planted it firm upon the waters beneath.*" The NJB translates verse 2 as "*it is he who laid its foundations on the seas, on the flowing waters fixed it firm.*" The NKJV translates verse 2 as "*For He has founded it upon the seas, And established it upon the waters.*"

Psalm 136:6 is another passage that makes it clear that the earth is positioned "upon" the waters of the great deep. That verse reads in part, "*To him that stretched out the earth above the waters....*" The NRSV translates it as "*who spread out the earth on the waters....*" The NJB translates it as "*He set the earth firm on the waters....*" The Hebrew word that is translated as "above" and "on" in these renderings is, again, *al*.

The Bible therefore indicates that the earth is planted or established firmly "upon" the waters of the great deep. The problem, of course, is that picture is contrary to what we know about the structure of the earth, for the earth is not "planted" upon waters underneath it. Some apologists would undoubtedly say that the waters refer to the earth's oceans. However, the earth is mostly rocky material, and its oceans are, in comparison with the bulk of the earth, scarcely a film of moisture on its surface. Moreover, these passages, and the ones we previously looked at, taken together make it clear that the earth is positioned "on" the waters of the cosmic ocean beneath it and that the waters of the cosmic ocean also extend "above" the firmament of heaven.

So, in context of the biblical cosmos, how is the disk of the flat earth held in place in the midst of the cosmic ocean? In particular, how is the earth "planted firm upon the waters beneath"?

The answer is found in 1 Samuel 2:8:

> ...*for the pillars of the earth are the Lord's, and he hath set the world upon them.*[1]

The Hebrew word translated as "pillars" is *matsuwq*, which has the meanings of "pillar" and "column." Therefore, the disk of the earth is placed upon pillars that go into the depths of the waters beneath the earth.

In the Bible, these pillars are also called the foundation or the foundations of the earth. A foundation is something that a structure is placed or built upon. In Micah 6:2 of the KJV we read:

> *Hear ye, O mountains, the Lord's controversy, and ye strong foundations of the earth....*

The NEB translation of Micah 6:2 uses the term "pillars" in place of "foundations" and calls them the "everlasting pillars that bear up the earth."

Job 9:6 in the KJV also mentions the pillars:

> [God] *shaketh the earth out of her place, and the pillars thereof tremble.*

The Hebrew word translated as pillars here is *'ammuwd*, which, like *matsuwq*, has the meaning of "pillar" or "column."

Psalm 82:5 makes a similar statement about the earth's supports: "... *all the foundations of the earth are out of course.*" The Hebrew word translated as "out of course" is *mowt*, which means "to shake" or "totter." The JPS Tanakh translates the passage as "*all the foundations of the earth totter.*" The language there implies several foundations—i.e., pillars—which, of course, would cause earthquakes when they totter or shake.

The Bible does not say what the pillars themselves are placed upon. In fact, in Job 38:4–6 of the KJV, God asks Job

> 4 *Where wast thou when I laid the foundations of the earth? declare, if thou hast understanding....*

> 6 *Whereupon are the foundations thereof fastened? Or who laid the cornerstone thereof;*

The word "foundations" appears twice in that passage, but two different words having different meanings are used in the original Hebrew. In the first occurrence, the Hebrew word translated as "foundations" is *yacad*, which, again, is a general term meaning "to set" or "to lay a foundation," as is indicated in the passage. However, the Hebrew word that is translated as "foundations" in the second occurrence is *'eden*, which has a more specific meaning of "a base" or "pedestal," as for "columns" or "pillars." BDB provides the following as an interpretation of *'eden*: "1. *Pedestal* ..., on wh. pillars of marble were set.... 2. *pedestals* of the earth on wh. its pillars were placed Jb 38:6." The NJB reflects this difference in meaning between the two words. Its translation of this passage is as follows:

> 4 *Where were you when I laid the earth's foundations? Tell me, since you are so well-informed! ...*
> 6 *What supports its pillars at their bases? Who laid its cornerstone—*

The NAB rendering of verse 6 is "*Into what were its pedestals sunk, and who laid the cornerstone.*" The implication of the passage is that only God knows what the bases of the earth's pillars are supported upon. The sense of the pillars in this passage certainly cannot be applied to a spherical, unsupported earth. It can, of course, be applied to a flat earth that is supported above the "great deep" by pillars.

Psalm 75:3 is another passage that mentions the pillars of the earth:

> *The earth and all the inhabitants thereof are dissolved: I bear up the pillars of it.*

The Hebrew word translated here as "pillars" is, again, *`ammuwd*, the same word used in Job 9:6. The Hebrew word translated as "dissolved" is *muwg*, which, according to KB, means "to sway backwards and forwards" and "undulate." The NRSV provides the following translation of the verse:

> *When the earth totters, with all its inhabitants, it is I who keep its pillars steady.*

The NIV translates it as:

> *When the earth and all its people quake, it is I who hold its pillars firm.*

This passage thus indicates that earthquakes are caused by a shaking of the earth's supporting pillars, and that God stops earthquakes by steadying the pillars.

In Jeremiah 31:37 we find:

> *Thus saith the LORD; If heaven above can be measured, and the foundations of the earth searched out beneath, I will also cast off all the seed of Israel for all that they have done, saith the LORD.*

This verse specifically mentions the foundations—that is, the pillars—that are under the earth. The Hebrew word translated as "searched out" is *chaqar*, which means "examine" or "explore." The implication of the passage is that the biblical God is saying that if anyone can go into the abyss under the earth and examine the pillars holding up the earth, he would reject the seed of Israel. In effect, he is saying he will never reject the seed of Israel because no mere mortal has the ability to explore the pillars under the earth.

There are several other references in the Bible to the pillars and the foundations of the earth. And since the vault of heaven is placed rim to rim with the disk of the earth, the pillars under the rim of the earth's disk also provide support for the vault of heaven. Thus, when the pillars of the earth shake, the pillars of heaven also shake. This is reflected in 2 Samuel 22:8, which states: *"Then the earth shook and trembled; the foundations of heaven moved and shook, because he was wroth."*

Because the flat disk of the biblical earth is placed on pillars, of necessity it is fixed in place and the only movement that the earth normally experiences occurs when the pillars tremble and cause earthquakes. For example, as we have already seen, Psalm 75:3 states (here, the NIV) *"When the earth and all its people quake, it is I who hold its pillars firm."* Other passages in the Bible state that the earth will be removed from its place in some future time during the Day of the Lord. That would necessarily mean that the biblical earth is normally fixed in place. Job 9:6, quoted earlier, provides an example:

> [God] *shaketh the earth out of her place, and the pillars thereof tremble.*

Isaiah 13:13, in describing the Day of the Lord, provides another:

> *Therefore I will shake the heavens, and the earth shall remove out of her place, in the wrath of the LORD of hosts, and in the day of his fierce anger.*

The NRSV translates the passage as:

> *Therefore I will make the heavens tremble, and the earth will be shaken out of its place, at the wrath of the LORD of hosts in the day of his fierce anger.*

By indicating that the earth will be shaken or removed out of its place in the Day of Judgment, these verses also indicate that the earth is normally fixed in place and will remain so until the Day of the Lord. These passages thus provide additional support for the understanding that the biblical earth does not move.

In addition to those verses, there are several other verses in the Bible that *do* specifically state that the earth does not move. Psalm 93:1 is one example: "... *the world also is established that it cannot be moved.*" The NEB gives an even clearer translation of the verse: "*Thou hast fixed the earth immovable and firm.*"

Psalm 96:10 makes a similar statement: "... *the world also shall be established that it shall not be moved....*" The NEB translation of that verse reads: "... *He has fixed the earth firm, immovable....*"

Another such statement is found in 1 Chronicles 16:30: "... *the world also shall be stable, that it be not moved.*" The NEB translation of the same verse reads: "... *He has fixed the earth firm, immovable.*" And Psalm 104:5 states "... [God] *laid the foundations of the earth, that it should not be removed for ever.*"

So, according to the Bible, the earth is fixed in place and does not move. Certainly, the Bible gives absolutely no inkling anywhere that the earth is actually rotating on its axis and orbiting the sun.

Nevertheless, the creationist author, Henry Morris, tried to argue that the rotation of the earth is implied in Job 38:14: "*It is turned as clay to the seal.*" The pronoun "it" refers to the earth, and Morris interprets this as the figure of a clay vessel being turned, or rotated, on a potter's wheel to receive an impression from a seal.[2] However, the Hebrew word that is interpreted as "turned" in this passage is *haphak*, which is more properly interpreted as "changed." The same word is used in Psalm 105:29: "*He turned their waters into blood,*" which certainly does not mean the waters were "rotated" into blood. Moreover, rather than a figure of a clay vessel on a wheel, the passage more likely alludes to a flat clay tablet such as those that scribes commonly used at that time for writing inscriptions upon. Such tablets were often impressed with the seal of the person who did the dictation. Thus, when the surface of the clay tablet is impressed with a seal, it is changed in form. That, in fact, is how the passage is interpreted in the RSV: "*It is changed like clay under the seal.*" The NIV likewise translates it in a similar way: "*The earth takes shape like clay under a seal.*" Rather than indicating that the earth is turning like a clay vessel on a wheel, the "turned" in the wording of the KJV would thus more properly mean that the surface of the earth is "changed" as if it were impressed by a seal. The passage therefore does not imply that the earth is a rotating sphere.

Because the ancient Hebrews believed the flat earth, along with its firmament of heaven above, to be totally surrounded by a cosmic ocean and supported by pillars, the idea of an unsupported, spherical earth moving through the extraterrestrial space of the modern view would have been totally alien and inconceivable to them. Therefore, in indicating that the biblical earth does not move, and in the context of the biblical cosmos, those passages noted above do *not* mean that the earth is not moving through outer space as in the modern view of the cosmos. Instead, those passages must relate to a concept of a non-moving earth that is compatible with the biblical view of the cosmos. It should thus be clear that we can understand the meaning of those passages only by looking at them in the context of the biblical cosmos. By doing so, the only understanding that we can properly derive from those passages is that the God of the Bible, by virtue of the pillars he placed under the earth, is preventing the earth from "moving"—falling, actually—into the abyss under the earth. Indeed, that is the only movement of the earth that the structure of the biblical cosmos would allow.

In fact, that understanding is supported by the original Hebrew word that is translated as "move" or as a related form in every one of the above passages that indicate that biblical earth does not move. That Hebrew word is *mowt*, the meanings of which include "to be moved," "totter," and "shake," but also "slip," "to dislodge," "let fall," and "drop." An example of its usage as "let fall" can be found in Psalm 140:10: "*Let burning coals fall upon them....*"

Now, those passages that say the earth does not move cannot be using *mowt* in the sense of a simple tottering or shaking of the pillars, for a simple tottering or shaking would cause only earthquakes, as suggested by Psalm 75:3, above. In the lands of the Bible, earthquakes are, and were, quite common, and, if those passages that indicate the earth does not move were using *mowt* in the sense of a simple totter or shake, they would be proved untrue every time an earthquake occurred. On the other hand, a tottering or shaking that was so severe that it would cause the earth to "remove out of her place" would imply that the pillars would collapse and cease holding up the earth. In those passages that indicate the earth does not move, then, *mowt* must have the more severe meaning of "to dislodge," "let fall," or "drop." With those meanings of *mowt*, as long as the biblical earth does not drop into the abyss, the passages would remain true even when earthquakes occur, and they would remain true until the "Day of the Lord" when God would "remove the earth out of her place."

We can therefore understand those passages that say that the earth does not move as meaning that the earth does not fall into the abyss under it because God has placed it firmly on its pillars. With that understanding, we could contextually interpret Psalm 104:5 as "God laid the foundations of the earth, that it should not fall." The other relevant verses could be translated in a similar fashion. For example, Psalm 93:1 would be translated as: ". . . the world also is established that it cannot fall."

That understanding of these passages is confirmed by a prophecy of the Judgment Day that is found in Isaiah 24:20 (here, the NKJV):

> *The earth shall reel to and fro like a drunkard, And shall totter like a hut; Its transgression shall be heavy upon it, And it will fall, and not rise again.*

That the biblical earth will reel to and fro, and totter like a hut, would, of course, be understandable if its supporting pillars shake. And, of course, if the shaking of the pillars were severe enough the biblical earth would dislodge from its place and fall into the abyss under the earth. In this passage, the original Hebrew word that is translated as "fall" is *naphal*, the primary meanings of which include "to fall" and "be cast down." In the context of the biblical cosmos and its end-times scenario, it is certainly quite easy to understand how the biblical earth could totter on its supporting pillars and fall. On the other hand, one would be hard put to apply Isaiah 24:20 to the modern view of the cosmos with its unsupported, spherical earth that is moving through the emptiness of outer space. (See Appendix C for a very graphic description in the Book of Enoch that also confirms this understanding about the movement of the earth in the biblical cosmos.)

Speaking of the end-times scenario, the Bible contradicts itself concerning the duration of the earth. Some passages, such as Psalm 104:5, which we looked at above, indicate that the earth is eternal. There are other passages that also state that the earth is eternal. For example, Psalm 78:69 states:

> *And he built his sanctuary like high* palaces, *like the earth which he hath established for ever.*

And in Ecclesiastes 1:4 we find:

> One *generation passeth away, and* another *generation cometh: but the earth abideth for ever.*

10. The Pillars of the Earth

However, other passages, such as Isaiah 13:13 and 24:20, which we looked at above, state that the earth is not eternal. And, of course, end-time prophecies concerning the destruction of the earth play an important part in the New Testament.

There is yet another consequence of a nonmoving earth in the biblical cosmos. Since, according to the modern view of the cosmos, the earth is a sphere that is rotating on its axis, its daily rotation periodically brings any sub-arctic location on its surface around to face toward the sun, causing that location to experience the light of day, and then around to face away from the sun, causing the location to experience the darkness of night. (In the arctic regions, of course, there are six months of daylight and six months of night as a result of the earth's axial tilt and orbit around the sun.) In addition, while one location on the surface of the earth is experiencing the light of day, another diametrically opposite location on the other side of the earth is experiencing the darkness of night.

Now, as we have seen, in contrast to the modern view of the spherical earth rotating on its axis and moving through space, the biblical earth is a flat disk fixed firmly in place upon its pillars. Because that arrangement prohibits the earth from moving, one would have to conclude that it must be the sun that moves in the biblical cosmos. And, in fact, there are several passages in the Bible which indicate that it is indeed the sun that moves. A notable example can be found in Ecclesiastes 1:5:

> *The sun also ariseth, and the sun goeth down, and hasteth to his place where he arose.*

According to this verse, as well as rising in the morning and setting in the evening, the sun must make "haste" to return to the place where it arose so it can rise from there again the next morning.[*] Now, we will often talk about the rising and setting of the sun; but, knowing that these apparent actions of the sun are actually caused by the rotation of the earth, we certainly would not say that the sun, having set, must make haste to return to the place where it is to arise again the next morning. Ecclesiastes 1:5 therefore makes it clear that the sun actually does move across the firmament of heaven in the biblical cosmos; and, because the sun *is* actually making a daily journey across the firmament of heaven, it must somehow make its way back to its starting point in order for it to arise there again

---

[*] A similar statement about the sun can be found in 1 Esdras 4:34 of the Apocrypha (this is from the RSV Apocrypha):

> *... The earth is vast, and heaven is high, and the sun is swift in its course, for it makes the circuit of the heavens and returns to its place in one day.*

the next day. Thus the sun must "hasteth to the place where he arose." (See Appendix C for related material in the Book of Enoch that describes how the sun "hasteth" to return to its place of arising after having set in the evening.)

In the last chapter we saw that Psalm 19:1–6 shows that the biblical earth is flat, but it also shows that it is the sun that moves in the biblical cosmos:

> 1 *The heavens declare the glory of God; and the firmament sheweth his handywork...*
> 4 *Their line is gone out through all the earth, and their words to the end of the world. In them he set a tabernacle for the sun,*
> 5 *Which is as a bridegroom coming out of his chamber, and rejoiceth as a strong man to run a race.*
> 6 *His going forth is from the end of the heaven, and his circuit unto the ends of it: and there is nothing hid from the heat thereof.*

That passage, like the others we looked at, describes an actual journey of the sun across the sky in the biblical cosmos and leaves no room for a rotating earth.

Psalm 104:19 provides another example of the moving sun in the biblical cosmos: "*He appointed the moon for seasons: the sun knoweth his going down.*" The NJB rendition of that verse states: "*He made the moon to mark the seasons, the sun knows when to set.*" Again, it is one thing to refer to the setting of the sun, but it is quite another thing to say "the sun knows when to set." That expression reflects a belief in a real movement of the sun and ascribes to the sun a volition in following its path across the sky and setting at the proper time each day. Again, the wording leaves no room for a rotating earth as the cause of the "apparent" motion of the sun across the sky.

That brings up what we learned in Chapter 7 about the nature of the light of day—specifically, the light of the daytime sky—in the biblical cosmos. We noted that several verses in the Bible indicate that the light of the daytime sky is not derived from the sun, but is an entity unto itself in the biblical cosmos. Genesis 1:3–5, remember, states:

> 3 *And God said, Let there be light: and there was light.*
> 4 *And God saw the light, that it was good: and God divided the light from the darkness.*
> 5 *And God called the light Day, and the darkness he called Night. And the evening and the morning were the first day.*

Here, God divided the light (*'owr*) from the darkness (*choshek* in the original Hebrew), which resulted in the first day, and we noted that the light of day came into being before the sun was created and even before God gave form to the earth. We also noted that Daniel 12:3 supports the view that the light of the daytime sky is an entity unto itself:

> *And they that be wise shall shine as the brightness of the firmament. . . .*

We also noted Ecclesiastes 12:2, which gives the "light"—which can only be the light of the daytime sky—equal standing with the other sources of light:

> *While the sun, or the light, or the moon, or the stars, be not darkened, nor the clouds return after the rain.*

And we also noted that Psalm 74:16 provided a similar view:

> *The day is thine, the night also is thine: thou hast prepared the light and the sun.*

Now, since the biblical earth is a flat disk, all parts of its surface must experience the light of day at the same time, and all parts of its surface must also experience the darkness of night at the same time. This means that, like the movement of the biblical sun, the light of the daytime sky, since it is an independent entity, must have a completely different means of bringing about its appearance than occurs on the actual earth of the modern view of the cosmos. Indeed, the same must be true for the dark of night since, in the biblical cosmos, it is contrasted with, and on the same level as, the light of day.

There are, in fact, other passages in the Bible that confirm there is such a different cosmological arrangement in the biblical cosmos. In Job 38, as we previously noted, God scolds Job by sternly questioning him about his knowledge, or lack thereof, concerning various aspects of creation and the cosmos. Within that chapter, Job 38:12, like the passages above, indicates that the light of day—as the first light of dawn—is independent from that of the sun:

> *Hast thou commanded the morning since thy days; and caused the dayspring to know his place*

The Hebrew word translated as "morning" there is *boqer*, which means "dawn," or the first light of day that appears before the rising of the sun. In the second part of the passage, the Hebrew word translated as "dayspring" is *shakar*, which also means "dawn." The "dayspring" of the second part therefore reiterates the "morning" of the first part. That passage thus indicates that God commands the dawn to come forth and shows it, as the first light of day, its place in the morning sky before the appearance of the sun. That ties in with Job 38:19–20, which is another passage in the same chapter that shows the independent nature of the biblical light of day—and of the darkness of night, as well. Here is the NIV rendering of the passage:

> 19 *What is the way to the abode of light? And where does darkness reside?*
> 20 *Can you take them to their places? Do you know the paths to their dwellings?*

There, the biblical God is indicating that, when it is night, the light (*'owr* again) of day must have a place where it dwells until it is ready to come forth at the beginning of the following new day; and that, when it is daytime, the darkness (*choshek* again) of night must likewise have a place where *it* dwells until it is ready to come forth at the beginning of the new night. In asking Job if he can show the light and the darkness the way to their separate abodes after their respective appearances each day, the biblical God is proclaiming that he himself performs that task. Job 38:19–20 thus attests that the light of day and the darkness of night, as well as being independent entities in the biblical cosmos, each have their own separate dwellings that they come out of when it is time for their respective appearances during the course of a day and return to when they have finished their tasks. This passage therefore clarifies the words of Job 38:12, above, in which it is asked, "*Hast thou commanded the morning . . . and caused the dayspring to know his place.*" Thus God commands the morning dawn (or dayspring), as the first light of day, to come forth from its dwelling.

That cosmological arrangement might appear to be ludicrous to us today, but it is fully consistent with the other aspects of the biblical cosmos that we have discerned, and it leaves no room for the continuous nature of both night and day that exists on what we know is a rotating earth. Indeed, like its prescribed movement of its sun, the very nature of the biblical cosmos requires some such arrangement, and Job 38:12 and 19–20, in describing that arrangement, indicate that the biblical earth does not need to rotate in order to bring about the light of day and the dark of night.

The passages reflecting that arrangement, and also those that describe the independent nature of the light of the daytime sky, thus provide added evidence in support of our understanding that the biblical earth does not move. If, as occurs on the actual earth, different locations on the biblical earth were separately experiencing daylight and night at the same time, then the light of day and the darkness of night would be constantly on duty—so to speak—rather than having to retire to their separate abodes upon finishing their daily tasks.

10. The Pillars of the Earth

Some biblical apologists might argue that these passages from Job are mere poetry and are not intended to be taken literally. But this is supposed to be God speaking and expounding upon his divine creative and commanding powers, and doing so in a stern and serious tone. Would the writer of Job have deliberately had his God expound on his divine powers by using an example that is so far out of touch with reality as to be so ludicrous? That is not likely. The writers of the Bible stood in awe of their God, and they would not have put words in his mouth that would have given the appearance of demeaning his abilities. Therefore, though it might seem ludicrous to us today, the view of the cosmos that is expounded on in these passages from Job must have been fully consistent with the cosmological view that the author and his fellow Hebrews had—a view that is also fully compatible with the cosmos of the Bible as we have discerned it.

Moreover, if one holds that Job 38:12 and 19–20 are mere poetry and are not intended to be taken literally, how does one draw the line between what is meant to be taken literally in other parts of the Bible and what is not? If we, today, say that these passages are not intended to be taken literally simply because they contradict our understanding of the cosmos, are we justified in saying that any other part of the Bible *is* intended to be taken literally. After all, we might not be aware that some other part of the Bible that is accepted literally today actually contradicts certain facts that we are not aware of, just as the ancient Hebrews accepted these passages from Job as being literal because they were not aware of the true nature of the cosmos. Moreover, saying that these passages from Job, or any other passages in the Bible, are not intended to be taken literally because they do not conform to modern knowledge turns the Bible into a book that is different from what its authors intended.

From what we have seen in this chapter, it should be clear that the biblical earth is fixed in place and does not move (except when the God of the Bible will shake the earth out of its place in the "Last Days").

It is because the Bible is so clear about the immovability of the earth that Galileo got into trouble with the Catholic Church when he tried to show that the earth rotates on its axis and moves around the sun according to the cosmological system that Copernicus devised. For promoting that cosmological system, Galileo was hauled before the Inquisition in Rome in the year 1615. The verdict of the Church inquisitors was as follows:

> The first proposition, that the sun is the center and does not revolve about the earth, is foolish, absurd, false in theology, and heretical, because it is expressly contrary to Holy Scripture.... The second proposition, that the earth is not the center but revolves about the

sun, is absurd, false in philosophy, and, from a theological point of view at least, opposed to the true faith.[3]

Under threat of being thrown into the dungeons of the Inquisition, Galileo acquiesced and agreed not to promote the Copernican theory any further. The Congregation of the Index subsequently decreed that "... the doctrine of the double motion of the earth about its axis and about the sun is false, and entirely contrary to Holy Scripture." The same decree condemned the writings of Copernicus and "all writings which affirm the motion of the earth." Thus the faithful were prohibited from reading all books that promoted the Copernican view of a moving earth.

In 1632, thinking that the religious atmosphere had changed sufficiently, Galileo published the *Dialogue of the Two Great World Systems*, a work that presented the Copernican view as a play on the imagination. The response of the Church was not one of amusement. Galileo was again hauled before the Inquisition and again threatened with imprisonment and torture unless he recanted. Galileo recanted and was placed under house arrest for the rest of his life. It was not until the year 1835 that the Catholic Church removed from the Index the condemnation of those works that promoted the idea of the double motion of the earth.

Galileo actually got off relatively lightly; he was only threatened with torture, forced to recant, and placed under house arrest for the rest of his life for saying, in contradiction to the Bible, that the earth rotates on its axis and moves around the sun. In the year 1600, the Church had Giordano Bruno, who refused to recant, burned at the stake for saying the same thing (and also for saying, among other things, that the earth is not the only world).

Because of what the Bible has to say about the immovability of the earth, the Protestant reformers—Luther, Calvin, Wesley, and several others—also rejected the Copernican system with its moving and rotating earth. Of Copernicus, Luther said the following:

> People gave ear to an upstart astrologer who strove to show that the earth revolves, not the heavens or the firmament, the sun and the moon.... This fool wishes to reverse the entire science of astronomy; but sacred Scripture tells us that Joshua commanded the sun to stand still, and not the earth.[4]

Calvin, in condemning those who accepted the Copernican system, referred to Psalm 93:1 (which we looked at earlier in this chapter) and said, "Who will venture to place the authority of Copernicus above that of the Holy Spirit?"[5]

In fact, belief in an immovable, and even flat, earth has had a long history in Christianity. Citing the Bible, many early leaders of Christianity, such as Eusebius, John Chrysostom, Lactantius, Tertullian, Theophilus of Antioch, and Clement of Alexandria, expressed their flat earth belief at a time when most scholars understood that the earth is a sphere.[6] Later, in the sixth century, Cosmas Indicopleustes wrote a book called *Christian Topology*, which advocated a flat-earth system in opposition to the Ptolemaic system with its spherical earth.

Of course, there were some Christian scholars who were more enlightened. In the eighth century, for example, Bede made it clear in his *Ecclesiastical History of the English People* that the earth is a sphere. And, in the thirteenth century, Thomas Aquinas likewise affirmed the sphericity of the earth.

Eventually, most Christian leaders and theologians did come to accept that the earth is a sphere. Nevertheless, most of those same Christian leaders and theologians, as previously noted, continued for some time to reject Copernicanism and held to the biblical view of an immovable earth. It was not until within the eighteenth century that most of the Christian leaders gradually and grudgingly accepted the Copernican view of a moving earth. Still, even as late as the nineteenth century, several Christian leaders and theologians, both Catholic and Protestant, continued to hold to the Ptolemaic system and to reject Copernicanism. For example, in a book published by the Lutheran Synod of Missouri in 1873, a former president of a Lutheran teachers' seminary stated the following concerning which one of the two systems is correct:

> But the wise and truthful God has expressed himself on this matter in the Bible. The entire Holy Scripture settles the question that the earth is the principal body of the universe, that it stands fixed, and that the sun and moon only serve to light it.

He went on to say:

> Let no one understand me as inquiring first where truth is to be found—in the Bible or with the astronomers. No; I know beforehand—that my God never lies, never makes a mistake; out of his mouth comes only truth, when he speaks of the structure of the universe, of the earth, sun, moon, and stars....[7]

Because the Bible is so clear in its assertions that the earth does not move, there are some Bible believers even today who reject the conventional modern view that the earth moves in space. These Bible believers, who are called geocentrists, adhere to the biblical view of an immovable earth. The Genesis

Institute, for example, is a creationist organization that has promoted the geocentric viewpoint. The Association for Biblical Astronomy (ABA), formerly called the Tychonian Society, is another organization devoted to spreading the word about the "truth" of the geocentric viewpoint of the Bible. Yet another is the Cercle Scientifique et Historique (CHSHE) which has offices in France and Belgium. In the preface of *Geocentricity*, a book put out by the ABA, we find the following:

> The ABA favors a large universe rotating about the earth once per day with the earth located at the dynamic center of the universe, whereas the CESHE advocates a small universe centered on the earth with a rotating earth. . . . Basically, the issue is one of the inerrancy and preservation of scripture, especially in light of the fickle pronouncements of science. At stake is the authority of the Bible in all realms, including in the realm of science.[8]

Another geocentric organization is the Fair Education Foundation. In *The Earth is Not Moving*, a book published by that organization, we find the following:

> Indeed, not only are no excuses made herein for the Bible's commitment to geocentricity (a non-moving Earth and an Earth-centered universe), the whole purpose of this book is to show that the Scriptures tell the Truth on this subject (and, by extension, *all* subjects!) even though the entire world has been taught otherwise.
>
> In other words, the topics that follow are an expose of the most universally taught and believed lie in the world today, namely, the lie that "science" has proven that the Earth is rotating on an axis daily and orbiting the sun annually.[9]

The "geocentrist" Bible believers often declare that the conventional modern view of astronomy is a "satanic lie," just as the "scientific" creationists likewise frequently declare that evolution is a "satanic lie." The geocentrists generally believe that the sun and moon go around the immovable earth, while the other planets go around the sun. They also generally believe that the stars are on a spherical shell (the concept of which they derived from the biblical firmament) that encloses the system.

For that matter, belief in the flat earth cosmology of the Bible also did not completely die out. The flat-earth viewpoint had a forceful resurgence in the nineteenth century; it began in England in 1849 with the publication of a pamphlet called *Zetetic Astronomy* and the subsequent publication of a book called *Earth Not a Globe*, both by Samuel Rowbotham, who wrote under the pseudonym of "Parallax." Rowbotham was frequently able to win debates against his spherical-earth opponents and gained a large number of converts. (This has a counterpart in present-day debates between creationists and evolutionists. In these debates, the "scientific" creationists, quoting out of context and resorting to false and deceptive use of "evidence," are often able, on rhetorical grounds, to hold sway over their evolutionist opponents. In a fast-paced debate, the debater who resorts to deception has a clear advantage over his opponent. After all, it may take only thirty seconds for a debater to make a deceptive statement, but several minutes for the opponent to show why that statement is deceptive—even if, in the unlikely case, he had the literary resources immediately on hand to refer to in order to refute the deceptive statement. What it boils down to is that the purpose of a debate is to win the debate, not to show where truth can be found.)

Even today, though the Flat Earth Society has been made the butt of many jokes, it is (or was, at least until recently) a very real organization and its members consider the spherical-earth view to be a hoax that is meant to discredit the Bible.[10] Though the membership of the society waned in numbers up through the latter part of the twentieth century, it continued to promote biblical cosmology. According to the Flat Earth Society, the Apollo landings on the moon were faked, as were the photos of the spherical earth taken from outer space. (As of this writing, there is a Web site called "The Flat Earth Society," but it is a parody and has no connection with the actual Flat Earth Society.)

The devoutly fundamentalistic geocentrists and flat-earthers thus give adequate testimony to the fact that one does not need to be a critical unbeliever to come to the conclusion that the biblical earth is immovable, or even flat.

Chapter 11
The Pit of Sheol

Although, as was pointed out in the last chapter, some Bible believers still cling to the non-moving and geocentric, and even flat earth, aspects of the biblical cosmos, most modern-day believers in the Bible ignore its clear statements indicating that the earth is fixed in place and does not move. Instead, they try to show that the cosmology of the Bible is compatible with the commonly accepted and incontrovertible findings of science concerning the earth in space. For example, they will often point to Job 26:7 as evidence of that view. In the KJV, that verse states:

> *He stretcheth out the north over the empty place, and hangeth the earth upon nothing.*

Believers in the Bible frequently say that passage reflects the modern view of an earth that is unsupported in the emptiness of extraterrestrial space. They also say that the passage proves that the Bible is the word of God because, at the time it was written, only God could have known what we now know about the earth in space.

But that argument is fallacious on several counts. In the first place, as will be shown in Chapter 13, some of the ancient Greek scientists not only knew that the earth is a sphere (as we previously noted), but also proposed that the earth spins on its axis and revolves around the sun along with the other planets. Why then should the writer of Job 26:7 be considered to have been able to have had an understanding of the earth's place in the cosmos only through a revelation from God? That passage, therefore, can hardly be taken as proof that the Bible must be a revelation from God because of its *supposed* allusion to an earth "hanging" in extraterrestrial space.

In the second place, in saying that the earth "hangeth" upon nothing, Job 26:7—like several other biblical passages that we examined in the previous chapter—implies that the earth does not move. Yet we know the earth is actually orbiting the sun at 18 miles per second, or 66,700 miles per hour.

In the third place, if the intent of Job 26:7 is to indicate that the earth is located in extraterrestrial space, it should have stated something like "God surrounded the earth with nothingness," or "God established the earth within nothingness." But, in fact, the passage actually indicates something quite different. In fact, a close study of the verse indicates that it is not what believers in the Bible make it out to be. What, for example, does the first part of the passage mean when it states that God "stretcheth out the north over the empty place"? What is meant by "the north"? Does it refer to some location on the earth? Or does it mean the direction? And what is meant by the "empty place"? If the contextual meaning of "the north" in the passage is not readily apparent, the meaning of the "empty place" under it cannot be readily apparent either.

Going on to the second part of the verse, what is meant when it states that God "hangeth the earth upon nothing"? The Hebrew word that is translated as "hangeth" is *talah*, which means "to hang (up)." Most of the occurrences of the word *talah* in the Bible are in reference to execution by hanging, as, for example, in Joshua 8:29: *"And the king of Ai he hanged on a tree until the eventide...."* In a few places, the word *talah* is used in reference to hanging up some object, as in Ezekiel 15:3: *"Shall wood be taken thereof to make any work? or will men take a pin of it to hang any vessel thereon?"*

The Hebrew word that is translated as the preposition "upon" in the second part of Job 26:7 is *al*, the meanings of which include "above," "over," "on," or "upon." So there is the question of which meaning should be applied to *al* in the passage. If the meaning is "upon," it would refer to what the earth is hanging upon or down from (as, for example, a coat hangs upon or down from a coat hook). In that case, the passage would apparently be saying that the earth is hanging (or suspended) *from* nothing; that is, it is hanging down from some celestial "peg" that is "nothing." That is what the KJV actually seems to be indicating when it states that God *"hangeth the earth upon nothing."* The NRSV rendition of the passage is almost identical, for it states that God *"hangs the earth upon nothing."* On the other hand, if "above" or "over" is the proper translation of *al*, the apparent meaning of the passage is that the earth is hanging or is suspended *over* or *above* "nothing," and that is how some other versions of the Bible translate it. For example, the NIV rendition of the passage is *"... he suspends the earth over nothing."* The JPS Tanakh also translates the word as "over" instead of "upon": ". . . [God] *suspended earth over emptiness.*" These two different interpretations of *al* therefore have quite different contextual meanings. In the one case, the "nothing" that is being referred to is what the earth is hanging down from, whereas in the other case the "nothing" is what is underneath the earth.

Regardless of which preposition ("above," "over," "on," or "upon.") is used in Job 26:7, that passage implies that the biblical earth has an upper side and an under side, for those prepositions imply directionality. Therefore, Job 26:7 does not negate what we have so far determined about the shape of the biblical earth. Certainly, there is nothing in the passage to indicate that the earth is a sphere, and the passage could just as well apply to a flat, disk-shaped earth that is covered by a solid vault of heaven and "hanging" upon, or over, "nothing."

Because the passage implies that the biblical earth has an upper side and an under side, it also implies that the biblical cosmos has an overall inherent top-to-bottom directional arrangement that has at least three tiers: The earth would be the middle tier, what is above the earth would be the upper tier, and what is below the earth would be the lower tier. That, of course, substantiates the view of the biblical cosmos we have already discerned. That is, it is a structure with several tiers, including the firmament of heaven, the "circle" of the flat earth below the firmament of heaven, and the abyss under the earth.

That view, of course, stands in stark contrast to the modern view of the cosmos. In the modern view, the earth does not have an upper side and an under side, and the cosmos does not have an inherent top-to-bottom directional arrangement. For someone standing at the earth's north pole, "up" is actually pointing—with respect to the background star field surrounding the earth—in a direction that is opposite from the "up" that someone standing at the south pole perceives. Indeed, every point on the entire surface of the earth has an "up" that is opposite to its counterpart point on the opposite side of the earth. Hence, the earth has no inherent upper side or under side. An astronaut drifting in extraterrestrial space also cannot point to a particular star and say that star is "up" and then point to another star and say that star is "down." Nor can he say that the earth is "under" one part of the cosmos and is "above" another part. An astronaut orbiting the earth could, of course, say from force of habit that the earth is "down" from his position in space. But that "down" direction constantly changes with respect to the stellar background as his position in earth orbit changes. In the actual cosmos, then, it is meaningless to say the earth is "upon," "over," or "above" anything or, again, "nothing." That underscores the difference between the modern view of the cosmos and the inherent top-to-bottom layered arrangement of the biblical cosmos.

11. The Pit of Sheol

Job 26:7 therefore provides added support to our understanding that the biblical earth is flat, for only a flat earth can have an upper side and an under side. And only an earth with an upper side and an under side can be hanging "upon" something, or be "over" or "above" something, or, again, "nothing."

It seems clear, then, that the description of the earth in Job 26:7 is not compatible with the modern view of the earth in space. But then, what is the real meaning of Job 26:7? What is meant by "the north" and the "empty place" of the first part of the verse? And what does the "nothing" of the second part of the verse represent?

For starters, the reference to "the north" cannot simply mean the direction; in order for "the north" to be "stretched" over the "empty place," it would have to be something material. A location in the northern part of the earth does not seem to fit either, for there is nothing in the Bible to suggest that the northern part of the disk of the earth is uniquely stretched over an "empty place" under it. It would therefore appear that the contextual use of the north in the passage is as a figure of speech that stands for something else, and that it would have been understood as such by the people who lived at the time the Book of Job was written.

How, then, do we determine the contextual meaning of "the north" in Job 26:7? One obvious way would be to try to find other passages in the Bible that could help in determining its meaning. But first, let's take a look at the original Hebrew word that is translated as "north" in the passage. That word is *tsaphown*, which, in fact, has a meaning of "north" or "northward." It was also the name of a mountain to the north in Syria that the locals believed to be the home of the gods. However, it is not likely that that mountain is what is being referred to in Job 26:7. What would it mean in saying that the mountain was "stretched over the empty place"? And why would the verse jump from referring to a mountain and then to the whole earth as hanging over or upon nothing?

Because the contextual meaning of the word as it is used in the passage is not readily clear, some translations of the Bible, instead of using the word "north," transliterate *tsaphown* and leave it at that. The NRSV, for example, renders the passage as *"He stretches out Zaphon over the void."*

So, are there any clues elsewhere in the Bible that might help us in resolving the contextual meaning of "the north" in Job 26:7? In our search for clues, elsewhere in the book of Job would be a good place to start; its author knew what he meant when he made reference to "the north," and perhaps he used similar terminology in another passage.

As a matter of fact, the word "north" does appear in two other places in the KJV translation of Job: verses 37:9 and 37:22. Verse 37:9 states: *"Out of the south cometh the whirlwind: and cold out of the north.*" That verse can be eliminated, however, because the original Hebrew word translated there as "north" is *mezareh* rather than *tsaphown*. *Mezareh* actually means "scattering winds," and, in the context of the passage, is taken as meaning the scattering winds that bring the cold. We had previously noted in Chapter 8 that the original Hebrew word for "south" in that verse is *cheder*, which actually means "chamber," i.e., a "chamber" of heaven. Again, the NRSV gives a more accurate translation of the verse than does the KJV: "*From its chamber comes the whirlwind, and cold from the scattering winds.*"

That leaves Job 37:22, which states in the KJV:

Fair weather cometh out of the north: with God is *terrible majesty.*

This appears promising, for the original Hebrew word that is translated as "north" there is *tsaphown*. So what does it mean when it says that "fair weather cometh out of the north?" The Hebrew word that is translated as "fair weather" is *zahab*, which actually has "gold" as its primary meaning. Moreover, "fair weather" does *not* appear as a meaning for *zahab* in any of the referenced lexicons. For that matter, Job 37:22 is the *only* location in the entire KJV where *zahab* is translated as "fair weather," in more than 380 other locations in the KJV, it is translated as "gold" or "golden." In addition, several other versions of the Bible translate *zahab* in Job 37:22 as "golden splendor" or as a similar term. For example, the NIV reading of the passage is as follows:

Out of the north he comes in golden splendor; God comes in awesome majesty.

What, then, does the word "north" mean in this verse? The following note for the verse in the NAB sheds some light on the matter:

Now the storms of doubt and ignorance disappear, and from the North, used here as a symbol for God's mysterious abode, comes the splendor of the manifestation of God's majestic ways."

So, according to this note, the "north" is a symbol for God's "mysterious abode." However, in the context of the passage, "the north" is not a mere symbol. The passage appears to indicate that God comes in "golden splendor" from his "mysterious abode" in the north. That would seem to indicate that the ancient Hebrews believed that there was a special relationship between their God and "the north." That understanding is also confirmed by the BDB in its exposition on the Hebrew word for "north": "*remote parts of north* Is 14[13] (as divine abode)." So both the NAB and the BDB lexicon relate "the north" to God's abode.

11. The Pit of Sheol

Wherever it was considered to be, God's "abode" would also be where his throne would be located. That could not have been the mountain named "North" in Syria where the pagan gods of the locals were believed to live. Some have suggested that God's abode was considered to have been on that mountain, but the writers of the Bible would not have associated their God with such a place. No, the verse must have some other meaning for "north."

Moreover, numerous passages in the Bible place God's throne in heaven, as for example in Psalm 11:4:

> *The LORD is in his holy temple, the LORD'S throne is in heaven....*

And Psalm 103:19:

> *The LORD hath prepared his throne in the heavens....*

And also Isaiah 66:1

> *Thus saith the LORD, The heaven is my throne, and the earth is my footstool....*

The reference to "the north" in Job 37:22 would therefore appear to indicate the part of heaven where the ancient Hebrews specifically believed their God's "abode" and his throne to be located.

Ezekiel's vision, a part of which we looked at in Chapter 8, provides support for that conclusion. The vision begins in Ezekiel 1, and the relevant parts, here from the ASV, are as follows:

> 1 *Now it came to pass ... as I was among the captives by the river Chebar, that the heavens were opened, and I saw visions of God.*
> 4 *And I looked, and, behold, a stormy wind came out of the north, a great cloud, with a fire infolding itself, and a brightness round about it...*
> 5 *And out of the midst thereof came the likeness of four living creatures....*
> 22 *And over the head of the living creature there was the likeness of a firmament, like the terrible crystal to look upon, stretched forth over their heads above.*
> 25 *And there was a voice above the firmament that was over their heads...*
> 26 *And above the firmament that was over their heads was the likeness of a throne, as the appearance of a sapphire stone; and upon the likeness of the throne was a likeness as the appearance of a man upon it above.*

Note that Ezekiel saw the heavens "open" and the "visions of God" coming "out of the north." Here, the Hebrew word translated as "north" is *tsaphown*, the same word that is used in the original Hebrew of Job 26:7 and Job 37:22. Significantly, the vision had the appearance of the firmament of heaven with God's throne above it. This passage in Ezekiel therefore provides substantial support for the idea that we developed from Job 37:22 concerning the location of God's "secret abode" and throne in the northern part of the firmament of heaven.

Isaiah 14:12–15 is another passage that provides evidence for the idea that the God of the Bible has his throne in the northern part of the sky:

> 12 *How art thou fallen from heaven, O Lucifer, son of the morning! how art thou cut down to the ground, which didst weaken the nations!*
> 13 *For thou hast said in thine heart, I will ascend into heaven, I will exalt my throne above the stars of God: I will sit also upon the mount of the congregation, in the sides of the north:*
> 14 *I will ascend above the heights of the clouds; I will be like the most High.*
> 15 *Yet thou shalt be brought down to hell, to the sides of the pit.*

"Lucifer" (*heylel*, "morning star" in the Hebrew) desired to ascend into heaven, exalt his throne above the stars of God, and sit upon the mount of the congregation in the "sides of the north." The Hebrew word that is translated as "north" in that passage is *tsaphown*, the same word found in the original Hebrew of Job 37:18, Job 26:7, and Ezekiel 1:4. The Hebrew word translated as "mount" in the passage is *har*, and, in its interpretation of that word, *Strong's Concordance* notes that "mount" or "mountain" is sometimes used in a figurative sense in the Bible. Since the "morning star" wanted to ascend into heaven and exalt his throne above the stars of God, his throne would necessarily be placed upon the firmament of heaven, in which the stars are set. And, as it states in verse 14, in ascending "above the heights of the clouds," the morning star would "be like the most High," or again like God sitting on his throne above the firmament of heaven. And since the "morning star" would be sitting on his throne on the firmament of heaven, the "mount of the congregation" must therefore be an allusion to the firmament of heaven. That would be understandable, as the vault of heaven could, in a figurative sense, be considered a "mount" over the earth. In that case, the "congregation" would refer to the other stars of the night sky, or the "host of heaven," i.e., the angels.

The reference to the "sides of the north" in the passage would therefore refer to the north side of the firmament of heaven (the Hebrew word translated as "sides" in the passage is *yerekah*, which has "flank" or "sides" among its meanings). This passage thus provides added support to the conclusion that the biblical God's "abode" is in the northern part of the firmament of heaven.

With the added evidence provided in the verses that we have examined above, we can make the following provisional clarifying interpretation of Job 37:22:

> Out of his abode in the northern sky he comes in golden splendor; God comes in awesome majesty.

Now, we concluded from Job 37:22, Isaiah 14:12–15, and the first chapter of Ezekiel that God's "mysterious abode" is located in the northern part of the sky. But what, in the cosmological sense, is so special about the northern part of the sky?

Actually, that the ancient Hebrews would have considered the northern sky to have a special relationship to their God makes a certain amount of sense. When one observes the movement of the stars during the course of a night, it appears that they travel in a circle around a point in the northern sky. (That point is near where Polaris, the North Star, is located today, but, because of the precession of the earth's axis over time, there was no significant star that was visible near that point during the time the Old Testament was being written. Thus, at that time, all of the stars would have appeared to move in circles around that point.) As the stars were, in Old Testament terminology, considered to be the "host of heaven," it would be logical for the ancient Hebrews to have considered that the point around which the stars circled to have had a special significance insofar as their God was concerned. Perhaps it was even believed that the point around which the stars circled was where God's throne was located in God's "mysterious abode" above the firmament. (In the Book of Enoch there is a passage that describes the heavenly host as circling around a crystalline structure wherein God dwelled, which, by implication would also be where his throne was located. Though it does not explicitly state that the crystalline structure was located in the north part of heaven, the description of the circling host of heaven implies that it was. See Appendix C.)

If the ancient Hebrews did believe that their God's abode was located at that point in the northern sky, one might wonder why there are no passages in the Bible that clearly reflect that view. In answer to that, perhaps it was for the same reason that the ancient Hebrews avoided saying the name of their God. Because of their religious sensitivities, they possibly also avoided drawing attention to the exact location of their God's abode in the firmament of heaven. Providing general statements about his abode—such as referring to it as the north—might have been acceptable; but the writers of the Bible quite possibly would not have wanted to give the appearance of pinpointing the exact location of their God or of his secret abode. Doing so may well have been considered to be their equivalent of a taboo.

Regardless of whether or not the ancient Hebrews actually considered their God's abode to be specifically at the point in the northern night sky where the stars circled, it still appears likely that they considered his abode to be in the north part of the firmament of heaven. Because of the relative importance of "the north" in that respect, the term could well have been expanded in meaning to be a figure of speech that included, by association, the whole of the firmament of heaven. Job 37:22 would therefore have been understood as meaning *"Out of heaven he comes in golden splendor....,"* which would make perfect sense.

Since Job 37:22 lends support to the conclusion that, in certain contexts, "the north" is actually a figure of speech used as a reference to heaven, we would be warranted in concluding that the author of Job retained that connection in his reference to the "north" in Job 26:7: *"He stretcheth out the north over the empty place...."* That passage would therefore have been contextually understood as "He stretcheth out heaven over the empty place...." The NIV partially concurs with that understanding, for it provides the following translation of the passage: *"He spreads out the northern skies over empty space...."* However, since "the north" also appears to have been used as a figure of speech for the whole of heaven, it would not have been only the "northern skies" that were "spread" over "empty space," but rather all of the sky. The NIV version of the passage would therefore be more properly rendered as "He spreads the skies [or heaven] over empty space." Significantly, in its rendition of the first part of Job 26:7 the NEB translates *tsaphown*, "the north," as meaning just that: the whole of the sky, with no qualification: *"God spreads the canopy of the sky over chaos."*

So when Job 26:7 states that God "stretcheth out the north," it can be taken as meaning that God stretched out the firmament of heaven, or that God stretched out the sky. The original Hebrew word translated as "stretched" in the passage is *natah*, which does have the meaning of "to stretch out" or "to spread out." It is noteworthy that the idea of God's spreading out or stretching out the sky or heaven appears in numerous other passages in the Bible. For example, Job 9:8 states: "[God] *alone spreadeth out* [natah] *the heavens, and treadeth upon the waves of the sea."* Isaiah 45:12 provides another example: *"I, even I, have made the earth, and created man upon it; I, even My hands, have stretched out* [natah] *the heavens...."* There are several other such passages in the Bible.

Of particular note, the Hebrew word *natah*, with "to spread out" as one of its meanings, brings to mind the Hebrew word *raqa`*, which—as well as "to beat out"—also means "to spread out." We saw in Chapter 8 that same word was used in Job 37:18: *"Hast thou with him spread out the sky, which is strong, and as a molten looking glass?"* That verse, remember, indicates that God "spread out," or "beat out" the solid firmament of heaven. Thus, the use of the word "stretched" in Job 26:7 follows the same usage as in several other passages in the Bible in relation to God's stretching out, or spreading out, the heavens.

So we can now understand that "the north" of Job 26:7 is actually a figure of speech for the firmament of heaven. The NAB, in fact, has a note for Job 26:7 which bears out that understanding. That note states: "The North: used here as a synonym for the firmament, the heavens."

It follows, then, that the words *"He stretcheth out the north over the empty place"* of Job 26:7 mean that God stretched out the firmament of heaven over the "empty place" below it. But what is significant is that the Hebrew word that is translated as "the empty place" in the passage is *tohuw*, the meanings of which include "formlessness," "emptiness," and "nothingness." Elsewhere in the Bible, and depending on the version, *tohuw* is translated as "confusion," "nothingness," "chaos," "emptiness," "thing of naught," or "broken down"—e.g., in Isaiah 24:10, 34:11, and 41:29. But most significantly, that same Hebrew word is also found in Genesis 1:2, where it is translated in the various versions of the Bible as "without form," "formless," or "unformed," as, for example, here in the KJV: *"And the earth was without form, and void; and darkness was upon the face of the deep. . . ."*

But that is not all. As we saw in Chapter 7, the Hebrew word that is translated as "the deep" in Genesis 1:2—*". . . and darkness was upon the face of the deep. . . ."*—is *tehowm*, which also means "the abyss." Moreover, in the cosmological sense, *tohuw* means "watery chaos,"[1] which reflects the initial condition of "the deep" before God began the creative process. Thus, in Genesis 1:2, the "watery chaos" of the unformed earth was equated with *tehowm*, the abyss of the "deep." Job 26:7 therefore indicates that, when God stretched out the "north" (i.e., the firmament of heaven), he stretched it out "over" the "watery chaos"—*tohuw*—of the deep. In fact, as we saw above, the NEB translation of Job 26:7 reflects that understanding: *"God spreads the canopy of the sky over chaos."* The JPS Tanakh also agrees with that interpretation of *tohuw*: *"He it is who stretched out Zaphon* [the "north," heaven] *over chaos. . . ."* It seems clear, then, that the first part of Job 26:7 is a restatement of the creation of the firmament of heaven over the waters of the deep as described in Genesis 1:7: *"And God made the firmament, and divided the waters which were* under the firmament from the waters which were above the firmament." The "waters which were under the firmament" would be those of the deep, or *tehowm*, which are equated with the "watery chaos" of *tohuw*.

So *tohuw*—translated as "without form" in Genesis 1:2 and as "the empty place" or "chaos" in Job 26:7—refers to the chaotic state of "the deep" over which the biblical god created the firmament of heaven. This indicates that, in certain biblical contexts, "the north" is indeed meant as a figure of speech for the firmament of heaven as a whole, for the biblical God created the whole firmament of heaven, and not just the northern part of it.

We can now analyze the remainder of Job 26:7, which, together with the first part of the verse, provides an example of the literary parallelism that is frequently found in the Bible. We'll use the JPS 1917 Tanakh version of the verse here since, instead of using the term "upon nothing," it uses "over nothing," which makes more sense in the context.

> *He stretcheth out the north over the empty space, and hangeth the earth over nothing.*

In this type of parallelism, the first part of a passage has a parallel in the second part, which has a conceptual relationship and a grammatical structure that is similar to the first part. In some cases, the second part restates the reference to the object of the first part. In other cases, the second part refers to something that is different from the object of the first part, but which has a close relationship to that object. Job 26:7, in fact, provides an example of the second type of parallelism—that is, the first part of the verse describes the creation of the firmament of heaven over the watery chaos of the deep, and the second part of the verse describes the creation of the earth over "nothing."

We saw in Chapter 10 that there are numerous mentions in the Bible to the waters of the deep under the earth, as, for example, in Psalm 24:1–2 (here, the NEB):

> *The earth is the Lord's. . . . For it was he who founded it upon the seas and planted it firm upon the waters beneath.*

The implication here is that when God created the earth under the firmament of heaven, he positioned it over the "watery chaos"— *tohuw*—of the deep that had been previously divided by the creation of the firmament.

And that is also what the author of Job also seems to saying in verse 26:7: *He stretcheth out the north* [the firmament of heaven] *over the empty space* [tohuw, "nothingness," "watery chaos"], *and hangeth the earth over nothing."* The implication is that the author intended the "nothing" of the second part of the verse to parallel and to reiterate "the empty space" (or

11. The Pit of Sheol

"empty place" in other versions of the Bible) of the first part. However, instead of using *tohuw*, which was the word he used in the first part of the verse, he used *beliymah* in the second part, which several versions of the Bible translate as "nothing." There are, however, several Hebrew words that mean "nothing," and they appear numerous times in the Bible. But, curiously, Job 26:7 provides the *only* occurrence of *beliymah* in the Bible—which brings up the possibility that, since there are no other passages with which to compare usage of the word, it may have had nuances that present-day translators are not aware of. The Hebrew word *beliymah* is, in fact, a combination of two words: *beliy* (or *belii*), which has various meanings, including a basic meaning of "wearing out," and *mah*, which has meanings that include "what," "aught," and "anything." *Beliy* is sometimes used as the negative "not," so combined with "mah" as *beliymah*, it can be taken as having a literal meaning of "not anything," or "nothingness."

That brings us back to the JPS Tanakh, which translates *beliymah* as "emptiness" in Job 26:7:

> *He it is who stretched out Zaphon over chaos, Who suspended earth over emptiness.*

And that in turn brings us back to *tohuw*, which is translated as "chaos" in that passage, but which also has "emptiness" among its various meanings. The JPS Tanakh thus, in poetic reiteration, equates *beliymah* of the second part of the passage with *tohuw* of the first part. Even more significantly, as previously noted, another meaning of *tohuw* is "nothingness." In Job 6:18 of the KJV, for example, *tohuw* is translated as "nothing.": "*The paths of their way are turned aside; they go to nothing, and perish.*" Thus, it seems clear that *beliymah*, translated as "nothing" in Job 26:7 of most versions of the Bible, parallels, and is taken as meaning the same thing as, or reiterates, *tohuw*—which can also be translated as "nothing"—in the first part of the verse.

So, although he used different Hebrew words for what was under the firmament of heaven in the first part of the passage, and what was under the earth in the second part, it would appear that the author of Job, in using poetic reiteration, had the same thing in mind for both parts of verse 26:7. In the first part of the passage, God stretched the firmament of heaven over *tohuw*—the watery chaos of the abyss of the deep. In the second part, God in turn hanged the earth over *beliymah*, which would, of necessity, also refer to the watery chaos of the abyss of the deep.

But there appears to be more to it than just that, for *beliymah* seems to have had another nuance of meaning that can be applied to yet another aspect of the biblical cosmos. The Hebrew word *beliy*, which forms the first part of *beliymah*, besides having a basic meaning of "a wearing out," also means "destruction" or "corruption," as exemplified in Isaiah 38:17-18 (here the RSV):

> 17 *Lo, it was for my welfare that I had great bitterness; but thou hast held back my life from the pit of destruction* [beliy], *for thou hast cast all my sins behind thy back.*
> 18 *For Sheol cannot thank thee, death cannot praise thee; those who go down to the pit cannot hope for thy faithfulness.*

"The pit of destruction" is, of course, a reference to Sheol (*she'owl* in the Hebrew, which is usually translated as "hell" in the KJV), and is, in fact, reiterated as "Sheol" in verse 18 of the above passage.

Strong's Concordance points out that Sheol, during most of the Old Testament time, ". . . was not understood to be a place of punishment, but was simply the ultimate resting place for all mankind." The sense in the Old Testament, then, is that *everyone* who dies, not just evildoers, would go down into Sheol. For example, in Genesis 37:35 we find Jacob lamenting his son Joseph, who he believes is dead (this from the RSV):

> *All his sons and all his daughters sought to comfort him; but he refused to be comforted, and said, "No, I shall go down to Sheol to my son, mourning." Thus his father bewailed him.*

Sheol, hell, was not generally considered to be a place of eternal, fiery torment until later when it played a prominent part in Christian belief. (The Book of Enoch, which was written within two centuries before the Christian era, was apparently instrumental in the development of the idea of Sheol, or hell, as a place of eternal, fiery torment; see Appendix C). In contrast, the references to Sheol in the Old Testament usually indicate that it was a dark place where the dead existed in a shadowy state. Job 10:20–22 illustrates this quite well (here from the ASV):

> 20 *Are not my days few? cease then, And let me alone, that I may take comfort a little,*
> 21 *Before I go whence I shall not return, Even to the land of darkness and of the shadow of death;*
> 22 *The land dark as midnight, The land of the shadow of death, without any order, And where the light is as midnight*

The Bible is not consistent in its placement of Sheol. Some verses seem to indicate that Sheol is located within the earth, but others indicate that it is in the abyss of the deep under the earth. Of course, there was no requirement that the ancient Hebrews over the hundreds of years of their existence would have exactly the same view about the location of Sheol. It

11. The Pit of Sheol

is possible that some of the ancient Hebrews considered Sheol to be within the earth, while others considered it to be in the abyss under the earth.

Job 10:21–22, above, seems take the latter view, for it describes the land of darkness—i.e., Sheol—as being a place "without any order." That would be a reference to the watery chaos of *tohuw*, the abyss of the deep under the earth. Job 26:5–6, provides further evidence of that view (here from the ASV):

> 5 *They that are deceased tremble Beneath the waters and the inhabitants thereof.*
> 6 *Sheol is naked before God, And Abaddon hath no covering.*

The "waters" that are mentioned in this passage would necessarily need to be the waters of the abyss. The verse also appears to indicate that the "deceased" reside in a place that is within the deepest depths of the "waters."

Psalm 69:14–15 also confirms this understanding (here, the KJV):

> 14 *Deliver me out of the mire, and let me not sink: let me be delivered from them that hate me, and out of the deep waters.*
> 15 *Let not the waterflood overflow me, neither let the deep swallow me up, and let not the pit shut her mouth upon me.*

The Hebrew word translated as "waters' in verse 14 is *mayim*, the same Hebrew word that is translated as "waterflood" in verse 15. In verse 15, "let not the pit shut her mouth upon me" reiterates "neither let the deep swallow me up," thus equating the "pit" (Sheol) with the deep. These verses thus reinforce the conclusion that Sheol is associated with the deep under the earth.

A further example of this view is found in Ezekiel 31:15, which states (here the ASV):

> *Thus saith the Lord Jehovah: In the day when he went down to Sheol I caused a mourning: I covered the deep for him, and I restrained the rivers thereof; and the great waters were stayed....*

The original Hebrew word translated as "deep" in that passage is *tehowm*, i.e., the abyss under the earth, and the Hebrew word translated as "waters" is, again *mayim*. Again, Sheol is associated with the deep.

Jonah 2:2–6 provides another example, for it states (here, the NIV):

> 2 *And he said: "I cried out to the LORD because of my affliction, And He answered me. "Out of the belly of Sheol I cried, And You heard my voice.*
> 3 *For You cast me into the deep, Into the heart of the seas, And the floods surrounded me; All Your billows and Your waves passed over me....*

> 5 *The waters surrounded me, even to my soul; The deep closed around me; Weeds were wrapped around my head.*
> 6 *To the roots of the mountains I sank down; the earth beneath barred me in forever. But you brought my life up from the pit, O LORD my God.*

The word translated as "the deep" in verse 5 is, again, *tehowm*. In this passage the deep is equated with the "belly of Sheol," or "the pit." The passage is also clear in indicating that Sheol is located in the watery abyss under the earth. Jonah's references to the waters that surrounded him and his sinking down to the roots of the mountains and being barred by the earth indicate that the watery abyss he was in was located under the earth.

Regardless of whether the ancient Hebrews considered Sheol to be within the earth or under the earth, they considered it to be a real place that was a physical part of their cosmology, and not some otherworldly realm beyond human ken (just as we concluded in Chapter 9 that the ancient Hebrews considered heaven to be a real, physical place and not some ethereal abode beyond human ken). Amos 9:1–2 provides an example of physical nearness of both of these parts of their cosmology:

> 1 *I saw the Lord standing beside the altar: and he said, ... I will slay the last of them with the sword: there shall not one of them flee away, and there shall not one of them escape.*
> 2 *Though they dig into Sheol, thence shall my hand take them; and though they climb up to heaven, thence will I bring them down.*

Even those passages which seem to indicate that Sheol is located within the earth do not necessarily contravene the idea that it is actually located in the abyss under the earth. One such passage is Numbers 16. In that chapter, Korah and his followers had set themselves against Moses. At God's direction, Moses tells the assembly to leave the tents of the rebels. God subsequently causes the ground to open up to send Korah and his followers, and their property as well, into Sheol. Here are the relevant verses from Numbers 16 in the NRSV:

> 31 *As soon as he finished speaking all these words, the ground under them was split apart.*
> 32 *The earth opened its mouth and swallowed them up, along with their households— everyone who belonged to Korah and all their goods*
> 33 *So they with all that belonged to them went down alive into Sheol; the earth closed over them, and they perished from the midst of the assembly.*

It is significant that Korah and his followers were swallowed up alive and, along with their possessions, went down into Sheol. The implication is that Sheol was considered to be a real, physical place under the surface of the earth and not some otherworldly domain. The question is, then, how far down within the earth was Sheol considered to be? There is nothing in the passage to indicate that the splitting of the earth did not go all the way down to the depths of the abyss under the earth. Indeed, in the above passage from the book of Jonah we noted that the watery deep—the pit of Sheol—that Jonah was in was located under the earth. We also noted that Jonah had repeatedly referred to the waters that surrounded him, and he said that he had sunk down to the roots of the mountains and was barred by the earth. That would indicate that the earth was above him, and he therefore would have been in the abyss under the earth. So too, could the splitting of the earth have carried Korah and his followers down past the "roots of the mountains" into the abyss under the earth.

Another passage indicating that the pit of Sheol is a watery place is Psalm 88:3–7, which states (again, the ASV):

> 3 *For my soul is full of troubles, And my life draweth nigh unto Sheol.*
> 4 *I am reckoned with them that go down into the pit; I am as a man that hath no help. . . .*
> 5 *Cast off among the dead, Like the slain that lie in the grave, Whom thou rememberest no more, And they are cut off from thy hand.*
> 6 *Thou hast laid me in the lowest pit, In dark places, in the deeps.*
> 7 *Thy wrath lieth hard upon me, and thou hast afflicted me with all thy waves.*

There, "the deeps" and "the waves" refer to the waters of the abyss under the earth, and the text is plain in showing that the "lowest pit," or Sheol, is located in the waters of "the deeps."

In 2 Samuel 22:5–7 and 16–17 we find David giving voice to a song that provides another reference to the watery nature of Sheol (here, from the JPS 1917 Tanakh):

> 5 *For the waves of Death compassed me. The floods of Belial assailed me.*
> 6 *The cords of Sheol surrounded me; the snares of Death confronted me.*
> 7 *In my distress I called upon the LORD, yea, I called unto my God; and out of His temple He heard my voice, and my cry did enter into His ears.*

> 16 *And the channels of the sea appeared, the foundations of the world were laid bare by the rebuke of the LORD, at the blast of the breath of His nostrils.*
> 17 *He sent from on high, He took me; He drew me out of many waters;*

The "waves," "floods," and "waters" in that passage are references to the abyss of the deep under the earth and are equated with the "cords of Sheol." Note also that the passage states "the foundations of the world were laid bare." The foundations would have to refer to the pillars that support the earth above the abyss under it, and God laid bare the pillars so that he could draw David "out of many waters." Note also that the verse refers to the "floods of Belial." The original Hebrew word that is transliterated as "Belial" in the passage is *beliyyaal*, the first part of which is the Hebrew word *beliy*, the same word that we noted is the initial part of *beliymah*, the term that was translated as "nothing" in Job 26:7. The second part of the word, *yaal*, has a meaning of profit or gain, but in a negative sense as it is commonly used in the Old Testament; an example is found in Proverbs 11:4: *"Riches profit not in the day of wrath...."* Thus, *beliyyaal*—consisting of *beliy*, "not," and *yaal*, "profit"—has a meaning of "worthlessness."

The expression, "the floods of Belial," is also found in Psalm 18:4–5 (here again from the JPS 1917 Tanakh):

> 4 *The cords of Death compassed me, and the floods of Belial assailed me.*
> 5 *The cords of Sheol surrounded me; the snares of Death confronted me.*

Thus this passage and 2 Samuel 22:5–6 associate "the floods of Belial" with Sheol, the watery abyss of the deep under the earth.

It is also interesting to note that "Belial" eventually came to be used as another name for Satan, who, in New Testament times was considered to be the overseer of Sheol. The development is noticeable in Nahum 1:11 of the NJB: *"From you has emerged someone plotting evil against Yahweh, one of Belial's counsellors...."* Satan is frequently called Belial in the Dead Sea scrolls, and the term is also found in the New Testament. In 2 Corinthians 6:15, for example, we find: *"And what concord hath Christ with Belial?..."*

Since, in Old Testament times, *beliyyaal* (a variant of *beliy*), was used in association with, and was another name for, Sheol, it is reasonable to conclude that the ancient Hebrews used *beliymah* (another variant of *beliy*) as yet another name for Sheol or "hell." That would thus provide further evidence for the connection between Sheol, the abyss of the deep, and *beliymah* in Job 26:7.

And that brings up Job 26:5–6, which we previously referred to, and, of course, immediately precedes Job 26:7, the verse that is the subject of our discussion. Again, those verses state (here again from the ASV):

> 5 *They that are deceased tremble Beneath the waters and the inhabitants thereof.*
> 6 *Sheol is naked before God, And Abaddon hath no covering.*

The Hebrew word that is translated as "Abaddon" is *'abaddown*, which has a meaning of "place of destruction" or "destruction." In fact, several versions of the Bible translate *'abaddown* as "destruction" in that verse. Here, for example, is the KJV rendition of the verse:

> *Hell* is *naked before him, and destruction hath no covering.*

The verse provides another example of biblical poetic reiteration, in which *'abaddown* is used as another word for Sheol, or "hell." Thus, the second part means the same thing as the first part and both parts indicate that Sheol—which is the same as *'abaddown*, also called "destruction—is naked or uncovered before God. But, as previously noted, the Hebrew word *beliy* also includes "destruction" in its meanings, as in Isaiah 38:17 (here from the NIV): "... *In your love you kept me from the pit of destruction* [beliy]...." The "pit of destruction" is, again, a reference to Sheol, which is translated as "hell" in the KJV. The equating of Sheol with "destruction" is found in several places in the Bible, and thus provides further evidence that *beliymah* had a connotation that linked it to Sheol. Psalm 103:4 provides an example:

> *Who redeemeth thy life from destruction; who crowneth thee with lovingkindness and tender mercies*

In this case the Hebrew word translated as "destruction" is *shachath*, which has, according to *Strong's Concordance* has the meaning of "a pit." Several versions of the Bible, in fact, translate *shachath* as "pit" in that passage; for example, the RSV: "*who redeems your life from the Pit....*" That same word is also translated as "destruction" in Psalm 55:23 where its meaning quite clearly refers to Sheol:

> *But thou, O God, shalt bring them down into the pit of destruction: bloody and deceitful men shall not live out half their days; but I will trust in thee.*

It would follow, then, that the Hebrew word *beliymah*, as well as meaning "nothing," as it is usually rendered in Job 26:7, could also mean "destruction." That understanding could easily follow from the meaning of *beliy*, which as we previously noted, has a basic meaning of "wearing out." If something becomes excessively worn out, it becomes for all intents and purposes "destroyed." It would thus not be a stretch to consider that Job 26:7 could be translated as "He ... hanged the earth over "destruction." However, "destruction" is another word for the pit of Sheol, and "destruction" also indicates a chaotic condition, thus again bringing to mind the chaos of *tohuw*, the watery abyss of the deep under the earth.

It follows, then, if *beliymah* was used as another term for Sheol in Old Testament times, and given what we have learned about the first part of the passage, Job 26:7 could be contextually translated as follows:

> He stretched out the firmament of heaven over the chaotic waters of the abyss, and suspended the earth over the pit of Sheol.

In Job 26:6 the "is" in "*Sheol is naked before God*" does not appear in the original Hebrew, so it could just as well be read as "Sheol *was* naked before God." Job 26:6 should therefore be taken as describing the original state of Sheol before the creation of heaven and the earth. Thus Sheol, the watery chaos of the abyss—*tehowm*, which is equated with *tohuw*—that initially was naked and had no covering before God, was subsequently "covered," first with the firmament of heaven, and then with the disk of the earth, just as it is indicated in Genesis.

On the basis of our analysis, we can now complete our contextual interpretation of Job 26:5–7:

> 5 They that are deceased tremble beneath the waters of the abyss and its inhabitants.
> 6 Sheol was naked before God, and destruction had no covering.
> 7 He stretched out the firmament of heaven over the watery chaos, and suspended the earth over the pit of Sheol.

That passage brings another point of similarity between the abyss of the deep (that is, *tehowm*) and Sheol; just as the Bible does not describe the abyss of the deep as having been created by God, but to have been pre-existing, so too it does not describe Sheol as having been created by God. Thus, Sheol and the abyss of the deep can also be equated on that similarity.

There is yet more evidence that the abyss of the deep under the earth was considered to be Sheol. For example, in Old Testament times, *'abad-down*, as noted above, was used as another word for Sheol. However, it later became yet another name for Satan, as exemplified in Revelation 9:11 (here, the ASV):

> *They have over them as king the angel of the abyss: his name in Hebrew is Abaddon, and in the Greek* tongue *he hath the name Apollyon.*

Note that Abaddon, "Satan," is the angel, or king, of the abyss. The Greek word translated as "abyss" is *abussos*, which has as its meanings "bottomless," "the abyss," and "the pit." Some other versions of the Bible translate *abussos* as "the abyss," while yet others translate it as "the bottomless pit." The KJV rendition of Revelation 9:11, for example, has the following:

> *And they had a king over them, which is the angel of the bottomless pit, whose name in the Hebrew tongue is Abaddon, but in the Greek tongue hath his name Apollyon.*

A pit completely enclosed by the earth would not be "bottomless." However, in the biblical cosmos, the abyss under the earth could easily have been considered bottomless.

There are several references to the abyss in the Book of Revelation. Here is Revelation 9:1 of the ASV:

> *And the fifth angel sounded, and I saw a star from heaven fallen unto the earth: and there was given to him the key of the pit of the abyss.*

In New Testament times the abyss was believed to be the home of demons, and in Luke 8:30–31 the demons who possessed a man asked Jesus not to send them back to the abyss. Here, in the KJV rendition of that passage, *abussos*, meaning the abyss, is translated as the "deep," the same word the KJV translators used for the translation of *tehowm*, the Hebrew word for the watery abyss in Genesis 1:2:

> 30 *And Jesus asked him, saying, What is thy name? And he said, Legion: because many devils were entered into him.*
> 31 *And they besought him that he would not command them to go out into the deep.*

Several other versions of the Bible translate *abussos* in this passage as "abyss"—here, the NRSV:

> *They begged him not to order them to go back into the abyss.*

In Romans 10:7 of the KJV we find the following, which equates the abyss—*abussos*, translated as "the deep"—with the place of the dead, or Sheol, as it was called in the Old Testament:

> *Or, Who shall descend into the deep? (that is, to bring up Christ again from the dead.)*

The NKJV translates the verse as follows:

> *or, " 'Who will descend into the abyss?' " (that is, to bring Christ up from the dead).*

That passage therefore provides yet further evidence that the place of the dead was considered to be within the waters of "the deep."

Revelation 20:1–3 also refers to the abyss (here the NJB):

> 1 *Then I saw an angel come down from heaven with the key of the Abyss in his hand and an enormous chain.*
> 2 *He overpowered the dragon, that primeval serpent which is the devil and Satan....*
> 3 *He hurled him into the Abyss and shut the entrance and sealed it over him....*

In addition, the Greek Septuagint uses *abussos* as the Greek translation of the Hebrew *tehowm* in Genesis 1:2. A note to Luke 8:31 in the NAB further clarifies the connection between the abyss mentioned in the above passages and the "watery deep" of the creation story in Genesis:

> Abyss: the place of the dead (Romans 10:7) or the prison of Satan (Rev 20:3) or the subterranean "watery deep" that symbolizes the chaos before the order imposed by creation (Genesis 1:2). (NAB)

Though the note gives three different definitions for "abyss," all three, as the evidence indicates, refer to the same thing: the abyss of the deep, or Sheol, under the earth.

It seems clear, then, that *abussos*, the original Greek word that is translated as the "abyss" or "deep" in the New Testament, refers to the same thing as *tehowm*, the original Hebrew word for "abyss" or "deep" in Genesis. So, again, when Job 26:6 is added to Job 26:7, it completes the abbreviated version of the creation of heaven and the earth that is described in Genesis 1. However, the version in Job emphasizes the relationship that the pit of Sheol, the abyss, has with each part of the creation process.

All of the material presented in this chapter thus drives the conclusion that Job 26:7—*He ... hangeth the earth upon nothing,* or *He ... suspends the earth over nothing* as the NIV states—has nothing to do with the modern concept of the earth in extraterrestrial space. Rather, in the biblical cosmos, the "nothing" over which the earth hangs or is suspended is a reference to the abyss, the pit of Sheol, under the earth.

Chapter 12

The Lights in the Firmament

Now let's get back to the creation story in Genesis 1. The cosmological portion of that story continues in verses 14 through 19.

> *14 And God said, Let there be lights in the firmament of the heaven to divide the day from the night; and let them be for signs, and for seasons, and for days, and years:*
> *15 And let them be for lights in the firmament of the heaven to give light upon the earth: and it was so.*
> *16 And God made two great lights; the greater light to rule the day, and the lesser light to rule the night: he made the stars also.*
> *17 And God set them in the firmament of the heaven to give light upon the earth,*
> *18 And to rule over the day and over the night, to divide the light from the darkness: and God saw that it was good.*
> *19 And the evening and the morning were the fourth day.*

According to this passage, the sun, moon, and stars are merely lights set in the solid firmament of heaven; they have no purpose but to give light to the earth and to provide the earth with signs and seasons.

So here in Genesis the creation of the sun, moon, and stars is simply an adjunct to the creation of the earth. This passage also makes it clear that the sun, moon, and stars are subordinate to the earth; however, if there is one thing that modern cosmology has revealed to us concerning the relationship of the earth to the rest of the cosmos, it is that the earth has no such favored position in the universe. Thus, the sun, moon, and stars were not "made" to shine on the earth and to provide the earth with "signs and seasons."

There is also no indication in this passage, or anywhere else in the Bible, that the sun—the "greater light"—is actually a massive body with a diameter more than 100 times that of the earth. Nevertheless, this passage from Genesis does provide some additional insight concerning the question of whether it is the sun or the earth that moves in the biblical cosmos. According to Genesis 1:9, God gave form to the earth on the third day of creation; Genesis 1:16–19 then goes on to state that God created the sun on the fourth day. However, from the modern scientific perspective we know that the earth (as well as every other planet in the solar system) orbits the sun, which is in the center of, and is the dominant body in, the solar system.

The question therefore arises: If the earth was created the day before the sun was created, what was the earth doing until the sun was created? Was it orbiting the place where the sun would appear later? Or was it parked in space waiting for the sun to appear so it could commence orbiting around that body? If the earth was orbiting the place where the sun would later appear, the sun's gravitational field would have needed to have been created at the same time the earth was created (that is, the day before the sun itself was created) so that the earth would have a proper orbit. If the earth had been parked in space when it was created, it would have needed to be accelerated from 0 to 66,700 miles per hour when the sun was created on the next day; in fact, it would have needed to reach its final velocity in a very short time and with the right orientation in order to attain the proper orbit. If it did not, its orbit would have been highly elliptical instead of being nearly circular.

From the perspective of the biblical cosmos, of course, those questions and the problems relating to them are not even a consideration. In Chapter 10 we found a large amount of evidence in the Bible to indicate that the biblical earth is fixed in place and does not move, which means, of course, that in the biblical cosmos it must be the sun that moves. Thus, the earth could be created first, and the sun could be created later in its subservient biblical role as a mere light that moves across the firmament of heaven and shines on the earth.

Therefore, since it indicates that the earth was created before the sun, the first chapter of Genesis fits in with, and provides added support for, the conclusion we came to in Chapter 10 from other passages in the Bible that the biblical earth does not move.

Genesis 1:16–19 also states that the moon—the "lesser light"—is a light that God made to "rule" over the night. But if that is so, why then does the moon not stay in the night sky in the same way the sun stays in the day sky? In actuality, there are several moonless nights during each month, and, during each month, the moon spends just as much time in the daytime sky as it does in the night sky. When the moon is in the daytime sky, it is always less than full (except when the full moon is on one horizon while the sun is on the opposite horizon) and its reflected light is washed out against the bright sky, so we do not notice it very often. That brings up another point: The moon has no light of its own, but instead reflects light from the sun. That, of course, contradicts the biblical view in which the moon is a light just as the sun and stars are.

Speaking of the stars, the above verses from Genesis raise another problem. We saw in Chapter 6 that there must be something in the neighborhood of 2×10^{22} stars in the universe, which is probably a quite conservative estimate. That is at least 20,000,000,000,000,000,000,000 stars. If we accept the biblical timeline and apply it to the modern view of the cosmos, all of those stars—which, as we now know, are actually other suns, many of which have planets of their own—would have had to wait for their creation until the day after the earth was created. In the meantime, the earth would have existed in the incomprehensibly immense cosmos as an insignificant mote all by itself.

To accept such a view of the earth in space and time would be the ultimate in geocentric absurdity. Again, as we saw in Chapter 6, the existence of the earth is utterly of no significance in the cosmic scheme of things, and the earth is lost in the massive numbers of stars and planets that exist in the universe. Moreover, we can see galaxies of stars that are billions of light years from the earth, which means that it must have taken billions of years for the light coming from them to have reached us. How then, could those stars have been created only a few thousand years ago? From a rational standpoint, there would be absolutely no justification or sense in saying that the earth was created the day before those 2×10^{22} stars were created.

Devout Bible believers, of course, do not look at it that way. To them, since the Bible is the word of God, there must be an explanation to account for the facts of the situation. They therefore come up with various scenarios, such as proposing that the speed of light was considerably faster during the creation week, or that the light from the newly created stars and galaxies was itself created in passage to the earth, thus negating the need for the otherwise required time. But all such scenarios have absolutely no scientific foundation and, in fact, contradict the evidence as well as established principles of science.

For example, in Einstein's equation, $E = mc^2$, the component "c^2" is the speed of light squared—that is, the speed of light multiplied by itself. Since the speed of light is such a large number to begin with, multiplying it by itself results in an enormously large number. When that number is multiplied by "m," which denotes the mass of the material that the equation is being applied to, the result is the energy—the "E" component of the equation—that is available in the mass. According to the equation, then, even a very small amount of mass can be converted to an enormous amount of energy. That equation, in fact, provides the basis for the devastating power of atomic bombs, for the energy produced by nuclear power plants, and for the source of the heat and light put out by the stars in the universe, including the sun.

Now, if the speed of light during the creation week had been increased by the required amount as proposed by the Bible believers, Einstein's equation dictates that the amount of energy released in nuclear reactions would have been of incalculably greater magnitude at that time than is currently the case. As a result, the intensely accelerated fusion processes taking place within all of the stars in the universe, including the sun, would have caused them blow up and perhaps even have converted their total mass into pure energy. In addition, the radioactive material in the earth and the other planets would have spontaneously and completely fissioned, which would have turned them into planet-sized atomic bombs.

Of course, if they were confronted with this information about the results of their proposal about the increase of the speed of light during the creation week, the Bible believers would undoubtedly say that everything was under God's control and he prevented those results from happening. But then, why should their whole scenario not be taken as being equivalent to a child's "just so" story?

The suggestion that the light from the newly created stars and galaxies was itself created in passage fares no better. Through telescopes we can see events taking place in stars and galaxies that are hundreds of thousands, millions, and even billions of light years away. Is God a deceiver, then, in providing the appearance of non-existent events in the light that was created in passage to the earth?

But, of course, in the cosmos of the Bible there is no need or place for the scenarios put forth by the creationists, for there are no such vast distances in the biblical cosmos. The biblical cosmos *is* geocentric, and it *is* considerably smaller than is the actual cosmos. Therefore, in the biblical cosmos there is no problem relating to the passage of time for the arrival of the light from the stars following their creation, for the stars are mere lights set in the solid firmament of heaven not so far above the disk of the earth.

The Bible repeatedly shows the ancient Hebrew view of the earth-centered cosmos in which the sun, moon, and stars are merely timekeeping lights that move across the solid vault of heaven above the earth—lights that are subordinate to the earth and to God's whims. An example that we previously noted can be found in Psalm 104:19 (here the NJB):

He made the moon to mark the seasons, the sun knows when to set.

And we also noted Ecclesiastes 1:5:

The sun also ariseth, and the sun goeth down, and hasteth to his place where he arose.

The God of the Bible, of course, has control of the heavenly lights. Job 9:7, for example, states that God "... *commandeth the sun, and it riseth not; and sealeth up the stars.*" God's direct "command" to the sun not to rise shows again that, in the biblical cosmos, it is the sun that moves, rather than the earth. Like Psalm 104:19, Ecclesiastes 1:5, and several other passages in the Bible, that verse leaves no room for a rotating earth.

Job 9:7 also states that God "... *sealeth up the stars.*" The Hebrew word translated as "sealeth" is *chatham*, which means "to seal up," or "to lock up." But what does that passage mean in saying that God seals up or locks up the stars. One would be hard put to apply that verse to the stars of the modern view of the cosmos; however, as some descriptions in the Book of Enoch show, Job 9:7 is quite understandable in the context of the biblical cosmos. (See Appendix C for information about this.)

Isaiah 13:10, which describes "the Day of the Lord," is another passage that shows the subordination of the stars to the earth:

For the stars of heaven and the constellations thereof shall not give their light; the sun shall be darkened in his going forth, and the moon shall not cause her light to shine.

If one believes that the multitudinous stars of the modern view of the cosmos—stars that are actually other suns, many of which have their own planetary systems—are the same ones described in that passage, one would have to believe that God created those suns to be subordinate to the earth and to cease giving out their light when the God of the Bible wreaks his divine judgment upon the earth.

Since, in the biblical cosmos the earth is not a rotating sphere and the sun is merely a relatively small light moving across the solid vault of heaven, it is a simple matter for the God of the Bible to make the sun move across the sky in whatever way he wills. In Isaiah 38:8, for example, we read:

Behold, I will bring again the shadow of the degrees, which is gone down the sun dial of Ahaz, ten degrees backward. So the sun returned ten degrees, by which it was gone down.

That passage specifically states that the sun moved backwards across the sky by ten degrees under the command of the biblical God. Again, that passage indicates that in the biblical cosmos it is the sun that moves, rather than the earth.

And we find in Amos 8:9:

> *And it shall come to pass in that day, saith the Lord God, that I will cause the sun to go down at noon, and I will darken the earth in the clear day.*

If the biblical earth were the rotating sphere of the modern view, God would have to make the earth's rotation speed up considerably in order to make the sun go down at noon. But it is not necessary for God to modify the earth's rotation in the biblical cosmos, for in the biblical cosmos the sun is a mere light moving across the firmament of heaven and God can change its course on his whim. Again, the verse clearly indicates that it is the sun that moves and not the earth.

Another such passage is Habakkuk 3:11:

> *The sun and moon stood still in their habitation: at the light of thine arrows they went, and at the shining of thy glittering spear.*

Here, both the sun and the moon stand still in their "habitation" (the original Hebrew word is *zebul*, which means "lofty residence"). The implication of the passage is that the sun and moon actually do move across the heaven in the biblical cosmos and that the God of the Bible can make them stop in their courses.

These verses thus make it clear that, in the biblical cosmos, the sun and the moon are mere lights that move across the vault of heaven, and that the God of the Bible can change the normal paths of those lights to suit his whim. That is well illustrated by the most well-known of the passages reflecting that view, Joshua 10:12–13:

> *12 Then spake Joshua to the Lord . . ., and he said in the sight of Israel, Sun, stand thou still upon Gibeon; and thou, Moon, in the valley of Ajalon.*
> *13 And the sun stood still, and the moon stayed, until the people had avenged themselves upon their enemies. Is not this written in the book of Jasher? So the sun stood still in the midst of heaven, and hasted not to go down about a whole day.*

The writer of that passage did not say that he himself had observed the sun and moon stand still. Nor did he say that any of his readers should remember having seen the sun and moon stand still. Nor did he even say

12. The Lights in the Firmament

that the story of the sun and moon standing still was handed down among the people from the time of its supposed occurrence. All he did to provide his readers with a note of authenticity for the story was to refer to the "book of Jasher"—a nonbiblical book. (The author of Habakkuk 3:11, quoted above, also likely used the book of Jasher as the basis for saying "The sun and moon stood still in their habitation.")

Nevertheless, Joshua 10:12–13 makes it clear that the biblical God was supposed to have made the sun, rather than the earth, stand still. If one does not accept that the Bible teaches that the sun and the moon are mere lights moving across the vault of heaven under the control of God, one would have to believe that God stopped the earth from rotating for a whole day (or that God performed some optical sight-of-hand, as some Bible believers have suggested) just so the Israelites could slaughter the people whose land they wanted to take.

The passage also states that both the sun and the moon stood still above the land. That would imply that the moon was supposed to have added its light to that of the sun to make the day brighter; otherwise, in the context of the "miracle," it would have been pointless for both of them to have remained motionless in the sky. In the actual cosmos, however, the moon has no light of its own, but rather reflects the light of the sun. Moreover, if the sun and the moon—that is, the actual sun and moon rather than the biblical sun and moon—were positioned in the sky as described in the passage in Joshua, the moon would not be in position to reflect *any* of the sun's light. After all, the moon is closer to the earth than is the sun, and, with their positioning as described in the passage, the sun would be shining on the other side of the moon, not the side facing the earth.

Furthermore, in the context of our knowledge about the actual physical sizes and distances of the sun and the moon, the description given in Joshua 10:12–13 represents a physical impossibility if it is taken to mean that the sun and the moon were both in the sky above the locations described in the passage and the land was in full daylight. Although the KJV and several other versions of the Bible use different prepositions for the positioning of the sun and the moon in this passage—the sun "upon" Gibeon and the moon "in" the valley of Ajalon—the same preposition is actually used in both places in the original Hebrew. The NIV and the NJB accordingly use the same preposition in their respective translations. The NIV rendition is "*O sun, stand still over Gibeon, O moon, over the Valley of Aijalon*" and the NJB is "*Sun, stand still over Gibeon, and, moon, you too, over the Vale of Aijalon!*" But Gibeon and the valley of Ajalon are only a few miles apart on the earth. Therefore, if the sun and the moon (that is, the celestial bodies that we know them to be, rather than the relatively small, nearby lights the ancient Hebrews believed them to be) were located over Gibeon and the valley of Ajalon, an eclipse of the sun would have occurred, for the moon would have covered the sun. The land would therefore have been in darkness for the biblical day, rather than in full sunlight.

Joshua 10:12–13 therefore clearly shows that, in the biblical cosmos, the sun and the moon are relatively small and near the earth's surface—that is, they are merely lights moving across the solid vault of heaven not all that far above the earth (and certainly not, as in the modern view, 230,000 miles to the moon or 93 million miles to the sun).

This passage in the book of Joshua also shows that the ancient Hebrews had no inkling of the true nature of the sun and moon, and, since that lack of knowledge is reflected in the Bible, one must conclude that the authors of the Bible had no special inspiration from God concerning the natural cosmological order of things. (Again, the ancient Greek scientists near the latter part of the Old Testament era knew more about the nature of the cosmos than did the "inspired" authors of the Bible. Aristarchus, in fact, calculated the size and distance of the moon and came up with measurements very close to the actual ones, and he determined that the sun is much larger than the earth and much further away from the earth than is the moon. This will be described in more detail in Chapter 13.)

There is one other problem with the event described in this passage. At the time it supposedly occurred, there were several civilizations—such as the Babylonian, the Egyptian, and the Chinese—that carefully kept written records and would have made note of unusual astronomical occurrences. If such a celestial event as described in the book of Joshua had actually occurred when it was supposed to have occurred, it certainly would have caused considerable consternation and controversy, and it surely would have been specifically dated and clearly reported in the records that those civilizations produced. Yet no such record is known.[1]

As well as the creation of the sun and moon, Genesis 1:14–19 also describes the creation of the stars. Significantly, however, there is not the slightest hint in that passage from Genesis or anywhere else in the Bible that the stars are actually other suns that are at an enormous distance from the earth. The only stars the God of the Bible knows of are the stars that can be seen with the naked eye, and those stars are mere lights whose only purpose is to shine down on the earth and to be for "signs and seasons." There is nothing in the Bible to indicate that there is a whole universe of stars out there or that the stars that can be seen with the naked eye are but an infinitely tiny fraction of all the stars in the cosmos. If the stars were made to provide the earth with "signs and seasons," why then are the vast, incomprehensibly immense majority of the stars too far away to be seen with the naked eye? The answer is, of course, that the visible stars—the stars set in the solid firmament of the biblical heaven—are the only stars in the biblical cosmos. Genesis 1:14–19 thus makes it clear that, in the biblical cosmos, the stars are subordinate to the earth and are mere lights that have no purpose but to give light to the earth and to provide the earth with signs and seasons.

To be sure, some biblical apologists like to quote certain passages from the Bible—passages which seem to indicate that those who wrote them understood something about the vast nature of the cosmos and therefore must have been inspired by God. For example, one passage that is frequently quoted by Bible believers is Jeremiah 33:22, which states that "the host of heaven cannot be numbered." In chapter 6 we took a look at the vastness of the universe and the enormous number of stars that it contains; the Jeremiah passage might therefore seem to reflect an understanding of the vastness of the cosmos. However, a look at the complete verse reveals the numbering of the host of heaven to be more a result of hyperbole than from an actual understanding of the true number of stars in the cosmos:

> *As the host of heaven cannot be numbered, neither the sand of the sea measured: so will I multiply the seed of David my servant, and the Levites that minister unto me.*

Certainly the seed of David and that of the Levites was not multiplied beyond measure as the Jeremiah passage would indicate. In Genesis 22:17, the God of the Bible makes a similar promise to Abraham: "*. . . I will multiply thy seed as the stars of the heaven, and as the sand which is upon the sea shore. . . .*" Those and other, similar passages in the Bible show that such language merely refers to a large, ill-defined number. For example, 1 Samuel 13:5 numbers the Philistines who gathered to fight with Israel "as the sand which is on the sea shore in multitude." Several other passages in the Bible also number masses of people as "the sand of the sea."[2] Certainly, the peoples referred to in these passages did not actually have numbers that were equivalent to the sands of the sea.

Furthermore, such passages must be looked at in light of the time and place in which they were written. The ancient Hebrews lived in a relatively dry climate and at a time when the light of the stars was not attenuated by the dust, chemical, and artificial light pollution that prevents most of us from viewing the night sky in all its glory. Thus, those living at that time were able to see considerably more stars than most of us could ever hope to see in our present-day urban and suburban settings.

In any case, even though only about 3,000 stars can be seen at one time under the best of conditions, one gets the impression that there are a great many more when viewing an unobscured and moonless night sky. Besides, 3,000 stars are still a large number to count at one sitting, particularly when there is no way of tagging the stars that one has already counted. Furthermore, while one is trying to count the stars over the course of a night, those stars that are in the western sky set and disappear below the horizon, while still other stars rise above the horizon in the eastern sky and come into view. In addition, as the seasons progress through the year, whole new constellations of stars appear in the nighttime sky. The person who wrote the Jeremiah passage could therefore have easily believed that the stars visible in the night sky are uncountable, and yet he would not have had any awareness of the true magnitude of the numbers of stars in the cosmos.

In Genesis 15:5, we find that God himself is seemingly unaware of the actual number of stars that can be viewed with the naked eye. In that passage he is speaking to Abram (here, the NKJV):

> *Then He brought him outside and said, "Look now toward heaven, and count the stars if you are able to number them." And He said to him, "So shall your descendants be."*

Abram's descendants were supposed to number considerably more than the 3,000 stars that can be viewed with the naked eye. Thus, in Deuteronomy 1:10, the God of the Bible subsequently says to the Israelites through Moses:

> *The LORD your God has multiplied you, and here you* are *today, as the stars of heaven in multitude.*

Several other passages in the Bible make similar statements.[3] If anything, such passages severely limit the number of stars in the cosmos, for they make the number of stars comparable to the number of the people spoken about in the passages. It is thus clear that the writers of the Bible had no understanding of the vast number of stars in the cosmos or even of the limited number of stars visible to the naked eye from the earth. Certainly Moses, as portrayed in Deuteronomy 1:10, thought that the number of the stars visible in the sky was comparable to the number of people standing before him. That crowd of people was supposed to have numbered much more than the three thousand visible stars they were compared with, but obviously considerably less than the actual number of stars in the cosmos.

Even if these passages did imply an understanding that a vast number of stars exist, it would still not mean that the Bible is the word of God. In the fifth century B.C.E., the ancient Greek philosopher Democritus put forth the idea that the Milky Way consists of untold numbers of stars that cannot be seen individually with the naked eye. Why then should the writer of the Jeremiah passage, or any other such passage in the Bible, be considered to have been able to have had such an understanding only through a revelation from God?

12. The Lights in the Firmament

The creation account in Genesis does not even mention the creation of the other planets in our solar system. Indeed, the Bible does not acknowledge the fact that there are other worlds in our solar system, much less in the systems of countless other suns. The only place the word "planets" occurs in the KJV is in 2 Kings 23:5:

> *And he put down the idolatrous priests ...; them also that burned incense unto Baal, to the sun, and to the moon, and to the planets, and to all the host of heaven.*

The passage could be taken as referring to the planets as the moving star-like points of light as they are seen from the earth, but the original Hebrew word translated as "planets" there is *mazzalah*, which more properly means "constellations" or "signs of the Zodiac." Several versions of the Bible, in fact, use one or the other of those translations of *mazzalah*, as for example the NKJV: "*Then he removed the idolatrous priests ..., and those who burned incense to Baal, to the sun, to the moon, to the constellations, and to all the host of heaven.*"

One might ask Bible believers why planets, as other worlds in our solar system or elsewhere in the cosmos, should even exist. Although the Bible does not explicitly state that the earth is the only world, that view is implicit in all of the biblical descriptions relating to the cosmos, from the creation to the End Times. Indeed, that view is mandated by the cosmology of the Bible; nowhere does the Bible describe the creation of other worlds, and the question of whether other worlds exist is not even a consideration. Given the specific steps in the creation of heaven and the earth as described in Genesis, as well as the eventual destruction of heaven and the earth as described elsewhere in the Bible, one must conclude that the earth is a unique creation in the biblical cosmos and is the only world in that cosmos. For that matter, since the biblical cosmos consists of a disk-shaped earth covered by a solid vault of heaven, there is no room for any other worlds.

Some Bible believers might refer to the Epistle to the Hebrews to prove that the Bible knows of other worlds. In Hebrews 1:1–2 in the KJV, we find:

> *God ... Hath in these last days spoken unto us by his Son, whom he hath appointed heir of all things, by whom also he made the worlds.*

And in Hebrews 11:3 we find:

> *Through faith we understand that the worlds were framed by the word of God, so that things which are seen were not made of things which do appear.*

Under closer analysis, however, an assertion that other worlds are being referred to in these verses would not hold up.

For starters, the word "world" appears more than 200 times in the New Testament of the KJV, and in the majority of those occurrences the original Greek word is *kosmos*, which specifically refers to the physical world. In both of the referenced verses in the Epistle to the Hebrews, however, the original Greek word for "worlds" is *aionas*, which is the plural of *aion*; the English word "aeon" or "eon," meaning "age," is derived from that same Greek word. Moreover, *Strong's Concordance* defines *aion* as follows: "prop. an *age*; by extension, *perpetuity* (also past); by implication the *world*; spec. (Jewish) a Messianic period (present or future)." *Harper's Bible Dictionary* (under "World") states that *aion* "has temporal connotations," meaning "this world/the world to come." BDAG defines *aion* as "a segment of time as a particular unit of history, *age*." The Greek word *aion* therefore essentially means an "age," and *aionas* would mean "ages." In its rendition of Hebrews 1:1–2 the NJB has correctly translated the word:

> *... God ... in our time, the final days, he has spoken to us in the person of his Son, whom he appointed heir of all things and through whom he made the ages.*

The NJB also renders *aionas* as "ages" in Hebrews 11:3.

It is clear, then, that *aionas* of Hebrews 1:1–2 and 11:3 has a meaning that more properly refers to the ages of this world—specifically, the temporal existence of this present world and that of the world to come. A closer look at Hebrews 1:1–2 supports that meaning, for it refers to the "last days" of this world (the end of the present age) and alludes to the next world (the next age) as the heritage of Jesus; both worlds (or ages) are made by God. This idea is also found in 2 Peter 3:10–12:

> 10 *But the day of the Lord will come as a thief in the which the heavens shall pass away with a great noise, ... the earth also. ...*
> 12 *Nevertheless we ... look for new heavens and a new earth, wherein dwelleth righteousness.*

That is confirmed by the Old Testament passages that Peter apparently used as the source for his "prophecy." One is Isaiah 66:22:

> *For as the new heavens and the new earth, which I will make, shall remain before me, saith the Lord, so shall our name remain.*

Another is Isaiah 65:17:

> *For, behold, I create new heavens and a new earth: and the former shall not be remembered, nor come into mind.*

12. The Lights in the Firmament

The same idea is found in Revelation 21:1:

> *And I saw a new heaven and a new earth: for the first heaven and the first earth were passed away; and there was no more sea.*

These passages make it clear that the "worlds" being referred to in Hebrews are the ages, or states, of the earth itself. Therefore, the Greek word *aionas*, even if it is translated as "worlds" as it is in the KJV, cannot be taken as referring to other worlds in the solar system, much less elsewhere in the universe.

As far as the Bible is concerned, then, there are no other worlds. What we know to be other planets in our modern view of the solar system are, in the biblical cosmos, merely points of light that are lumped in with the other lights that are the stars set in the solid firmament of the biblical heaven. Therefore, the only world that the Bible knows of is the flat, immovable, disk-shaped earth that the God of the Bible created as the centerpiece of his creation—that is, the only world the ancient Hebrews knew about.

In contrast, according to the modern view, the earth is but one of the nine planets (or eight planets, if one accepts the recent demotion of Pluto) in the solar system and was formed at the same time and by the same process as the other planets. The differences between the earth and the other planets in the solar system are due merely to variations in the composition of the dust disk that surrounded the newly formed sun, as well as to the differing distances of the planets from the sun as they accreted from the dust disk. Moreover, given the incredible number of stars in the universe, one must conclude that similar processes have formed planetary systems around a considerable number of those other stars. Indeed, as we noted in chapter 6, numerous stars that appear to have planetary systems in the process of forming around them have been found, and astronomers have actually detected scores of planets orbiting other stars.

Though the planets as other worlds are not mentioned in the Bible, a reference to their apparent nature as "wandering stars" can be found in the Epistle of Jude, verses 13–14, where we find the following:

> 13 . . . *wandering stars, to whom is reserved the blackness of darkness for ever.*
> 14 *And Enoch also, the seventh from Adam, prophesied of these.* . . .

The Greek word that is translated as "wandering" in this passage is *planetes*, which means "a wanderer." The word *planetai* was used by the ancient Greeks to refer to those "stars" that move, or "wander," against the backdrop of the fixed stars. Our word "planet," of course, is derived from that Greek word.

Jude verses 13–14 therefore appear to indicate that its author was referring to the planets as those "stars" that move or wander in paths that are different from those of the background stars. Still, to properly understand the contextual meaning of the passage, one must understand that in verse 14 Jude cited the pseudepigraphic Book of Enoch to justify what he said about the "wandering stars": "*And Enoch also, the seventh from Adam, prophesied of these. . . .*" Jude also made reference to, and even quoted, other passages from that same book as if it were inspired scripture; in fact, several early Christian leaders did consider the Book of Enoch to be inspired literature, and many of its words are echoed in other books of the New Testament. This results in some serious problems for Bible believers, for the Book of Enoch provides a great many highly detailed descriptions of the cosmos. Those descriptions, many of which echo the material that we have found in the Bible and which supplement what we have already learned about the biblical cosmos, are, from the modern standpoint, pure nonsense.

That is demonstrated by the words in the above passage from Jude, "*. . . wandering stars, to whom is reserved the blackness of darkness for ever.*" There, Jude is referring to those stars that are to be punished—as related in the Book of Enoch—by being imprisoned in a "place of darkness" for not keeping their appointed timekeeping circuits in heaven. This means that, according to the Bible (since the Epistle of Jude specifically cites the authority of the Book of Enoch in this matter), the stars are entities who have a volition of their own; they are therefore responsible for their actions and can be punished for not properly performing their timekeeping functions. (See Appendix C for further information about the relationship of the Epistle of Jude to the Book of Enoch.)

The idea that the stars are such entities, or "angels," appears elsewhere in the Bible. Several passages mention the "host of heaven," a term that appears to have been applied either to the stars of heaven themselves or to entities, or angels, who control the stars. For example, we find in 2 Chronicles 18:18:

> *Again he said, Therefore hear the word of the LORD; I saw the LORD sitting upon his throne, and all the host of heaven standing on his right hand and on his left.*

We also find the following in Judges 5:20: "*They fought from heaven; the stars in their courses fought against Sisera.*" And Job 38:7 states: "*When the morning stars sang together, and all the sons of God shouted for joy?*" That verse provides an example of biblical parallelism in which the second part of the verse reiterates the first part; thus "the sons of God" (i.e., the angels) restates "stars." The verse, then, can be taken as indicating that the stars are, or have an association with, the angels.

12. The Lights in the Firmament

Isaiah 14:12–15, which we looked at in Chapter 11, also indicates that the stars in the biblical cosmos are entities:

> 12 *How art thou fallen from heaven, O Lucifer, son of the morning! how art thou cut down to the ground, which didst weaken the nations!*
> 13 *For thou hast said in thine heart, I will ascend into heaven, I will exalt my throne above the stars of God: I will sit also upon the mount of the congregation, in the sides of the north:*
> 14 *I will ascend above the heights of the clouds; I will be like the most High.*
> 15 *Yet thou shalt be brought down to hell, to the sides of the pit.*

The Hebrew word translated as "Lucifer" is *heylel*, which means "morning star," or "shining one," and several versions of the Bible, in fact, use the term "morning star" or similar terms rather than "Lucifer" in the passage.[4] The NJB, for example, translates verse 12 as follows:

> *How did you come to fall from the heavens, Daystar, son of Dawn? How did you come to be thrown to the ground, conqueror of nations?*

According to the Book of Enoch, which, again, Jude apparently thought of as inspired scripture, these entities, or "angels," control or hold, and are sometimes identified as actually being, the "lights" in heaven that are the sun, moon, and stars. As such, they have the responsibility of keeping the lights in the sky on their proper courses across the firmament of heaven. (See Appendix C for more information about this.)

"Lucifer," the "morning star," in Isaiah 14:12–15 has been equated with Satan. That identification is implied in Revelation 9:1:

> . . . *and I saw a star fall from heaven unto the earth: and to him was given the key of the bottomless pit.*

This passage from Revelation indicates that the author considered the star that fell to the earth to be a conscious entity, which provides further evidence for our conclusion about the nature of the "host of heaven."

Also concerning the identification of stars with angels, we find the following in Revelation 1:20:

> *The mystery of the seven stars which thou sawest in my right hand, and the seven golden candlesticks. The seven stars are the angels of the seven churches: and the seven candlesticks which thou sawest are the seven churches.*

In that passage, the seven stars are specifically described as the angels of the seven churches, and that description reflects the belief that the stars are synonymous with, or are identified with, angels. This is, of course, completely at odds with the nature of the stars in the modern view of the cosmos, and it completely relegates the biblical view of the stars to ancient Hebrew mythology.

We have seen that, as well as indicating that the sun physically moves across the solid vault of heaven, the Bible also indicates that the sun is relatively close to the earth. Revelation 19:17 provides another example of this near-earth view of the sun:

And I saw an angel standing in the sun; and he cried with a loud voice, saying to all the fowls that fly in the midst of heaven, Come and gather yourselves together unto the supper of the great God.

The angel who was standing in the sun was likely considered to be one of those entities, or angels, who guide the lights of heaven—the sun, moon, and stars—in their circuits across the firmament of heaven. This passage thus provides further evidence that the writers of the Bible viewed the sun as something quite different from the intensely hot, 860,000-mile-in-diameter and 93-million-mile-distant body that science describes. The passage in fact indicates that the biblical sun is relatively small and is within shouting distance of the fowls in the sky. (Again, see Appendix C for related material in the Book of Enoch about the entity who controls the sun, as well as those who control the moon and stars.)

Another example showing the subordination of the stars to the earth is provided in a vision described in Daniel 8:10:

And it [the horn] *waxed great, even to the host of heaven; and it cast down some of the host and of the stars to the ground, and stamped upon them.*

This passage shows that, according to the Bible, the stars are lights that can be cast down to the earth. It also mentions the host of heaven in association with the stars. Again, there is not the slightest hint in the Bible that the stars are actually other suns. In the biblical cosmos, rather than being suns that are immensely larger than the earth, the stars must be considerably smaller in order for them to fall to the earth.

The subordination of the cosmos to the earth is carried on through the Bible in its descriptions of the events of the "Last Days." In Matthew 24:29, we therefore find Jesus saying this about the stars:

Immediately after the tribulation of those days shall the sun be darkened, and the moon shall not give her light, and the stars shall fall from heaven, and the powers of the heavens shall be shaken.

12. The Lights in the Firmament

That passage clearly shows the subordination of the biblical cosmos to events here on earth. Furthermore, because it states that the stars will fall from heaven, it clearly shows, as does Daniel 8:10, above, that biblical cosmology is glaringly different from modern cosmology. From the standpoint of modern cosmology, it is nonsensical to talk about stars falling from heaven.

But what is most significant about Matthew 24:29 is that it shows that the biblical Jesus did not know that the stars are actually other suns, but believed instead that they are lights that can fall to the earth from the firmament of heaven. This ties in with Matthew 24:31, which, as we noted in Chapter 9, indicates that the biblical Jesus believed that heaven is a solid vault placed over a flat earth. From these passages, one must conclude that Jesus had no divinely inspired knowledge about the true nature of the cosmos. This, of course, has significant implications for any claims of divinity or of any special revelatory knowledge that one might make for Jesus.

In Revelation 6:13–14, the assertion that the stars are lights that can fall from heaven is repeated in the vision of the Last Days:

And the stars of heaven fell unto the earth, even as a fig tree casteth her untimely figs, when she is shaken of a mighty wind. And the heaven departed as a scroll when it is rolled together....

In addition to describing the stars as falling to the earth from heaven, the above passage from the Book of Revelation likens heaven to a scroll that can be rolled up. Those concepts are also found in Isaiah 34:4:

And all the host of heaven shall be dissolved, and the heavens shall be rolled together as a scroll: and all their host shall fall down, as the leaf falleth off from the vine, and as a falling fig from the fig tree.

There Isaiah states that the stars—the host of heaven—will fall from the sky.

Those passages have a bearing on what is stated in 2 Peter 3:10–13:

10 But the day of the Lord will come as a thief in the night; in the which the heavens shall pass away with a great noise, and the elements shall melt with fervent heat, the earth also and the works that are therein shall be burned up.
11 Seeing then that all these things shall be dissolved, what manner of persons ought ye to be in all holy conversation and godliness,

> 12 *Looking for and hasting unto the coming of the day of God, wherein the heavens being on fire shall be dissolved, and the elements shall melt with fervent heat?*
> 13 *Nevertheless we, according to his promise, look for new heavens and a new earth, wherein dwelleth righteousness.*

This is another passage that clearly shows the subordination of the cosmos to the events of the Last Days here on the earth. Note that the passage states: "*. . . we, according to his promise, look for new heavens and a new earth, wherein dwelleth righteousness.*" From those words, it is apparent that Peter believed in the immediacy of the coming Day of the Lord in which the contemporary cosmos will be overturned and a new heaven and a new earth will be established. Other passages in the New Testament, some of which we referred to above, also indicate that the early Christians believed that they were living in the Last Days and that the Day of the Lord was at hand. (This will be looked at in more detail in Appendix B.)

Note that 2 Peter 3:10–13 states "the heavens shall pass away with a great noise." In biblical times, as we noted in Chapter 8, scrolls were occasionally made of thin sheets of metal. Since, in the biblical cosmos, heaven is a thin vault of metal, it surely would make a "great noise" if it were "rolled together as a scroll," much as a sheet of metal makes a noise like thunder when it is shaken.

These passages therefore make it clear that the sun and moon and stars, according to the beliefs of the ancient Hebrews—and of the early Christians who came after them—are mere lights that their tribal god, Yahweh, set in the firmament of heaven solely for the benefit of the earth, an earth that, in their view, is the centerpiece of creation.

From the biblical perspective then, it naturally follows that, because the sun, moon, and stars have no other purpose than to be a part of God's plan for the earth, their eventual destruction is also tied to the prophesied destruction of the earth in the Last Days. When the Last Days come, the sun and moon will not give their light, the firmament of heaven will be rolled up like a scroll, the stars will fall to the earth, and everything will be consigned to the flames of God's judgment. God will then create a new heaven and a new earth that will be cleansed of the unrighteousness and injustices of the old.

That view of the Last Days perhaps reflects an understandable yearning by those who expressed it, but it is a view that is no more grounded in reality than is the cosmology upon which it is based.

Part Four
The Cosmological Answer

Chapter 13

The Biblical Cosmos:

Metaphor or Mythology?

In Part Two of this book, we took a look at the enormous cosmos that science has uncovered. We found a cosmos that has billions of galaxies, each with billions of stars that are other suns, and we saw that an incredible number of other planets must exist in that cosmos. We also found that the cosmos has been expanding for more than 13 billion years and has prospects of continuing its expansion for time beyond comprehension.

In light of our cognizance of that vast cosmos, it would be appropriate to repeat the questions that appeared near the opening of Chapter 7. In the context of the biblical worldview, why would the God of the Bible create a vast universe such as that which science has uncovered, yet subordinate its very existence to an insignificant speck of dust such as is the earth? And why would the God of the Bible create such a universe—an incomprehensibly immense universe with an unfathomable capacity for a bewildering number of galaxies, suns, and planets—only ultimately to destroy it all when the events of the Last Days play out here on the earth?

In answer to those questions, many Bible believers would doubtlessly say that God can do anything he wants, and, if he wanted to create the cosmos that science has revealed, that was his business. If there is a God, that may be so, but that does not mean that that God is the God of the Bible, for the God of the Bible knows nothing of the cosmos that science has revealed. As we have seen, the cosmos that science has revealed is totally at odds with the cosmos that the Bible describes—the cosmos created by Yahweh, the tribal god of the ancient Hebrews.

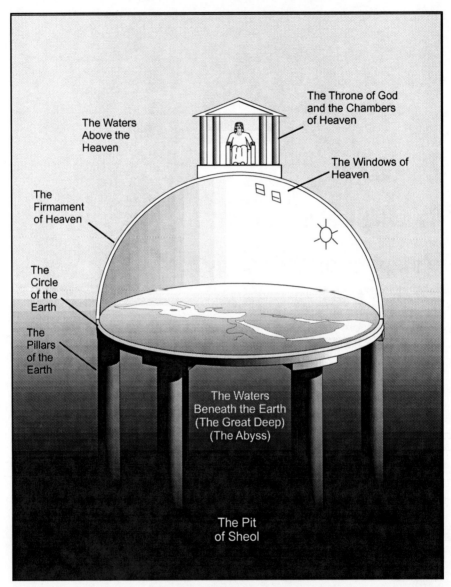

Figure 5. The Biblical Cosmos

That is made clear by Figure 5, which graphically shows the results of our extensive analysis of the biblical cosmos in Part Three of this book. The cosmos of the Bible is a small (when compared to the expansive cosmos that science has revealed) earth-centered abode, the sole purpose of which is to provide a stage for unfolding the plan that the biblical God has for mankind. In the biblical cosmos, the earth is a flat, immovable disk set on pillars that support it in the waters of the

13. The Biblical Cosmos: Metaphor or Mythology?   191

great deep. Placed rim to rim with the disk of the earth is the solid firmament, or vault, of heaven in which are set the lights that are the sun, moon, and stars. The solid firmament holds back the waters that are above the firmament, and on top of the firmament sit the chambers of heaven and the throne of God. Finally, in the abyss of the great deep under the earth is the pit of Sheol.

By constructing this view of the cosmos from a large number of biblical passages, we have demonstrated quite thoroughly that it is explicit in the Bible and that it permeates the Bible from beginning to end. But what is particularly significant, this view of the biblical cosmos is obviously not a chaotic assemblage; it is a coherent picture with a distinct order, even if it is nothing like the modern view of the cosmos. If the writers of the Bible did not have a clear idea of what they perceived their cosmos to be, we would not have been able to derive a clear, self-consistent picture from what they wrote. And the fact that we can derive a clear, self-consistent picture from the cosmological descriptions that are found in the Bible is evidence that the view of the biblical cosmos that we have derived from our analysis is, in fact, the view that is predicated in the Bible.

But what is most significant, this is the only view of the cosmos that can be constructed from the plain, unaltered words of the Bible, and that perceived view makes perfect sense in light of the geocentric, cosmologically limited perspective of the Bible. Certainly, there is nothing in the Bible to indicate that the earth is a sphere that is rotating on its axis and orbiting the sun. Nor is there anything in the Bible to indicate that the sun, moon, and stars are anything more than lights set in the solid firmament of heaven.

The biblical view of the cosmos is completely at odds with our contemporary understanding of the nature of the cosmos—an understanding that has come about through centuries of study and observations. That understanding therefore provides the perspective by which to judge the credibility of the Bible. We, today, know that the stars are immense suns that are similar to our own sun, and that the vast majority of the stars are so far away that they can be seen only with sensitive instruments. On the other hand, the only stars that the God of the Bible knows of are those that can be seen with the naked eye from the earth, the stars that he created as mere lights set in a solid firmament to shine on the earth and to provide it with signs and seasons. Moreover, since the stars are mere lights set in the solid firmament of the biblical heaven, they can fall to the earth at the whim of the biblical God. In the modern view of the cosmos, however, the idea that the stars could fall from the sky to the earth is sheer nonsense. Therefore, if the stars are other suns, the only reasonable conclusion must be that the geocentric cosmology of the Bible has no basis in reality.

From the evidence, then, we must conclude there is absolutely nothing in the Bible to even hint that the biblical God created a universe such as that which science has uncovered. In the cosmos of the Bible there is no need or place for such a universe, for the cosmos that is revealed in the Bible is not the cosmos that science has revealed. Rather, the cosmos that is revealed in the Bible is based upon the geocentric, naïve, primitive, mythologically oriented beliefs of the ancient Hebrews.

That this is the correct view of the biblical cosmos is supported by the pseudepigraphic Book of Enoch. As shown in Appendix C, the Book of Enoch describes in considerable detail the structure of the cosmos and its relationship to biblical eschatology, and that structure clearly reflects the cosmological view that we have discerned from our study in Part Three of this book. Moreover, the fact that the author of the Epistle of Jude believed the Book of Enoch to be authoritative provides additional support for our conclusions about the view of the cosmos that is found in the Bible.

This view of the biblical cosmos is, of course, not really new. Many non-believers, and even some believers, have stated that the biblical earth is flat and that its heaven is a solid vault above the earth. More often than not, though, those who posit that view usually quote only a verse here or a verse there in support of it, and those who reject the idea that this cosmological view can be found in the Bible shrug off those verses as being merely poetic expressions. We have seen, however, that there are far more than just a few such verses, poetic and otherwise, in the Bible, and that they show that this view of the biblical cosmos permeates the Bible from beginning to end and is, in fact, an integral part of the biblical worldview.

Even today, some Bible believers, specifically the flat earthers and geocentrists, accept certain aspects of this view of the biblical cosmos. Moreover, a few scholarly books have been published in which this view of biblical cosmology has been explored and documented.[1] Some standard reference works, such as certain Bible dictionaries and encyclopedias, also occasionally indicate that the Bible posits this view of the cosmos. Still, most Bible believers of today do not understand that the Bible in fact presupposes such a view, and they merely assume that the God of the Bible created the universe that science has uncovered.

To counteract suggestions that biblical cosmology posits this view, some apologists have tried to interpret the language of the Bible in light of the incontrovertible and presently accepted facts concerning the modern view of the cosmos and have tried to show that the biblical cosmos conforms to those facts. But forcing such interpretations upon the text of the

13. The Biblical Cosmos: Metaphor or Mythology?

Bible does not make those interpretations valid, nor does it prove the intent of the writers of the Bible.

Other believers in the Bible take an opposite tack and say that the Bible is not meant to be a book of science, and therefore its statements relating to areas of science are not meant to be taken literally. If that is so, one might then ask why should any of its statements relating to other areas be taken literally. Some Bible believers, particularly creationists, see the problem that results from taking the position that the Bible is not meant to be a book of science. They therefore say that the Bible *is* a book of trustworthy science and is to be believed even when the findings made by modern scientists are in conflict with it. Where they can, or feel they must, these creationists "interpret" the Bible to make it appear to conform to the findings of science (for example, interpreting the biblical "circle of the earth" to mean a spherical earth). On the other hand, in certain cases they "interpret" the findings of science to make them appear to conform to the Bible (for example, interpreting certain geological depositions as being evidence of the biblical Flood, even though such interpretations have been repeatedly shown to be without any scientific merit). If they cannot make such interpretations from either a biblical or scientific standpoint, they totally reject the findings of science as being unfounded speculations or outright falsehoods.

Some biblical apologists argue that the cosmological language of the Bible merely reflects a use of metaphoric and poetic expressions. For that matter, of course, some of the "poetic" or "metaphoric" cosmological expressions that are used today, such as "the ends of the earth," are a result of a carryover from their usage in the Bible. Certainly, when we use such expressions today we understand that we are using them in a metaphoric or poetic sense because we know better than to use them to express an actual description of the cosmos (or at least those of us who are not geocentric creationists or flat-earthers know better). To the ancient Hebrews, however, those expressions were not necessarily metaphoric or poetic. Is the statement *"Can you beat out the vault of the skies, as he does, hard as a mirror of cast metal"* simply a poetic statement? Or does it express an actual belief about the physical nature of the vault of heaven? Certainly, if the idea of a god beating out the vault of the sky were put forth in a non-biblical belief system, we would consider it to be pure mythology. Similarly, is the statement, *"The sun also ariseth, and the sun goeth down, and hasteth to his place where he arose,"* merely metaphoric or poetic, or does it express a belief that the sun actually does physically move across the sky and therefore must "hasteth" to the place where it arose so it can arise from there again the following morning? If such a statement appeared in the literature of a non-biblical belief system, we would assign that literature to the mythology section of the library.

However, to the writers of the Bible, such cosmological expressions did in fact describe their perception of the cosmos. The cosmological view that we have gleaned from the Bible is supported by every single biblical passage that gives insight concerning the shape of the biblical earth, the structure of the biblical cosmos, and the nature of the "lights" in the firmament of heaven. In the Bible there is not one single passage that presupposes a spherical earth. There is not one passage in the Bible that says, for example, "All the nations around the world." Nor is there any statement indicating that God "molded" the "ball" of the earth. Nor is there any statement indicating that the earth is rotating on its axis or orbiting the sun, or that the stars are other suns.

Why is that? Did God have to give in "poetic" terms every single biblical expression that describes the shape of the earth or the earth's place in the cosmos? In addition to using these "poetic" expressions, why did God not provide at least some factual statements that would have revealed to the ancient Hebrews the true nature of the shape of the earth and its relation to the rest of the cosmos? If he had done so, it would have shown his great creative power and the magnificence of his creation. Because of what the Bible has to say about cosmology, are we to believe that God deliberately misled the ancient Hebrews concerning the structure of the cosmos? Or would it be more reasonable to conclude that no God had anything to do with the authoring of the Bible and that the cosmos that is described in the Bible came solely from the naïve cosmological beliefs of the ancient Hebrews? If the Bible is actually the word of God, would it make false statements, even if those statements are in the form of poetry? If God is truth, as many Bible believers assert, would God make untrue statements—in the form of poetry or not—in his "inspired" book? If God does not make untrue statements, but if untrue statements (such as *"The sun also ariseth, and the sun goeth down, and hasteth to his place where he arose"*) are found in the Bible, then by the rules of logic the Bible cannot be considered the word of God.

If one still insists that all such cosmological statements that are found in the Bible, instead of reflecting reality, are merely meant to be metaphoric or poetic, then another question must be raised: Why should not other descriptive expressions or passages in the Bible also be considered poetic or metaphoric? Why should only the cosmological statements be taken as figurative or metaphoric? Why should not the account of the biblical Flood also be considered a metaphoric fiction? Or, why should not the account of the death and resurrection of Jesus be considered merely metaphoric? Indeed, why should not anything else in the Bible be consid-

13. The Biblical Cosmos: Metaphor or Mythology?

ered mere poetry or metaphor and not actually be descriptive of any kind of reality?

If one therefore argues that all of the biblical expressions and statements indicating that the structure of the biblical cosmos is as we have discerned it to be are merely metaphoric or poetic, and are hence figurative, one cannot then maintain with any confidence that the same cannot be said of any other expression or statement found in the Bible.

In answer, apologists might say that it should be obvious which statements in the Bible should be taken figuratively and which statements should be taken literally. From this standpoint, they would argue that, since the Bible is the word of God, any statements that do not conform to some aspect of the real world must be meant to be taken figuratively. Therefore, they would say, the cosmology of the Bible should be taken figuratively because of the findings of modern science. But that line of argument would be unwarranted, for it ignores the intent and beliefs of the original writers of the Bible, and it makes interpretation of the Bible dependent upon the findings of science. Moreover, how can one know what is "obviously" meant to be taken figuratively or what is meant to be taken literally? Just as the cosmology of the Bible would have been taken literally when the Bible was written, but would now be taken figuratively, some other aspect of the Bible that is now taken literally might in the future be taken figuratively because of some future finding of science or of ancient history.

But even more significantly, what good is a "revelation" from God if it cannot stand on its own statements and be uniformly interpreted throughout history?

Some Bible believers argue that the societies of the time in which the Bible was written did not have the sophistication to understand the cosmos as we do, and that God therefore merely provided descriptions of the shape of the earth and the structure of the cosmos in terms that those early societies could understand and that conformed to the prevalent mythology.

But that argument is no better than the previous one. If the God of the Bible merely used expressions that conformed to the prevalent myths of the times concerning the cosmos, then why could he not have used expressions that conformed to the prevalent myths in other matters as well? But if the Bible conforms to the myths of the times in which it was written, why then should the Bible be considered the true, infallible word of God and not simply a collection of myths?

In any case, all such arguments fall short when we consider that the ancient Greeks, using their intellect and simple observations and means of measurement, had come to at least a partial understanding of the cosmos as we presently perceive it.

To be sure, like their Hebrew contemporaries, the ancient Greeks early on believed the earth to be flat. Moreover, many of the ancient Greeks believed that the planets and the sun, moon, and stars were attached to transparent crystalline spheres, much in the same way the ancient Hebrews believed that the sun, moon, and stars were set in a solid firmament.

However, as the culture and science of the ancient Greeks evolved, so did their ideas about the cosmos. For example, at a time when the ancient Hebrews still believed the earth to be a flat disk supported by pillars, the ancient Greeks had come to understand it to be an unsupported sphere. Previously, Greek philosophers had believed for mystical reasons that the earth is a sphere; however, other Greek philosophers eventually came along and provided more rationalistic arguments for a spherical earth—arguments that were based on scientific reasoning and observations. In the fourth century B.C.E., Aristotle summarized three of those arguments.

First, if the earth were flat, all of the stars that are visible from one point on the earth's surface at a given time of the year would be visible from all other points. However, it was observed that, if one traveled far enough in a southerly direction, the stars near the northern horizon would eventually disappear below the horizon, while new stars would appear and rise up from the southern horizon. This could be best explained if one were traveling on a curved surface. Therefore, the earth is a sphere.

Second, ships sailing off into the open sea were observed to have their bottom portions disappear below the horizon before their upper portions. This effect took place regardless of the direction and at all places. This meant that the sea and therefore the earth curved equally in all directions. Therefore, the earth is a sphere.

Third, during eclipses of the moon, the earth's shadow on the moon is always a segment of a circle regardless of the position of the moon in the sky. A sphere is the only solid object that casts a circular shadow regardless of the profile that it presents. Therefore, the earth is a sphere.

Aristotle also reasoned that objects had a natural tendency to fall toward the center of the earth, and that, in itself, would tend to make the earth form into a sphere. Thus, even though he did not know about the law of gravitation, Aristotle was able to come to a conclusion that reflects one of its effects.

Other Greek philosophers and scientists also made contributions to the development of an understanding about the cosmos. We have already noted that the fifth century B.C.E. Greek philosopher, Democritus, put forth the idea that the Milky Way consists of untold numbers of stars that cannot be seen individually with the naked eye.

Another such scientist was Heraclides, who, about 350 B.C.E., put forth rational arguments in favor of the idea that the earth spins on its axis. By this time the Greeks understood that the stars are at an immense distance from the earth. Heraclides reasoned that, if the earth did not rotate, then the stars would have to move at an incredible rate of speed to make one revolution around the earth in one day. It was far simpler to consider the stars to be stationary while the earth rotated.

Heraclides also disposed of the idea of the celestial crystalline spheres and proposed that the planets moved in orbits. Though he kept the earth as the center of the cosmos, he had Mercury and Venus move in orbits, in the correct order, around the sun, which in turn orbited the earth. This was a good first step even if he did not follow through and develop a more accurate description of the solar system.

About 230 B.C.E., Eratosthenes measured the circumference of the earth. He knew that on the twenty-first day of every June the sun was not quite directly overhead at noon in Alexandria, Egypt, where he lived, but that it *was* directly overhead at noon on that same date in Syene, which was some distance to the south. Using the angular difference in the height of the sun between the two locations on that day, along with the distance between them, with simple geometry he calculated a value for the circumference of the earth that was very close to the value we know it to be, which is about 24,900 miles.

Another Greek scientist, Aristarchus, subsequently developed a correct description of the solar system, with the sun in the center and the earth as the third planet from the sun. He also found that the axis of the earth is tilted 23½ degrees from the ecliptic.

Aristarchus also made measurements of the distance between the earth and the moon and the earth and the sun. He knew that the sun is farther away from the earth than is the moon because the moon on occasion comes between the earth and the sun, causing an eclipse of the sun. He reasoned that when the moon is exactly half full it is at the apex of a right angle between it and the sun and it and the earth. All he had to do to calculate the distance to the sun was to measure the angle between the earth and the half-full moon and the earth and the sun. Although his reasoning was flawless, he was unable to make his measurements with the accuracy that was needed. He had to calculate when the moon was *exactly* half full, and he would have had to make the measurement of the angle to the moon and the sun with a high degree of accuracy. Unfortunately, his instruments were not accurate enough for the task. If, as is very likely, Aristarchus had made his observations when either the sun or the moon was close to the horizon, atmospheric refraction would also have caused a significant inaccuracy in the angle he measured.

In any case, according to his calculations the sun was about 20 times farther from the earth than was the moon. He subsequently determined the moon to be about 250,000 miles away—a distance that is almost right. This meant that, according to his measurements, the sun was about 5 million miles away. That figure is quite a bit short of the true distance, but it provided an inkling of the true size of the solar system, a size that was considerably greater than anyone had previously imagined.

With the results of his calculations of the distances to the sun and the moon, Aristarchus could also calculate their sizes. He determined the moon to be about one-third the diameter of the earth—a figure not very far from its true size—and the diameter of the sun to be seven times that of the earth. Because he had made an error in calculating the distance to the sun, his estimate of its diameter was also off. But his calculations did show the sun to be considerably larger than the earth.

It is likely that the Greek scientists and philosophers had learned much more than we are presently able to give them credit for. However, a considerable amount of what the Greek scientists and philosophers had learned has been lost to mankind, and much of that loss is the result of deliberate destruction, and much of that destruction was caused by followers of Christianity. For example, Theophilus, the Christian bishop of Alexandria in the fourth century C.E., had a mob of his followers set fire to the Serapeum library in Alexandria in order to destroy the classical learning it contained and to remove the intellectual threat it held for the growing Christian church. Because of such losses, much of what we do know about the achievements of the ancient Greek scientists came from secondary sources, rather than from the original writings.[2]

It is also unfortunate that many of the other Greek philosophers and scientists either did not know about those achievements or did not accept their validity. For that matter, Ptolemy came along in the second century C.E., put the earth back in the center of the planetary system, and brought back the idea of the celestial spheres. Though he did make some improvements in the system that allowed for better astronomical predictions, his incorrect views about the structure of the solar system prevailed until Copernicus put the sun back into the center 1400 years later.

In any case, all this shows that at least some of the ancient Greeks were sophisticated enough to learn about many of the principles of scientific cosmology. That being the case, why did the God of the Bible not let the ancient Hebrews in on some of that knowledge when they were writing the Bible? Why did the God of the Bible not provide at least a few specific facts about the true structure of the cosmos and the earth's place in it? If he had, the Church would not have been misled concerning the nature of

13. The Biblical Cosmos: Metaphor or Mythology?

the cosmos and would have been saved from some of the embarrassment that resulted from its condemnation of Copernicus and its persecution of Bruno and Galileo.

Furthermore, why did the God of the Bible conform his cosmological descriptions in the Bible to the common myths and naïve beliefs of the ancient Hebrews? Why did the God of the Bible not include one single statement in the Bible indicating that the earth is a sphere and that it is rotating on its axis and revolving around the sun? Did he consider that his "chosen people" were less able to understand those things than were the ancient Greeks who figured them out by themselves? Or did the God of the Bible deliberately mislead the writers of the Bible concerning the shape of the earth and its place in the cosmos? Or, did the God of the Bible simply not know he was creating a spherical earth?

The only alternative, and the only reasonable conclusion, is that the ancient Hebrews believed that the earth is flat, that it is supported by pillars, that it is immovable, and that it is covered by a solid vault in which the lights of the sun, moon, and stars are set. It thus follows that the ancient Hebrews incorporated that mythic view of the cosmos into their scriptures. But if the view of the cosmos that is found in the Bible came from their myths, there is no reason for believing that they did not likewise inject other aspects of their mythology into their scriptures.

Therefore, if the ancient Hebrew writers of the Old Testament did not have the guidance of a God concerning what they wrote about cosmology, why then should they be considered to have had the guidance of a God concerning what they wrote about anything else in their scriptures?

The same holds true for the writers of the New Testament. Those writers fully accepted the mythological views of the Old Testament, including its cosmology, and they patterned their New Testament writings on that mythology. If the Old Testament falls because of its cosmology, then the New Testament must also fall, particularly since that same cosmology can be found in the New Testament.

In addition, as is shown in Appendix C of this book, Jude's reference to the Book of Enoch is particularly damaging to the authority of the New Testament. From his words, it is apparent that Jude accepted the Book of Enoch as scripture, and he even made reference to the cosmology of the Book of Enoch. But the cosmology that is described in the Book of Enoch, though it reflects the cosmology of the Bible, is pure nonsense from a modern standpoint.

If the Bible literalists still insist in believing that every word in the Bible is true and that the Bible is the word of God, they must accept the flat-earth cosmology of the Bible and reject all that science has found out about the cosmos, for the Bible does not allow for the existence of a spherical, moving earth, nor for the existence of a vast, ancient, galaxy-filled universe. On the other hand, if the literalists do not accept as true the literal words of the Bible concerning the cosmos, then they must admit that they cannot really accept as literally true anything else in the Bible.

Die-hard believers in the Bible, of course, may completely ignore what we have found out about biblical cosmology and its implications concerning the credibility of the Bible. On the other hand, some believers in the Bible may be perplexed about what we have found out about biblical cosmology because they have always strongly felt that "fulfilled" prophecies and other "evidences" found in the pages of the Bible are a powerful testimony to its credibility as the word of God.

So we have the question: If the Bible is not credible because of its cosmology, how can it be credible in these other areas that seemingly point to its being the revealed word of God? Leaving that question unanswered would undoubtedly leave many readers confused, so we should try to answer it as an adjunct to our examination of biblical cosmology. In Appendixes A and B, we will therefore take a critical look at some of the "evidence" that has convinced many believers that the Bible is the word of God, and we will show some significant problems that can be found in the Bible. By doing so, we will round out our examination of the Bible and show that there is a consistent picture concerning its credibility.

Chapter 14
The Implications

There is no question that the two views of the cosmos—the biblical and the modern—are deeply disparate. The biblical view of the cosmos presents a primitive, naïve, earth-centered, mythic construct that has no basis in reality and is similar to the mythic cosmological views of certain other ancient societies. In contrast, the modern view of the cosmos reflects the natural order of things as revealed by centuries of observation and, as such, is basic to a proper understanding of our place in the universe. Those differences are, of course, significant in and of themselves, but the implications of these different views of the cosmos go far beyond cosmology. If the cosmos of the Bible is based on the mythological outlook of the ancient Hebrews and has no basis in reality, why then should not other extraordinary depictions in the Bible be considered to be likewise based on the myths of the ancient Hebrews?

If the Bible cannot be believed when it indicates that heaven is a solid vault covering a flat, disk-shaped earth, why should it be believed concerning its account of the creation of the first man from the dust of the earth or of the first woman from a rib? If the Bible cannot be believed when it indicates that pillars hold the flat earth in place above the waters of a great deep, why should it be believed when it tells a story about a snake that talks and a tree of knowledge of good and evil? If the Bible cannot be believed when it states that the stars will fall to the earth from heaven in the Last Days, why should it be believed when it describes the fall of man in the Garden of Eden? If the cosmos of the Bible is mythic, why should not those other depictions in the Bible be likewise considered mythic? If none of those things should be given any credence because they are mythic, why then should any other extra-ordinary biblical depiction be given any credence? Indeed, why should any of the supernaturally oriented stories or viewpoints described in the Bible be given any credence? If none of these things should be given any credence, why should any of the writers of the books of the Bible be considered to have been inspired of God when it is clear that they believed these things and that these things played a part in their "inspiration"?

In other words, if the Bible cannot be believed concerning what it says about an important aspect of the natural order of things, why should it be believed concerning what it says about the supernatural order of things?

What, then, of God?

The question of whether a God exists is not the subject of this book. All kinds of arguments have been proposed to prove the existence of a God, but all of those arguments have counter arguments showing they are invalid in one way or another. Questioning whether a God exists or not is, in a sense, pointless, for any statements or arguments that we might make about the existence or nonexistence of a God are based on ignorance and not on real knowledge.

The point to be made here is that, even if there is a God and that God did cause the creation of the cosmos and everything that came from it, it still would not mean that the Bible is the revealed word of that God. The fact of the matter is that the true nature of the cosmos goes squarely against the cosmos of the Bible. Thus, even if there is a creator God, the cosmos he created, as we know it to be, shows that Yahweh, the tribal god of the ancient Israelites, is not that God, any more than is the Greek god Zeus or the Roman god Jupiter.

Therefore, even if there is a God, there is no justification in assuming that the Bible is the word of that God. The Bible would thus remain a collection of myths, folklore, embellished histories, and superstitious assumptions and speculations. Indeed, our comprehension of the cosmos should in all honesty bring us to an understanding that the Bible expresses the mythic beliefs of a naïve and self-centered people.

What this means is that, if there is a God, we can understand more about him by studying his creation than we can by mindlessly believing a book of myths and superstitions passed down from an age of myth and superstition.

What is ironic about all this is that someone who understands, contemplates, and appreciates the grandeur and the immensity of the 13-billion-year-old universe and all that has come from it, but who may be an atheist, is closer to understanding God, if there is a God, than is the Christian fundamentalist who, because of being mentally chained to the myths of the Bible, rejects or is simply ignorant of much of what science has uncovered about the cosmos and the natural world.

But then, what of morality?

Believers in the Bible, and particularly Christian fundamentalists, frequently argue that all morality is derived from the Bible, and they often question the morality of those who do not accept the Bible as the word of God. They also argue that setting aside the Bible would result in setting aside the basis of all morality.

But is that the case? Does all morality truly originate from the Bible? For that matter, does the Bible provide acceptable standards of morality for today's world?

Not really! In fact, as human societies developed and evolved from early times, they also developed and evolved certain norms of behavior. Because they help to stabilize human social interactions, some, if not most, of those norms of behavior are found in all societies. In certain societies, such as in the ancient Hebrew culture and in the Christian community that was heir to that culture, those norms were given authority by being called God's laws; but that does not take away from the fact that those norms were actually products of human social evolution. Indeed, many societies, past and present, have been ignorant of the Bible, yet have been characterized by most of the same moral standards that Bible believers claim as unique to themselves. The code of Hammurabi, for example, predates Moses, but provides antecedents to the biblical Ten Commandments. In fact, it is likely that many biblical standards were derived from the code of Hammurabi and other pagan sources.

Believers in the Bible also assert that the Bible has had a civilizing and benevolent influence on society. But is that also the case?

As well as listing some of the commonly accepted norms of social behavior, the Old Testament details numerous rituals, edicts, and punishments that reveal a primitive, superstitious, and harsh view of life and religion. Most of these rituals, edicts, and punishments are totally foreign to the social norms of today and are generally ignored by most modern Christians.

For example, Numbers 5:12–31 specifies the procedure that is to be used to determine if a woman has committed adultery. A priest is to give her a mixture of dirt and water to drink and charge her with a curse that her thighs would rot and her belly would swell if she is guilty. If she does not subsequently become sick, she is judged innocent, but if she does become sick she is judged guilty.

The Bible also specifies death, usually by stoning, for adulterers, blasphemers, those who work on the Sabbath, brides who are found not to be virgins, children who are disrespectful to their parents, and others whose "transgressions" would not warrant a death penalty in today's society.

The "morality" and social "benevolence" of the Bible is such that it prohibits cooking the meat of a young goat in its mother's milk, but does not prohibit slavery. In fact, the Bible condones slavery. In the commandment given in Deuteronomy 5:21, slaves are considered to be the property of their owners and are not to be coveted by the owners' neighbors. The KJV uses the term "servant" in that verse, but the NAB gives a more accurate translation: "*'You shall not covet your neighbor's wife. 'You shall not desire your neighbor's house or field, nor his male or female slave, nor his ox or ass, nor anything that belongs to him.'*" Deuteronomy 15:12 specifies that a Hebrew slave owner must free, after six years of servitude, a slave who is a fellow Hebrew; non-Hebrew slaves presumably could have been enslaved for life. And, in Colossians 3:22, rather than speaking out against slavery, Paul tells slaves (in Greek, *doulos*, which means "slaves," rather than "servants," as it is translated in the KJV New Testament) to obey their masters.

No, the Bible is hardly the ultimate authority in morality and social benevolence, and to act morally, one does not need to believe that the Bible contains God-given precepts by which to live. Certainly, to believe that it is wrong to take something that belongs to someone else, one does not need to believe in a God who ordered the ancient Israelites to attack another people and take their land. And to believe that murder is wrong, one does not need to believe in a God who commanded Moses, as described in Numbers 31, to order the Israelites to murder all of the men, male children, and non-virgin women of the vanquished Midianites. (As for the virgins, Moses told the Israelites to take them for themselves.)

The Bible, in fact, is full of divinely mandated atrocities. The Book of Joshua in particular describes in further detail the bloodthirsty genocide that the Israelites committed against the peoples of the lands they were invading. Those divinely mandated acts of the ancient Israelites that are described in the Bible should be reprehensible to anyone with any true moral sense. If such acts were committed in the same manner today, the whole world would condemn them and would charge the perpetrators with committing crimes against humanity. Yet, because those acts described in the Bible were supposedly committed at the behest of the God of the Bible, believers in the Bible consider them to have been moral acts.

Furthermore, from the very beginning, the God of the Bible shows questionable moral judgment. When he created Adam and Eve in the Garden of Eden, he created them so that they lacked the knowledge of good and evil. Yet the God of the Bible temptingly placed the tree of knowledge of good and evil in the garden and forbade Adam and Eve to eat of its fruit. Lacking the knowledge of good and evil, Adam and Eve would not have been able to understand that it would be "evil" to disobey God; in fact, it was only after partaking of the fruit and acquiring the knowledge of good and evil that they would have understood that disobeying God was "evil." Consequently, they could not have been responsible for their actions in disobeying the command not to eat of the fruit. Nevertheless, the God of the Bible punished them for eating thereof.

In addition, the biblical God, as related elsewhere in the Bible, often acted in a manner that should be reprehensible to anyone with any true moral sense. The Book of Exodus, for example, describes the God of the Bible as repeatedly hardening the heart of Pharaoh against letting the Israelites go, and then repeatedly punishing the Egyptian people because their ruler did not let the Israelites go. Is such a petty and capricious deity really worthy of worship? Any rational person would say *no*!

However, to those who believe in the Bible, morality is defined by the dictates and actions of the biblical God. Thus, according to the Bible believer, all of those previously described actions and commandments of the biblical God were moral.

If, as Bible-believing Christians argue, the Bible provides the basis for morality and social benevolence, then, against the standards that Christians claim for themselves, one must weigh the history of Christianity. That history, in fact, reveals much about the nature of Christian belief and the effects of that belief on society. Indeed, an examination of that history reveals that belief in Bible-based Christianity has not always been the moral and benevolent social force that Christians have frequently made it out to be.

As soon as the followers of Christianity were able to gain political power in the lands of the eastern Mediterranean during the early fourth century, there was a portent of what was to come. The combination of absolute political power and absolute theological certainty resulted in a requirement for absolute religious conformity. That requirement began to be expressed during the reign of Constantine with a growing intolerance towards those who held unorthodox beliefs about the nature of Jesus and the Church, and it became more vehement during the reign of Constantine's son, Constantius, under whom the inhabitants of several towns were massacred because they adhered to one "heresy" or another.

The rise of Christianity also resulted in a concerted effort to rid society of the remnants of the pagan religions. Constantius closed the pagan temples and decreed that those individuals who continued to practice the worship of the pagan deities were to be put to death and to have their property confiscated.

The rise of Christianity also had another effect on society during that era. Because it was generally believed that the Second Coming of Jesus was close at hand, and with it the end of the world, the study of the natural world was considered useless and a distraction from the message revealed in the Bible. As a consequence, the schools that could be found in most of the large towns, and the academies that had long been fixtures in almost every large city, were gradually abandoned. An outright animosity toward "pagan" learning and scholarship began to develop, exemplified in the year 391 C.E. when Bishop Theophilus of Alexandria, under the order of Emperor Theodosius, destroyed the Serapeum library in Alexandria and with it many of the important literary works of the Greco-Roman world.

In the growing Christian influence in that era, all learning and science came to be looked upon with suspicion and was identified with paganism. Consequently, scholars and scientists who taught others and who practiced their occupations did so at great risk to themselves. One such scholar and scientist was Hypatia of Alexandria, who worked in the palace library in Alexandria. She was a woman of renowned intellect and beauty and, in an age when women were normally not held in much regard, she was a noted neoplatonic philosopher and one of the foremost mathematicians, physicists, and astronomers of the time. Because of what she represented, Cyril, the Archbishop of Alexandria, inveighed against her, and in the year 415 C.E. his followers attacked her as she was on her way to her scholarly duties. They tore off her garments, murdered her by cutting the flesh from her bones with abalone shells, and then burned her remains.[1]

As the growing Christian church continued to spread over a larger area, there was a concerted effort to convert nonbelievers. If the power of the spoken message was not sufficient to achieve the desired result, the power of the sword was used, and, during the reign of Justinian, tens of thousands of non-Christians were forcibly baptized. As Christianity expanded into Europe, forced conversions were common. Charlemagne, for example, ordered death to those who refused baptism.

After Christianity was established in Europe, the call went out for crusades to regain Palestine from the Moslems. While on their way to the "Holy Land," the Christian crusaders massacred whole villages, and then, upon reaching Jerusalem, murdered thousands of Jews and Moslems. Later on, in the early thirteenth century, a crusade was called to eradicate the Albigensen "heresy" in southern France. As a result, whole towns were destroyed and tens of thousands of people were murdered. When the crusaders attacked the town of Beziers in southern France, they asked their commander, the papal legate, how they would be able to distinguish the town's fellow Catholics from its Catharistic "heretics." His reply: "Kill them all. God knows his own." The Christian crusaders, in their religious fervor, were quite happy to oblige the order. But after growing bored with giving the town's inhabitants a quick death by the sword, they amused themselves by torturing their victims in unspeakable ways.

With the establishment of the Inquisition, the Church sought to stamp out all forms of heresy and "witchcraft." The inquisitors also discouraged and even severely punished those who pursued scientific inquiry. Torture was used to induce confessions, and uncounted numbers of people were, over the centuries, tortured and burned at the stake or killed in other horrendous ways.

Following the Protestant Reformation, the search for heretics and witches continued, and the Protestants themselves added more victims to the mounting toll. Both the Catholics and the Protestants spread their religious obsessions to the New World. The Salem witch trials were established in New England and the Inquisition was established in Mexico and Peru. In addition, many of the American colonies passed laws governing religious belief and expression. Maryland, for example, had a law that provided the death penalty for anyone who denied the reality of the Trinity.

Over time, and with the Enlightenment, many people began to question whether religion should continue to have the ability to dominate society in such an excessive manner. The temper of the times caused governments to become more secular, and the Christian religious institutions gradually lost much of the political power that they had previously held in most of the countries of the Western world.* As a result, those institutions could no longer use the governmental power structures to force their theology upon society, and they were constrained to become somewhat more tolerant of those who adhered to different viewpoints. Consequently, society became less religiously coercive and the tortures and executions waned in number; most notably, the Inquisition finally came to an end in the early nineteenth century.

No one will ever know the true figure for the total number of people who were killed and tortured over the centuries because of the religious fervor of believers in Bible-based Christianity, but the number was certainly in the millions. In addition, the religious fervor, with its concern for the next world rather than this, resulted in the loss of much of the classical learning of the ancient Greek and Roman civilizations. It also put a brake on a continuance of the traditions of scientific inquiry, technological development, civic improvements, and artistic achievements that those civilizations had engendered. Thus, the rise and spread of Christianity

---

*The government of the United States of America, of course, was established with the ideals of the Enlightenment and with the separation of church and state in mind.

played a substantial part in thrusting the western world into the Dark Ages and in setting human progress back for several hundred years.

Doubtless there are those who would say that the foregoing descriptions present a one-sided view of the historical conduct of believers in Bible-based Christianity. Perhaps that view *is* one sided, but it is no more so than are the views presented by those individuals who seek to promote belief in Christianity and in the Bible. If one were totally ignorant about the history of Christianity, one would be hard put to learn anything about the religious excesses described above by listening to the rhetoric engendered by those individuals. Indeed, from listening to that rhetoric one would get the impression that Christianity has, ever since its inception, been the epitome of moral practice and of societal benevolence.

As much as modern Christians tend to brush off the historical excesses of earlier believers in their religion—if they are even aware that such excesses had taken place—those excesses cannot be ignored in any proper evaluation of Bible-based Christianity. Indeed, an evaluation of the history of Bible-based Christianity shows that belief in the Bible does not guarantee that the believers will act in a true moral manner. The inquisitors of a few centuries ago certainly believed they were acting morally when they tortured and killed "heretics" and "witches." They had the Bible to justify their actions, for the Bible makes it clear that heretics and witches are to be punished and even killed. In the view of the inquisitors, the tortures that they caused "heretics" and "witches" to endure here on earth were but an insignificant taste of the tortures that God himself would cause them to endure for all eternity. The inquisitors, then, were merely emulating the "justice" of God on an infinitely smaller scale.

Moreover, even today, belief in the Bible portends a continuation of the past excesses that occurred during the history of Christianity. The Christian Reconstructionists, the Dominionists, the Christian nationalists, and others like them are working to impose their Bible-based beliefs upon present-day society. If any of these groups gain the ascendancy, the horrors of the events that occurred during the history of Christianity could well happen again. Our democratic institutions would fall by the wayside and society would again be under the thumb of religious oppressors who would curtail religious and personal freedom. Of even more concern are those who would like to help bring on the biblical Apocalypse by exacerbating religious conflicts throughout the world. The result could very well be wars that would end civilization as we know it—and without the salvation of the biblical kingdom of God that those Bible believers stridently hope for.

In the final analysis, there is nothing moral about the Bible, for, in the final analysis, the Bible makes it clear that morality counts for nothing.

14. The Implications

According to a strict Christian view of biblical doctrine, a person who, throughout his life, acts responsibly toward his fellow man, never knowingly commits an offense against another person, acts charitably and helps others in their misfortune, and accepts personal responsibility and tries to atone for any of his actions that may have inadvertently caused injury to another person, yet dies without accepting Jesus as his savior, is doomed to eternal torture.

On the other hand—again according to a strict Christian view of biblical doctrine—a thorough scoundrel who never in his life did one kind thing for anyone, robbed and murdered scores of people, never accepted responsibility for his crimes, and even framed innocent people for his own wrongdoings, will spend eternity in paradise if he accepts Jesus as his savior just before dying.

In the final analysis according to a strict interpretation of Bible-based Christian doctrine, morality counts for absolutely nothing. In the final analysis, according to Bible-based Christian doctrine, salvation is what counts, and one achieves salvation by believing without question that one's own wrongdoings can be atoned for by the suffering and death of someone else. Such is Christian morality, a morality based upon the Bible.

In contrast to the Christian view of morality, rather than coming from a blind obedience to the presumed dictates of an all-powerful God, true morality and ethical behavior come from reasoned personal convictions and a concern for one's fellow humans.

And that brings up another point. In holding to what they consider to be absolute "truth," believers in the Bible often look upon nonbelievers as being somehow spiritually deficient, morally depraved, and intellectually perverse for rejecting the "evidence" of that "truth." As a result of that attitude, believers in the Bible often feel free to inveigh against nonbelievers. From their church pulpits, and on radio and TV, Bible believers frequently belittle, deride, slander, and denounce nonbelievers. In holding to their position of having "absolute" truth, Bible believers give no thought to the idea that perhaps nonbelievers have honest and valid reasons for being nonbelievers.

Moreover, contrary to what Bible believers say, non-believers as a whole are no more apt to engage in criminal acts than are believers, and, in fact, are often less likely to engage in such acts. For example, in certain European countries that have a much higher percentage of non-believers than does the United States, the crime rate is much lower than in the United States. This is particularly true in the Scandinavian countries, which have very low rates of religious belief and very low crime rates. If non-believers are more apt to commit crimes, why is crime not rampant in those countries?

Of course, showing that belief in the Bible is a delusion would not end the problem of unwarranted religious belief. It would seem that many, if not most, followers of religion have been conditioned to need a belief in something beyond and greater than themselves. It is therefore likely that, if some Bible believers—perhaps because of being convinced by the evidence presented in this book—did become disillusioned with the Bible and with the religion based on it, they would try to find another religious system to fill that need.

But there is a lesson to be learned from what we have found out about the Bible: Truth cannot be found by wearing the twin blinders of wishful thinking and unquestioning belief. Just as the Bible is the word of men, and not the word of God, so can the same be said of any other "holy" book and the religion based on it.

In coming to disbelieve in Bible-based religion, then, one should not desperately look around for another system of religion to take its place. The danger is that if too many people who become disillusioned with the Bible convert to another religion, particularly one that is based on restrictive and primitive inculcations that are similar to those found in the Bible, that religion could in turn become a force in society, and by becoming a force in society it could in turn bring on the same problems and intolerance that belief in the Bible has brought about.

Appendixes

Appendix A
Problems in the Old Testament

Many of those who try to promote belief in the Bible have felt compelled to do more than merely make an appeal to faith in order to convince others that the Bible is the word of God. To that end, these individuals have written countless books and have engendered innumerable discourses from pulpits and through the electronic media in which they have used all sorts of approaches and arguments in trying to prove their case. To a great many people, the rhetoric put forth by these individuals is quite persuasive, and, in many cases, seemingly irrefutable.

So we have the question: If the Bible is not credible because of its cosmology, how can the Bible promoters put forth such compelling arguments relating to other aspects of the Bible which seemingly prove that the Bible is the revealed word of God? Leaving that question unanswered would undoubtedly leave many readers of this book confused, so in this and in the next appendix we will take a critical look at some of the "evidence" that has convinced many believers that the Bible is the word of God. We will also take a look at some significant problems that can be found in other aspects of the Bible besides cosmology. Taken together, all of this material will show that there is a consistent picture concerning the credibility of the Bible.

Though many of those who promote belief in the Bible may have, through their rhetoric, convinced large numbers of people that the Bible is the word of God, the standards of evidence that they have used to support their arguments leave a lot to be desired, and the conclusions that they draw are frequently unwarranted.

For example, many biblical apologists argue that the Bible has been proved true, and therefore that it is the word of God, because it contains certain accounts and descriptions that have been verified by archaeological finds. But that argument is a leap in logic as well as faith. If an archaeological find verifies that a city, say, described in the Bible did exist, or that a battle described in the Bible was fought, does that mean the supernatural embellishments that were also given in the referenced account actually happened, or that other parts of the Bible are also true?

Many biblical apologists will argue that it does. But by the same argument, one could prove that the Greek gods exist and that the supernatural events described in the *Iliad* actually took place. After all, archaeology has shown that the city of Troy did exist, which substantiates Homer's account of the battle for Troy. If the argument holds in the one case, then why not in the other?

The only thing that can be reasonably said about such archaeological finds is that the writers of the Bible, as well as the writers of other books from ancient times, often wrote about real cities and real wars. But that should not be considered unusual, and neither should it be considered evidence that the supernaturalistic embellishments those writers incorporated into their writings actually happened. In ancient times, after all, people frequently interpreted events in their lives according to their particular mythological and supernaturalistic beliefs.

Moreover, some archaeological findings conflict with the accounts of some of the events that are described in the Bible. For example, it appears that, contrary to the biblical account of the Israelite conquest of Canaan, many of the people who came to be known as Israelites actually settled in the land more or less peacefully over a lengthy period, and that others were native to the area. Archaeological studies have detailed the sequences of habitations in many of the cities of Canaan over several thousand years. Those studies indicate that there is little or no evidence for the genocidal bloodbaths that the invading Israelites were supposed to have visited upon the original inhabitants of the land as described in the Bible. In fact, contrary to the descriptions given in the Bible, many of the "conquered" cities did not have defensive walls at the time of the purported conquest, while many of the other cities did not even exist at the time.[1] These disparities can be understood as being the result of myth building and historical revisionism by biblical authors who lived long after the times and events they were describing.

Believers in the Bible have also frequently argued that many biblical prophecies have been fulfilled, which, they say, proves that the Bible is the word of God. For example, they often declare that the fate of certain nations and cities in biblical times had been "prophesied" beforehand by various Old Testament prophets.

Ezekiel's prophecy that the city of Tyre would be destroyed and would be made as bare as a rock is a case in point; believers in the Bible frequently refer to that prophecy when they try to prove that the biblical prophets were inspired by God. However, although many people have been swayed by that particular "proof," the facts are not as remarkable as they have been made out to be, and, in fact, they stand in direct opposition to the claims made by Bible believers. Because so much has been made of this prophecy, we shall examine it in detail.

Ancient Tyre was an important Phoenician city that was located on an island in the Mediterranean just off shore in what is now southern Lebanon. Its initial settlement during the third millennium B.C.E. was prompted by the existence of two natural harbors on the island, one on the south side and the other on the north side. The harbor on the north side—called the Sidonian port because it faced Sidon on the coast to the north—was, in fact, one of the best along the eastern end of the Mediterranean. The Tyrians took full advantage of the two harbors on the island, and, as a result, they became a renowned commercial seafaring power.[2]

Of course, living on an island in the sea meant that certain necessities were in short supply and had to be brought in; an associated settlement on the mainland near the island was therefore used to facilitate supplying the island city with water, timber, and other provisions. The Tyrians also dug wells and water storage cisterns on the island, which, along with the water that was ferried in, allowed the city to support a fair-sized population.

Over time, the Tyrians became quite wealthy because of the extensive commercial trade in which they engaged, including the sale of an expensive purple dye they made from mollusks that were to be found in the waters nearby. Tyre also established several colonies on the coasts of the Mediterranean, including, most notably, Carthage on the African coast.

The Tyrians, however, did not always have smooth sailing. As a result of their strategic location and their growth in commercial importance, the island city of Tyre and its associated mainland settlement were subject to attacks by their more militaristic and expansionist neighbors. To counteract such threats and to protect themselves, the inhabitants of both Tyre and the mainland city built walls around their respective settlements.

To support their arguments concerning Ezekiel's prophecy, biblical apologists often try to give the impression that Tyre consisted primarily of the mainland city when Ezekiel made his prophecy, and that the island had no particular significance until events forced its eventual settlement.[3] However, as noted above, the island had in fact been inhabited long before

Ezekiel's time, and it overshadowed the mainland settlement in importance because of its harbors and because it was easily defendable and thus better able to withstand the might of its militaristic neighbors. Moreover, the original Phoenician name of the city is *Sur*, which means "rock," for the city derived its name from the geological nature of the island upon which it stood. The mainland settlement, on the other hand, was called Ushu.[4] The Hebrew version of the name of the island city was *Tsor*, and that is the word that is used for the name of the city in the original Hebrew of Ezekiel. (The name "Tyre" comes from the later Greek name for the city.)

The events leading up to Ezekiel's prophecy concerning Tyre had their beginning when Babylonia, under Nebuchadrezzar its king, conquered Jerusalem in 598 B.C.E. and exiled its leading citizens to Babylon in 597 B.C.E. After having gained control of the lands of the eastern Mediterranean, Nebuchadrezzar next sought to complete some unfinished business, and he made plans to invade Egypt. That did not go without notice by those who had been subjugated by the Babylonians, and, feeling that Nebuchadrezzar would be distracted with his Egyptian campaign, Judah and Tyre subsequently revolted against Babylonian hegemony. The revolts were miscalculated, however, and Babylonia set forth to put down those who questioned its authority.

Ezekiel was one of those who had been exiled in the Babylonian captivity, and, of course, being in Babylon he would have known full well about the new Babylonian assault against Judah and its capital, Jerusalem. He, in fact, made reference to the beginning of the Babylonian siege against Jerusalem in chapter 24 of his book. It was near the end of the ninth year of the exile:

> 1. *Again in the ninth year, in the tenth month, in the tenth day of the month, the word of the Lord came unto me, saying,*
> 2. *Son of Man, write thee the name of the day, even of this same day: the king of Babylon set himself against Jerusalem this same day.*

Since the exile to Babylon occurred some time during the year 597 B.C.E., the first full year of the exile would have overlapped the remainder of the year 597 and some initial part of the year 596 B.C.E. By counting the years of the exile as continuing to overlap consecutive calendar years, this would bring Ezekiel's declaration concerning the point in time (that is, the tenth day of the tenth month of the ninth year of the exile) when "the king of Babylon set himself against Jerusalem" into 587 B.C.E., which is, in fact, the generally accepted year for the beginning of the siege. It did not take long for the Babylonians to again defeat Jerusalem. This time they razed the city to the ground.

Ezekiel subsequently made his prophecy against Tyre (referred to as "Tyrus" in some books of the KJV, including the book of Ezekiel; most other versions of the Bible use "Tyre" in referring to the city) in chapter 26. The prophecy begins with the following:

> 1 *And it came to pass in the eleventh year, in the first* day *of the month, that the word of the LORD came unto me, saying,*
> 2 *Son of man, because that Tyrus hath said against Jerusalem, Aha, she is broken* that was *the gates of the people: she is turned unto me: I shall be replenished,* now *she is laid waste:*

Ezekiel gives the date as *"the eleventh year, in the first day of the month,"* of the exile. Since he invariably gives the number of the month in numerous other references to the years of the exile, it would appear that the number of the month that would have been in this passage was lost some time later during the copying process, which left only the day of the month in the text. In any case, the eleventh year of the exile that Ezekiel gave as the time frame for his giving this prophecy concerning Tyre would have been during 586/5 B.C.E., not long after the destruction of Jerusalem.

Apparently the inhabitants of Tyre felt that the walls of their two cities made them secure against the Babylonian threat, and as verse 2 of the above passage indicates, they made light of Jerusalem's fate and sought to profit from the city's downfall. In essence, the inhabitants of Tyre apparently looked to take control of some of the territory that had belonged to Jerusalem as well as a share of the trade that it had before it was razed by Nebuchadrezzar's forces.[5] The NAB rendition of verse 2 provides wording that more clearly indicates what is being said:

> 2 *Son of man, because of what Tyre said of Jerusalem: "Aha! it is broken, the gateway to the peoples; now that it is ruined, its wealth reverts to me!"*

Ezekiel clearly took umbrage at the comments that the inhabitants of Tyre had made about the destruction of Jerusalem, for the thrust of his prophecy was to wish a similar fate upon their city. Ezekiel continued his prophecy against Tyre with the following (here, the KJV again):

> 3 *Therefore thus saith the Lord GOD; Behold, I am against thee, O Tyrus, and will cause many nations to come up against thee, as the sea causeth his waves to come up.*
> 4 *And they shall destroy the walls of Tyrus, and break down her towers: I will also scrape her dust from her, and make her like the top of a rock.*

> 5 *It shall be a place for the spreading of nets in the midst of the sea for I have spoken it, saith the Lord GOD: and it shall become a spoil to the nations.*
> 6 *And her daughters which are in the field shall be slain by the sword; and they shall know that I am the LORD.*

Taken by itself (as many believers in the Bible take it), this passage does not say which "many nations" would rise against Tyre, nor does it say when they would rise against Tyre. In any case, there was nothing very unlikely about what it describes. An alternate way of looking at the setting for this part of the prophecy would be to ask, "What would have been the likelihood that Tyre would *not*, within the subsequent several hundred years, be attacked in some war or other?" The answer is that likelihood would have been dubious at best. In fact, considering the times and the strategic and commercial importance of Tyre, the odds were reasonably good that the city *would* eventually be attacked and perhaps even be razed in some war at some time or other. Indeed, Tyre had previously been subject to attacks. About the year 724 B.C.E, for example, Shalmaneser V, the king of Assyria, attacked and captured the mainland city of Ushu. Though he failed to capture the island city of Tyre before his death, it did yield to his successor, Sargon.

The fact of the matter is that, in those times, wars and razed cities were hardly rare. The Babylonians, for example, would level the cities they conquered if they felt that doing so would provide an object lesson to others who might seek to resist Babylonian dominion. Indeed, as befitted the times, the Babylonians, as we previously noted, razed Jerusalem after again defeating that city when its inhabitants tried to throw off the yoke of Babylonian domination.

Concerning the "many nations" that are described as coming against Tyre in this part of Ezekiel's prophecy, one might therefore say, "Ho hum. So what else is new?" Simply put, in order for this part of Ezekiel's prophecy to be falsified, Tyre would have to have *not* been conquered and razed for a very extensive time after the prophecy was made. In that light, and considering the times, this part of Ezekiel's prophecy was hardly remarkable.

Believers in the Bible, of course, do not look at it that way. In fact, they regard the above passage as a remarkable example of divinely revealed prophecy because of events that occurred hundreds of years after Ezekiel wrote the passage.

However, that passage cannot really be taken by itself, for it is distinctly tied to the verses that follow it. In those verses, Ezekiel specifically "prophesied" that Tyre was to be conquered and laid waste by the Babylonians under Nebuchadrezzar. The following is found in verses 7–12 of Ezekiel 26:

> 7 *For thus saith the Lord GOD; Behold, I will bring upon Tyrus Nebuchadrezzar king of Babylon.* . . .

> 8 *He shall slay with the sword thy daughters in the field; and he shall . . . cast a mount against thee. . . .*
> 9 *And he shall set engines of war against thy walls, and with his axes he shall break down thy towers.*
> 10 *. . . thy walls shall shake at the noise of the horsemen, and of the wheels of the chariots, when he shall enter thy gates. . . .*
> 11 *. . . he shall slay thy people by the sword, and thy strong garrisons shall go down to the ground.*
> 12 *And they shall make a spoil of thy riches, and make a prey of thy merchandise: and they shall break down thy walls, and destroy thy pleasant houses: and they shall lay thy stones and thy timbers and thy dust in the midst of the water. . . .*

The introduction to the second part of the prophecy confirms that both parts belong together: The second part begins, "*For thus saith the Lord GOD; Behold, I will bring upon Tyrus Nebuchadrezzar king of Babylon. . . .*" The Hebrew word translated as "For" is *kiy*, which specifically expresses causality and can be translated as "because" as well as "for." Thus the word "For" in that verse clearly ties the two parts of the prophecy together and shows that the destruction that is described in the previous verses will be brought about by Nebuchadrezzar's forces as described in the succeeding verses.

In verses 14–17, Ezekiel goes on to describe how God would cause Nebuchadrezzar's destruction of Tyre to be total:

> 14 *And I will make thee like the top of a rock: thou shalt be a place to spread nets upon; thou shalt be built no more: for I the LORD have spoken it, saith the Lord GOD.*
> 15 *Thus saith the Lord GOD to Tyrus; Shall not the isles shake at the sound of thy fall, when the wounded cry, when the slaughter is made in the midst of thee?*
> 16 *Then all the princes of the sea. . .*
> 17 *. . . shall take up a lamentation for thee, and say to thee, How art thou destroyed, . . . the renowned city, which wast strong in the sea. . . .*

Then, in verses 19–21, Ezekiel emphasized that the city of Tyre would be completely lost to the world:

> 19 *For thus saith the Lord GOD . . . , I shall bring up the deep upon thee, and great waters shall cover thee;*
> 21 *. . . and . . . though thou be sought for, yet shalt thou never be found again, saith the Lord GOD.*

Thus, the second part of the prophecy makes it clear that Tyre was to be totally destroyed by Nebuchadrezzar's forces. According to the prophecy, the city of Tyre would be torn down and made bare like "the top of a rock" and would be covered by the "waters" of "the deep" and "built no more," and, though it would be "sought for," it would "never be found again." Therefore, from the perspective of the prophecy there could not have been any subsequent attacks against Tyre hundreds of years later by "many nations," for there would have been no city of Tyre for the "many nations" to attack. Hence, the "many nations" mentioned in the first part of the prophecy have to be taken as being allies, or perhaps mercenaries, of Nebuchadrezzar in his war against Tyre.

The reason for the prophesied total destruction of Tyre by Nebuchadrezzar is, after all, explicit in the context of Ezekiel's invective against Tyre in verse 26:3: *"Therefore thus saith the Lord GOD; Behold, I am against thee, O Tyrus...."* The word "therefore" specifically indicates that God would cause Tyre's destruction because the inhabitants of Tyre had belittled and sought to profit from the destruction of Jerusalem, which Ezekiel had mentioned in the previous verse. It certainly would have made no sense for Ezekiel to have meant that his prophesied destruction of Tyre was to take place after those who were the object of his prophetic invective had been dead for hundreds of years and would therefore not personally suffer for their having spoken ill of Jerusalem.

And that brings us to the events that were playing out when Ezekiel made his prophecy in the eleventh year of the exile. The Babylonians under Nebuchadrezzar, as noted above, had already defeated and had razed Jerusalem, and they were, in fact, preparing to lay siege against Tyre and the mainland settlement (if, indeed, they had not already begun the siege). From his position in the Babylonian exile, Ezekiel certainly would have been well aware of the plans that Nebuchadrezzar had for Tyre.

From his standpoint at the time, then, Ezekiel was basically "prophesying" of what would have appeared to be a foregone conclusion, for it seemed likely that the city of Tyre would soon be overpowered and sacked by Nebuchadrezzar's army just as Jerusalem had already been. (In his book, *Evidence that Demands a Verdict*, Josh McDowell overstated the length of time between the date of the prophecy and the date of the beginning of the Babylonian siege of Tyre, possibly to make it appear that Ezekiel's prophecy was not based on an observation of ongoing events.[6] There are many such statements in his book that could be considered misleading, and these statements have been frequently taken up and repeated by several other biblical apologists.)

Nevertheless, it turned out that such a conclusion about the likelihood of Tyre's imminent destruction was erroneous, for, contrary to Ezekiel's prophecy, Nebuchadrezzer's army failed to conquer Tyre and to lay it to waste. After a thirteen-year siege, a compromise was reached between the two warring parties, and under that compromise Tyre agreed to become a vassal of Babylon. It took some time for Tyre to recover economically from the many years of siege, but it subsequently existed in relative peace and prosperity for two and one-half centuries.

Significantly, in his prophecy about all that would happen to Tyre, Ezekiel made absolutely no mention of the fact that Tyre was ultimately able to withstand Nebuchadrezzar's siege. And, what is particularly noteworthy, Ezekiel made no mention at all of the fact that Tyre agreed to become a vassal of Babylon. If Ezekiel really was making an inspired and true prophecy against Tyre, that prophecy surely would have reflected the actual events and would have at least mentioned Tyre's degradation in becoming a vassal of Babylon. Instead, as noted above, Ezekiel prophesied that Tyre would be totally destroyed by Nebuchadrezzar's forces, that its inhabitants would be slaughtered, and that the city would be "built no more." Moreover, the prophecy declared that though the city of Tyre would be "sought for," it would "never be found again."

The facts provide a picture that is quite different from that given by most biblical apologists. As previously noted, the apologists frequently misrepresent the facts and try to give the impression that the city of Tyre consisted primarily of the mainland city and that the inhabitants of that city moved to the island—which, in their view was previously without much significance—and established a secure city there as a response to Nebuchadrezzar's siege.[7] The apologists will usually say that it was the island city that Nebuchadrezzar failed to conquer, but that he did conquer and destroy the mainland settlement, which, they say, fulfilled the second part of Ezekiel's prophecy. The problem is that the historical record concerning the siege of Tyre is sketchy, and there appears to be no account that specifically states that Nebuchadrezzar did in fact conquer and destroy the mainland settlement.[8]

In any case, unlike Jerusalem, which Nebuchadrezzar's forces would have been able to surround to prevent supplies from reaching the city, the mainland settlement would have had access to supplies carried by its fleet of ships to its seaward side, and would therefore have been able to hold out considerably longer than Jerusalem had been able to. Even if Nebuchadrezzar's forces did eventually breach the walls of the mainland city, those same ships would have been able to ferry its inhabitants to Tyre or to other locations. Indeed, it is possible that, if Nebuchadrezzar's forces did actually enter the mainland city, its populace had already completely abandoned it for the more secure—and long-time extant—island city of Tyre during the siege.

Regardless of whether or not Nebuchadrezzar's forces were able to enter the city on the mainland, the siege against that settlement and Tyre did not end with a military victory and a slaughter such as that which Ezekiel had prophesied. Perhaps the most effective evidence of that fact is provided by the words of Ezekiel himself. Several years after writing the above-quoted passages, Ezekiel off-handedly conceded that his "prophecy" concerning Tyre had failed to be fulfilled. As he stated in Ezekiel 29:17–18:

> 17 *And it came to pass in the seven and twentieth year, in the first month, in the first day of the month, the word of the Lord came to me, saying,*
> 18 *... Nebuchadrezzar king of Babylon caused his army to serve a great service against Tyrus: every head was made bald, and every shoulder was peeled; yet had he no wages, nor his army, for Tyrus, for the service that he had served against it.*

Ezekiel made that observation in the twenty-seventh year of the exile, which would have been about three years after the end of Nebuchadrezzar's failed thirteen-year siege against Tyre. Therefore, in contradiction to the prophecy that he had previously given, Ezekiel here indicates that not only was Nebuchadrezzar *not* victorious against Tyre, but also that he and his army failed to "make a spoil" of Tyre's "riches" or to "make a prey" of Tyre's "merchandise." Rather, Nebuchadrezzar had "no wages, nor his army," for his "service" against the city. During that service, the heads of his soldiers were worn bald from the helmets they wore, and their shoulders were peeled from carrying their armaments—all to little effect and no profit.

Though he previously had prophesied that Nebuchadrezzar's forces would slaughter the inhabitants of Tyre, Ezekiel made no mention of such a slaughter in the above passage. If such a slaughter had occurred, he surely would have mentioned it as a confirmation of his prophecy and as evidence of God's retribution. It is thus quite clear that Ezekiel's prophecy was a failed prophecy—in fact, a false prophecy.

However, as previously noted, in those times a city would be, as often as not, eventually attacked in one war or another. Such was the case with Tyre. Two and one-half centuries after Nebuchadrezzar's failed siege of Tyre, Alexander the Great had his army besiege the city because it thwarted his plans to conquer Egypt. To facilitate his assault on the island city, Alexander had his army use stones from the buildings of the mainland settlement to construct a causeway from the mainland to the island. After a seven-month siege, Tyre fell to Alexander's forces and most of its surviving inhabitants were sold into slavery.

Eventually, with an influx of new inhabitants, as well as the return of some of the original Tyrians who had abandoned it during Alexander's siege, Tyre once again became a power in the sea. In fact, only eighteen years after its conquest by Alexander, Tyre was able to resist for a time the armies of Antigonus, one of Alexander's generals who sought to set himself up as king of the conquered lands following Alexander's death. However, after fifteen months of siege, Antigonus's army finally conquered and "reduced" Tyre.

From these facts, we can understand why many biblical apologists tend to separate the first part of Ezekiel's prophecy from the remainder. Since, contrary to Ezekiel's prophecy, the Babylonians under Nebuchadrezzar failed to conquer and sack Tyre, the apologists have tried to salvage something from the prophecy by splitting it into two sections. According to these apologists, the part of the prophecy describing Nebuchadrezzar's assault on the city of Tyre actually referred to the mainland settlement and not to the island city. The other part of the prophecy—which mentioned the "many nations"—was supposed to have been fulfilled by Alexander's conquest of the island city and by its subsequent "reduction" by Antigonus.[9]

The wording in the prophecy, however, certainly does not support that interpretation. We have already seen that the two parts of the prophecy actually belong together and that it was Nebuchadrezzar who was supposed to cause the destruction of Tyre. Moreover, when he called Tyre by name, Ezekiel clearly meant the island city. In verse 26:5, for example, he prophesied that Tyre "shall be a place for the spreading of nets in the midst of the sea." The words "in the midst of the sea" can only refer to the island city and not to the mainland settlement. In verse 26:4 he also prophesied that God "will also scrape her dust from her, and make her like the top of a rock," and he repeated that prophecy in verse 14. In prophesying that fate, Ezekiel was making a play on the name of the city; as we noted, the Phoenician word translated as "Tyre" means "rock" and is a reference to the geological nature of the island that gave the city its name. Furthermore, in 26:17 he referred to Tyre as "the renowned city, which wast strong in the sea," and, in 27: 4–6, Ezekiel stated of Tyre: "Your borders are in the heart of the seas." Those verses again clearly refer to the island city.

In addition, speaking as if the words were coming directly from God, Ezekiel frequently addressed Tyre in the second person—i.e., as "you" or "thee"—when he described the calamities that would befall the city. In so doing, Ezekiel made a further differentiation between the island city of Tyre and the mainland settlements. In verse 26:8, for example, he stated:

"*He* [Nebuchadrezzar] *shall slay with the sword thy daughters in the field; and he shall . . . cast a mount against thee.*" KB states this concerning the passage: "(your) daughters in the fields **Ezk 26**$_{6\text{-}8}$ meaning the settlements dependent on Tyre . . . on the mainland . . . in contrast to the main city on the island." The RSV translates verse 26:8 as "*He will slay with the sword your daughters on the mainland; he will . . . throw up a mound against you.*" The NIV goes even further and translates the passage as "*He will ravage your settlements on the mainland with the sword; he will . . . build a ramp up to your walls.*" This wording thus shows that Ezekiel differentiated between the mainland settlements and the island city. In referring to Tyre's daughter settlements as being "on the mainland," Ezekiel clearly indicated that Tyre itself was not on the mainland. Ezekiel thus was prophesying that Nebuchadrezzar would destroy the mainland settlements ("your daughters on the mainland") and then would siege ("cast up a mound against you") and destroy the island city of Tyre.

In separating the first part of the prophecy from the remainder, the believers in the Bible thus do violence to the words of Ezekiel and make them say something they are not really saying. Moreover, if one holds that Alexander's conquest of Tyre and the subsequent reduction of the city by Antigonus fulfilled Ezekiel's prophecy, it would necessarily bring about an unjust conclusion: That is, though the inhabitants of Tyre in Ezekiel's time brought the wrath of God upon their city for their having spoken ill of Jerusalem, they got off relatively easy—but those who lived in the city two and one-half centuries later were made to suffer for it.

In any case, Alexander's attack on the city of Tyre underscores the fact that Ezekiel's prophecy was a false prophecy; if Ezekiel's prophecy had been a true prophecy, the Babylonians would have completely destroyed Tyre and there would have been no city for Alexander to attack. This shows that, to many of the believers in the Bible, reality means nothing. They ignore the facts and make whatever interpretations of both the Bible and history that are necessary in order to "prove" that the Bible is the word of God. Thus, somehow, Alexander's conquest of Tyre and Antigonus's "reduction" of it were a fulfillment of Ezekiel's prophecy that Nebuchadrezzar would lay waste to that city.

But that is not the end of the pronouncements that believers in the Bible have made about Alexander's conquest of Tyre. Many believers in the Bible have declared that Tyre has continued bare as a rock ever since, thus "proving" that the Bible is the word of God. They have even challenged nonbelievers to try to rebuild Tyre and thus falsify Ezekiel's prophecy, which, they say, can never be done because God has ordained that Tyre would never be rebuilt.

However, it is not necessary for nonbelievers to rebuild Tyre in order to falsify Ezekiel's prophecy and, by implication, the Bible. Soon after the "reduction" of the island city by Antigonus, and contrary to the statements made by many believers in the Bible, both Tyre and the mainland settlement were rebuilt. In later years, under Roman rule, Tyre was the capital of the province of Phoenicia and again became a rich commercial center. Even Jesus (in Matthew 11:21–22) spoke of Tyre as a living city and prophesied that it would be more tolerable for Tyre on the Judgment Day than for the cities of Palestine that rejected him; if Tyre had long since been made barren and was uninhabited, as the believers in the Bible aver, that prophecy was meaningless. In addition, the apostle Paul visited Tyre for a week and found disciples there (as described in Acts 21:3–5). A Christian cathedral was built in Tyre about 316 C.E. and several important councils of the Church were held there over the next two hundred years.

In the thirteenth century C.E., the Moslems conquered and razed both Tyre and the mainland settlement, which were at that time under the control of the Christian crusaders. Though the mainland settlement was not subsequently rebuilt, Tyre *was* eventually rebuilt and has continued to be inhabited on up to modern times. As H. Jacob Katzenstein states in his book, *The History of Tyre* (p. 9):

> The location of the city of Tyre is not in doubt, for it exists to this day on the same spot and is known as Sûr. . . . The character of the city has changed, however. In ancient times it was situated on an island, but from the time of Alexander the Great . . . the city has been linked to the mainland by a dike. . . .

Tyre's current place name of "Sûr" is derived from the ancient form of the name of the city. As we previously noted, the original Phoenician name of the city is *Sur*, and the Hebrew version of the word is *Tsor*. In most English-based maps it is also called "Tyre."

According to the Encyclopedia Britannica, Tyre had a population of twenty-three thousand as of 1982. The present-day island city covers most of the area covered by the original city on the island, though it is an island no longer, as silt has built up extensively along both sides of Alexander's causeway (or "dike," as Katzenstein called it) to form a peninsula. The silt also filled in the southern harbor, but the northern harbor is still usable.

Tyre suffered severe bomb damage in the late 1970s and early 1980s during the hostilities between Israel and the PLO. During the 2006 hostilities between Israel and Lebanon, Israel again bombed sites within the city. Nevertheless, the city lives on despite Ezekiel's "prophecy" that it would be made "like the top of a rock," and that it would be "built no more" after its "destruction" by Nebuchadrezzar.

In her book, *Tyre Through the Ages*, Nina Jidejian provides an aerial photo of Tyre that shows quite clearly the city on the peninsula. In his book, referred to above, H. Jacob Katzenstein provides another aerial photo of Tyre. You can use Google™ Maps™ to look at a satellite close-up photographic view of Tyre. Go to http://maps.google.com/maps, position the map view on Lebanon at the eastern end of the Mediterranean, and then use the slider to zoom in and enlarge the southernmost quarter of the country. (The map may not be available if you zoom in too close. If this is the case, zoom back out until it is visible again.) Tyre is located on the small finger of land that juts out from the mainland. When you locate Tyre, center it on the screen and click the button labeled "Satellite"; the map view will change to a photograph. Zoom in the image further to show the present-day city of Tyre.

In their conviction that the Bible is the word of God and that Ezekiel's prophecy must therefore have been fulfilled, many believers in the Bible simply ignore the historical importance of the island city of Tyre. They instead insist that Ezekiel's prophecy was fulfilled with the destruction of the mainland settlement, which they view as having been the city of Tyre rather than the island.[10] If that interpretation were correct, one would have to conclude that Ezekiel's prophecy against Tyre—a prophecy that Ezekiel had directed at the city because its inhabitants had spoken ill of Jerusalem—was not fulfilled until nearly two thousand years later when the mainland city was destroyed by the Moslems.

Nevertheless, as we saw, Ezekiel's prophecy was pointed clearly at the island city, and, according to the prophecy, Tyre was to be made bare as a rock, and the sea would cover it, and it would never be rebuilt. The present-day city of Tyre thus stands as clear evidence of the failure of Ezekiel's prophecy.

In any case, the area encompassing the Near East (the proper term for the area in question, rather than "Middle East," which the media wrongly uses) is littered with cities that have been abandoned by cause of war or otherwise. So even if Tyre had been totally destroyed by any of its conquerors and had never been rebuilt, it would hardly have been all that remarkable.

There is some related material concerning Ezekiel as a "prophet" of God that we should take a look at as well. In Ezekiel 29:1–13, Ezekiel gave a prophecy about Egypt:

> 1 *In the tenth year . . . the word of the LORD came unto me, saying,*
> 2 *Son of man, set thy face against Pharaoh king of Egypt, and prophesy against him, and against all Egypt.*
> 3 *. . . Speak, and say, Thus saith the Lord GOD; Behold, I am against thee, Pharaoh king of Egypt, . . .*

> 8 *Therefore, thus saith the LORD GOD; Behold, I will bring a sword upon thee, and cut off man and beast out of thee.*
> 10 *... and I will make the land of Egypt utterly waste and desolate, from the tower of Syene even unto the border of Ethiopia.*
> 11 *No foot of man shall pass through it, nor foot of beast shall pass through it, neither shall it be inhabited forty years.*
> 12 *... and I will scatter the Egyptians among the nations, and will disperse them through the countries.*
> 13 *Yet thus saith the LORD GOD; At the end of forty years will I gather the Egyptians from the people whither they were scattered.*

Ezekiel gave this prophecy during the tenth year of the exile, which was the year before he gave the prophecy against Tyre. He continued this prophecy by saying that Egypt would subsequently be the basest of nations and would never again be exalted above any other nation. At the time Ezekiel gave this prophecy, Nebuchadrezzar was preparing for his campaign against Egypt. However, as we noted, that campaign was sidetracked when Nebuchadrezzar initiated his siege against Tyre. It appears that the siege also effectively sidetracked Ezekiel's prophecy against Egypt.

Ezekiel returned to this prophecy in the twenty-seventh year of the exile. This was about the same time that Nebuchadrezzar was making his plans for his renewed campaign against Egypt after the unsatisfactory resolution of his thirteen-year siege against Tyre. Ezekiel continued the prophecy (which Ezekiel apparently thought of as another "sure bet") by prophesying that Nebuchadrezzar would be the instrument of Egypt's vanquishment. The continuation of the prophecy follows the passage in 29:17–18 that we previously looked at, in which Ezekiel off-handedly admitted that not only was Nebuchadrezzar *not* victorious against Tyre, but also that he had "no wages, nor his army," for his service against the city. Verses 19 and 20 continue as follows:

> 19 *Therefore thus saith the LORD GOD; Behold I will give the land of Egypt unto Nebuchadrezzar king of Babylon; and he shall take her multitude, and take her spoil, and take her prey; and it shall be the wages for his army.*
> 20 *I have given him the land of Egypt for his labour wherewith he served against it, because they wrought for me, saith the LORD GOD.*

Thus, according to this prophecy, though Nebuchadrezzar had failed to make a spoil of Tyre, he *would* make a spoil of Egypt as a reward for his unfulfilled labor against Tyre. Needless to say, this prophecy by Ezekiel was also an utter failure; Nebuchadrezzar did not conquer Egypt, and Egypt was never totally abandoned for forty years, nor were its inhabitants scattered among the nations—not at any time during the reign of the pharaohs or any time since. Moreover, contrary to Ezekiel's prophecy, Egypt remained a powerful nation for several hundred years after the time of Nebuchadrezzar.

Understandably, believers in the Bible tend to be silent about Ezekiel's failed prophecy concerning Egypt.

In the latter part of his book, Ezekiel made prophecies about "Gog, the land of Magog" that were of an apocalyptic nature. These prophecies were influential in the minds of believers during New Testament times when they were hearkened to in the apocalyptic Book of Revelation (Rev. 20:18–20). But one would be justified in wondering why any credence should be given to these apocalyptic prophecies made by Ezekiel when his prophecies concerning the contemporary fates of Tyre and Egypt turned out to be such dismal failures.

There are other prophecies in the Bible, of course, and believers in the Bible frequently say that many of those prophecies came to pass and thus have proved that the Bible is the word of God. But to properly analyze such assertions, one must understand that the peoples of those ancient times, and particularly the Hebrews, had a different approach to looking at history than we do today.

For one thing, as we have noted, ancient peoples often interpreted events in terms of their religious or supernaturalistic beliefs, and they frequently added religious or supernaturalistic embellishments to their accounts of events in order to make a point or to put across a conviction.

Moreover, in describing historical and even contemporaneous events, writers during biblical times sometimes put their words in the form of prophecies and wrote them as if they had been pronounced by some notable who had lived centuries before. By writing a commentary or an exhortation as if it were a prophecy by some ancient personage, and then subsequently causing it to be "found," a writer could have his own writings given some force of authority.

Such writings are called pseudepigraphic. During the latter part of the Old Testament era, several pseudepigraphic books were written—e.g., the Book of Enoch (also called 1 Enoch), the Book of the Secrets of Enoch, 4 Ezra, the Testaments of the Twelve Patriarchs, the Books of Adam and Eve, and many others.

There was another reason for writers of the time to take the pseudepigraphic route. During the third century B.C.E., the section of the Hebrew

Scriptures containing the books of the prophets was closed to new prophecies. As a result, anyone who felt moved by the prophetic impulse was, of necessity, forced to write his words as if they had been written by some personage who lived long before the scriptures were closed to contemporary "prophets." By taking such a route, a writer could hope that his writings would slip past the closure and would be accepted as scripture, or at least be received as having some authority.

There was yet another significant reason for writers of the time to take the pseudepigraphic route. Freedom of speech is a concept that was not generally accepted by certain, indeed most, regimes during those centuries, and if someone took it upon himself to write a harangue against those in power, it very likely would be at his own risk. However, by writing a pseudepigraphic "prophecy" about events of his own time, a writer could insulate himself from the potential consequences of his rhetoric, especially if the language of the prophecy was veiled so that only those "in the know" would understand its true meaning.

Certainly someone used that cautionary strategy in writing the Book of Daniel. It appears, in fact, that the Book of Daniel was written in response to a situation that existed during the second century B.C.E., long after the sixth-century B.C.E. time frame in which its purported author lived.

Several Bible scholars have pointed out many lines of evidence in support of that conclusion. For starters, the Book of Daniel is not found among the Books of the Prophets in the Hebrew canon, but rather is found among the Writings (which is the final section of the Hebrew canon). That placement would indicate that the Book of Daniel was written after the Books of the Prophets had been closed to further additions during the third century B.C.E. To go along with that, the literary style and language of the book also place it closer to the second century B.C.E. than to the sixth century B.C.E.

The external evidence also points to a late date for the composition of the Book of Daniel. Specifically, there is a glaring lack of mention of the Book of Daniel in any of the writings of the time until well after 165 B.C.E. About the year 190 B.C.E., for example, Jesus Ben Sirach enumerated and wrote praises of the former Israelite rulers and prophets, but made no mention of Daniel or of the Book of Daniel.[11] If the Book of Daniel had been in existence at the time he wrote his praises, and if Daniel had actually been a noteworthy prophet during the sixth century B.C.E., Ben Sirach surely would have included that book and that personage in those praises.

In addition, several studies have shown that the Book of Daniel has many anachronisms and errors in its descriptions of the period of the

Babylonian exile, which was when Daniel was supposedly writing, but that it has a regular correctness in its descriptions of the later Greek period, which was the period about which Daniel was supposedly prophesying. This would indicate that the actual author of the book was very familiar with the Greek period, but had hazy knowledge concerning the time of the Babylonian exile—which would be rather curious if the book had actually been written by someone who had lived during the exile.

One anachronism concerns the spelling of the name of the Babylonian ruler, Nebuchadrezzar. The spelling of the name with the internal "r," rather than with an "n," more correctly reflects the original spelling, which is *Nabu-kudurri-usar*. That is the spelling that the Babylonian ruler himself used in his inscriptions. On the other hand, the form of the name with an "n," *Nebuchadnezzar*, is based on the Greek spelling, *Nabuchodonosor*, which came into use following the Alexandrian conquest of the area more than two centuries after the Babylonian exile.[12] Thus, those books of the Bible that were written after the Alexandrian conquest usually use the Greek form of the name. It is notable that Ezekiel, who, in fact, *was* writing during the Babylonian exile, always used the more correct form in his writings. Significantly, *only* the Greek form of the name is found in the Book of Daniel, even though Daniel, like Ezekiel, was supposed to have written his book during the Babylonian exile, more than 200 years before the Greek conquest of the land. In Daniel 4:4, for example, even the Babylonian ruler uses the Greek form anachronistically in saying his own name: "*I Nebuchadnezzar was at rest in mine house....*"

Because of the textual evidence that can be found within the Book of Daniel, many Bible scholars feel that it was probably written between 167 and 165 B.C.E., and that its author wrote it to give encouragement to the Palestinian Jews who were being persecuted by Antiochus Epiphanes of Syria at that time. (Antiochus Epiphanes was one of the Seleucid rulers—i.e., the Hellenistic inheritors of a portion of the area conquered by Alexander the Great. Antiochus Epiphanes razed part of the city of Jerusalem, tried to wipe out the religion of the Hebrews, and caused the temple to be defiled—the "abomination that makes desolate" of Daniel 11:31.) The putative date of 165 B.C.E. for the completion of the writing of the Book of Daniel is based on the fact that its "prophecies" were historically accurate for events up to that date, but those for events subsequent to that date—that is, when the author of the book tried to make some actual predictions of the future from his standpoint in time—were off base.[13]

There is another piece of evidence that would appear to confirm the 167–165 B.C.E. time frame for the writing of the Book of Daniel.

Jeremiah had previously prophesied that it would be seventy years between the order to rebuild Jerusalem and the restoration of the Jews. However, in Daniel 9:24–25, the author of the book reinterpreted the prophecy to make the time span between the order to rebuild Jerusalem and the beginning of the religious restoration of the Jews encompass "seventy weeks." According to the conventional interpretation of the passage, the seventy weeks means seventy weeks of years, which would be 490 years. In the year 538 B.C.E., Cyrus had issued the decree allowing the Jews to rebuild the city and the temple, which, given the 490 years, would have brought the culmination of the prophecy to the year 48 B.C.E, a date of no particular significance in Jewish history. However, there is another way the calculation could be made. If the seventy "weeks" of Daniel 9:24–25 are taken as seventy "sevens" in adding up the years—that is, if the sevens refer to each number seven occurring within the count of the incremented consecutive years—a different result would occur. Thus, the seventh year from the decree would be first seven counted, the seventeenth year would be the second seven, etc., while the seventieth year, the seventy-first year, the seventy-second year, etc., would be the eighth, ninth, and tenth seven, etc., respectively. Those years having two sevens, such as seventy-seven, would count as two sevens each. By continuing this process on out for a total count of seventy sevens, we find that 371 actual years would accrue from the time of Cyprus's decree. That brings us to the year 167 B.C.E.

It is particularly significant that, with this interpretation, the time frame for the calculation began with the decree for the rebuilding of the temple and ended with what was supposed to be the beginning of the cleansing of the temple after its defiling by Antiochus Epiphanes. It would thus appear that the author of the book of Daniel happened to notice that Jeremiah's prophecy could be reinterpreted in such a way as to bring about its culmination in his own time. His use of that calculated date of 167 B.C.E. further supports the conclusion that the author of the book of Daniel wrote it to encourage his fellow Jews by manifesting to them that the restoration process was about to begin. Given the other substantial evidence for the approximate time frame of between 167 and 165 B.C.E for the composition of the Book of Daniel, the fact that the "seventy weeks" calculation can be so easily interpreted to produce a date of 167 B.C.E. can hardly be a coincidence.

Rather than being genuine prophetic books, then, writings such as the Book of Daniel were actually commentaries on events that were taking place at the time the true author was writing. In the case of the Book of Daniel, because its veiled language disguised its real meaning, it continued to be looked upon as prophetic scripture long after the actual events about which it "prophesied" had become a part of the past. With the coming of new generations, the "prophecies" in the Book of Daniel were simply reapplied to a future time.

In addition to the Book of Daniel, several other books in the Old Testament contain internal evidence that they were written, or at least revised, long after their supposed authors lived.

For example, the first five books in the Bible were, according to tradition, supposed to have been written by Moses. However, there are three passages in Genesis (11:28, 11:31, and 15:7) in which Abram (Abraham) is described as living in "Ur of the Chaldees" (or "Ur of the Chaldeans"). At the time when Abraham was supposed to have lived, the Chaldean Dynasty did not exist and would not exist until more than 1,200 years later. The Chaldean Dynasty, in fact, was established during the new Babylonian Empire by Nabopolassar, the father of Nebuchadrezzar, who was Ezekiel's contemporary. When Abraham was supposed to have lived, Mesopotamia, where Ur was located, was under the old Babylonian Empire, and, when Moses was supposed to have written Genesis, it was a part of the Assyrian Empire.

Therefore, the actual author of the book of Genesis had to have been writing after the Chaldean Dynasty was established in Babylonia, which was nearly a thousand years after the time of Moses. The author of the book of Genesis named Abram's city "Ur of the Chaldees" so his readers—who were more familiar with the Chaldean Dynasty than with the old Babylonian Empire—would know where Ur was located. But his having done so resulted in an anachronism. It would be as if someone of today "found" a manuscript that was supposed to have been written by a tenth-century traveler, but in which the traveler states he attended the coronation of Otto I by the archbishop of Mainz of the "Federal Republic of Germany." Such a statement would indicate that the actual author of the manuscript had written it no earlier than the latter half of the twentieth century—a thousand years after the time of Otto I.

References to Philistines in Genesis and Exodus provide yet another anachronism. The Philistines were Aegean Sea Peoples who invaded the lands of the eastern Mediterranean in the twelfth century B.C.E. and settled there. Though the term "Philistines" could not have been applied to the inhabitants of that area before that time, Genesis 21:34 states that "Abraham sojourned in the Philistines' land many days," which would have been some 800 years before there were any Philistines in the land. Several such references to the Philistines in Genesis and Exodus indicate that the author of those books lived long after the Philistines had arrived in the area and that he apparently assumed they had always been there.[14]

Thus, the references to the Philistines indicate that those books reached their final form a long time after the time of Moses.

Because of such anachronisms and numerous other pieces of evidence, many Bible scholars maintain that the Book of Genesis and several other books of the Bible dealing with the early history of the Hebrew people were written, or at least heavily edited, much later than commonly accepted. In fact, it appears that a large number of the supposedly pre-exilic books of the Old Testament actually reached their final form during or after the period of the Babylonian exile in the sixth century B.C.E. When the Persian ruler, Cyrus, gained control of the lands of Mesopotamia, he allowed the exiles in Babylon to return to their homeland, which motivated the returnees to renew and re-establish their former national identity and religious heritage. To help in bringing about the renewal process, certain of the religious leaders of the time apparently "found" certain scriptural documents and books—some of which were supposedly long lost—that would provide the needed inspiration. These books subsequently became part of the Hebrew Scriptures.

During the past two-hundred years, scholarly analysis of these books has revealed that several of them were formed by redactors, or editors, who combined, and added to, earlier traditions and documents that gave variant versions of certain stories, events, and religious viewpoints. Some of these variant traditions and documents reflected the differing views that were prevalent in the different geographical areas, specifically Judea and Israel, in which they originated, and their incorporation into the biblical text resulted in the numerous contradictions and inconsistencies that can found in the Bible. But the real significance of the post-exilic date for these redacted books of the Bible is that they present a late development in the religious thought of the Hebrew people and introduce a revised view of their history.

The scholars found that there were four main earlier source documents that were used in creating these books. Each of these source documents had distinctive peculiarities in style, wording, and subject matter that enabled the scholars to separate them out from the combined texts in which they appeared. To designate the individual source documents, they were assigned separate alphabetic letters according to certain characteristics they exhibited. One document was labeled J, for "Jehovah" (or Yahweh) because it consistently uses God's name when referring to God. Another document was labeled E because it uses the term "El" or "Elohim" in referring to God up until when Moses learns God's name in the account of the burning bush (in contrast, the J document uses God's name from the beginning). The largest document was labeled P because much of it deals with priestly matters. The fourth document was called D because it is found only in the book of Deuteronomy.

For example, the first four books of the Bible—the books of Genesis, Exodus, Leviticus, and Numbers—are clearly a result of the integration of various combinations of the J, E, and P source documents. These books, along with Deuteronomy, constitute the Torah. In his book *Who Wrote the Bible?*, Professor Richard Elliott Friedman shows in detail the process by which the Torah was put together, and, from various lines of evidence, he proposes that the redactor was the prophet Ezra.[15] Ezra had left Babylon for Jerusalem about eighty years after the first of the exiles in Babylon had made their return, and he carried with him a copy of the Torah that he apparently had redacted from the earlier source documents. Friedman suggests that Ezra integrated the various source documents to create the unified Torah in order to satisfy the different groups who considered the different source documents to be the written words of Moses. The ostensible purpose of the redacted Torah was to help bring together the different groups and help to create a new unified House of Israel.

In Chapter 7, remember, we saw that two such source documents provided the basis for the two different accounts of creation in Genesis. Genesis 1:1 through 2:3, in fact, had its source in the P document, whereas Genesis 2:4–25 had its source in the J document (the story of the subsequent events in the Garden of Eden also had its source in the J document).

To further demonstrate the use of the source documents in the construction of Genesis, let's take a look at the biblical story of the Flood. An analysis of that story shows that it actually consists of two separate and differing versions of the Flood story that were derived from the J document and the P document. It appears that the redactor cut up and interspersed together the texts of those two source documents to create a single story that integrated the details of the two original stories of the Flood. What is particularly significant about this is that it is possible to reconstruct the P document and the J document by separately extracting the verses that were derived from them and yet still have them each tell the two somewhat different but coherent versions of the Flood story.

The table below provides in two columns the source documents as I have extracted them from the Flood story according to the source breakdown that Friedman provided in his book. The column labeled P provides the verses that have their origins in the P document, and the column labeled J provides the verses that have their origins in the J document. If you read straight down the P column, you will find that it tells a coherent story. Likewise, if you read straight down the J column you will find that it also tells a coherent but somewhat different version of the story. Note the differences in the details between the two versions. The P version consistently uses "Elohim" to refer to God (references to both Elohim and Yahweh can be found in other parts of the P document, but here only

"Elohim" is used), whereas the J version consistently uses Yahweh, the name of God. P has only one pair of each of the different kinds of animals that go on the ark; J has seven pairs of the clean animals and one pair of the unclean. P has the waters of the Flood coming from the opened windows of heaven and the fountains of the deep; J has the waters of the Flood coming from rain only. P has a year as the total time from the beginning of the Flood to when the earth is dry enough to set foot on; J has 54 days as the total for that time. In P, Noah sends out a raven; in J, Noah sends out doves.

In order to help track the comparisons of the story lines in the two versions, I have aligned some of the parallel passages in the columns. Note that the J column does not include a description of the building of the ark as the P column does. There is probably a good reason for this omission. The other differences between the two versions make it quite likely that the descriptions of the construction of the ark were also different in the two source documents—perhaps so much so that if they were both kept in the combined story it might have seemed that Noah had built two quite different arks, whereas only one ark was supposed to have been built. It appears that the redactor tolerated the other differences in the two versions, but felt that he simply could not accommodate the differences in the descriptions of the construction of the ark. To resolve the problem, he therefore deleted the description of the construction of the ark that appeared in the J source and used only the description that appeared in the P document. But even with that, a trace of the description of the ark from J apparently appears in verse 8:13b, which mentions that Noah removed the "covering" of the ark when the Flood was over and the dry land appeared. The description of the construction of the ark is fairly detailed in P, but there is no mention of a covering in that description. As the covering is mentioned in the J extract of 8:13b, it is very likely that it was previously mentioned in the deleted text from J that described its version of the construction of the ark. If so, the "covering" was quite likely one of the significant differences between the descriptions of the ark's construction in the two source documents. It appears that, when the redactor deleted the description of the ark's construction from the J source, he neglected to follow through and delete the mention of the covering in verse 8:13b.

Incidentally, that brings up another point. The breakdown of the Bible into numbered verses did not occur in the original text of the Bible when it was written, but was added much later to assist in referencing passages. The assignment of the verse numbering was somewhat arbitrary and results in parts of the two source documents appearing together in a few of the verses.

It should also be pointed out that the redactor was free to make changes in the wording of the source documents in order to smooth out the transitions from one to the other; still, he kept such changes to a minimum.

P	J
6:9 *These are the generations of Noah: Noah was a just man and perfect in his generations, and Noah walked with God [Elohim].*	
6:10 *And Noah begat three sons, Shem, Ham, and Japheth.*	
6:11 *The earth also was corrupt before God [Elohim], and the earth was filled with violence.*	6:5 *And GOD [Yahweh] saw that the wickedness of man was great in the earth, and that every imagination of the thoughts of his heart was only evil continually.*
6:12 *And God [Elohim] looked upon the earth, and, behold, it was corrupt; for all flesh had corrupted his way upon the earth.*	6:6 *And it repented the Lord [Yahweh] that he had made man on the earth, and it grieved him at his heart.*
6:13 *And God [Elohim] said unto Noah, The end of all flesh is come before me; for the earth is filled with violence through them; and, behold, I will destroy them with the earth.*	6:7 *And the Lord [Yahweh] said, I will destroy man whom I have created from the face of the earth; both man, and beast, and the creeping thing, and the fowls of the air; for it repenteth me that I have made them.*
6:14 *Make thee an ark of gopher wood; rooms shalt thou make in the ark, and shalt pitch it within and without with pitch.*	
6:15 *And this is the fashion which thou shalt make it of: The length of the ark shall be three hundred cubits, the breadth of it fifty cubits, and the height of it thirty cubits.*	
6:16 *A window shalt thou make to the ark, and in a cubit shalt thou finish it above; and the door of the*	

P	J
ark shalt thou set in the side thereof; with lower, second, and third stories shalt thou make it. 6:17 *And, behold, I, even I, do bring a flood of waters upon the earth, to destroy all flesh, wherein is the breath of life, from under heaven; and every thing that is in the earth shall die.* 6:18 *But with thee will I establish my covenant; and thou shalt come into the ark, thou, and thy sons, and thy wife, and thy sons' wives with thee.*	
6:19 *And of every living thing of all flesh, two of every sort shalt thou bring into the ark, to keep them alive with thee; they shall be male and female.* 6:20 *Of fowls after their kind, and of cattle after their kind, of every creeping thing of the earth after his kind, two of every sort shall come unto thee, to keep them alive.* 6:21 *And take thou unto thee of all food that is eaten, and thou shalt gather it to thee; and it shall be for food for thee, and for them.*	6:8 *But Noah found grace in the eyes of the Lord* [Yahweh]. 7:1 *And the Lord* [Yahweh] *said unto Noah, Come thou and all thy house into the ark; for thee have I seen righteous before me in this generation.* 7:2 *Of every clean beast thou shalt take to thee by sevens, the male and his female: and of beasts that are not clean by two, the male and his female.* 7:3 *Of fowls also of the air by sevens, the male and the female; to keep seed alive upon the face of all the earth.*
6:22 *Thus did Noah; according to*	7:4 *For yet seven days, and I will cause it to rain upon the earth forty days and forty nights; and every living substance that I have made will I destroy from off the face of the earth.* 7:5 *And Noah did according unto*

P	J
all that God [Elohim] commanded him, so did he.	*all that the Lord [Yahweh] commanded him.*
7:6 And Noah was six hundred years old when the flood of waters was upon the earth.	
7:8 Of clean beasts, and of beasts that are not clean, and of fowls, and of every thing that creepeth upon the earth,	
7:9 There went in two and two unto Noah into the ark, the male and the female, as God [Elohim] had commanded Noah.	
7:11 In the six hundredth year of Noah's life, in the second month, the seventeenth day of the month, the same day were all the fountains of the great deep broken up, and the windows of heaven were opened.	
7:13 In the selfsame day entered Noah, and Shem, and Ham, and Japheth, the sons of Noah, and Noah's wife, and the three wives of his sons with them, into the ark;	*7:7 And Noah went in, and his sons, and his wife, and his sons' wives with him, into the ark, because of the waters of the flood.*
7:14 They, and every beast after his kind, and all the cattle after their kind, and every creeping thing that creepeth upon the earth after his kind, and every fowl after his kind, every bird of every sort.	
7:15 And they went in unto Noah into the ark, two and two of all flesh, wherein is the breath of life.	
7:16a And they that went in, went in male and female of all flesh, as God [Elohim] had commanded him.	

P	J
	7:10 *And it came to pass after seven days, that the waters of the flood were upon the earth.*
7:21 *And all flesh died that moved upon the earth, both of fowl, and of cattle, and of beast, and of every creeping thing that creepeth upon the earth, and every man:*	
7:24 *And the waters prevailed upon the earth an hundred and fifty days.*	7:12 *And the rain was upon the earth forty days and forty nights.*
	7:16b *... and the Lord* [Yahweh] *shut him in.*
	7:17 *And the flood was forty days upon the earth; and the waters increased, and bare up the ark, and it was lift up above the earth.*
	7:18 *And the waters prevailed, and were increased greatly upon the earth; and the ark went upon the face of the waters.*
	7:19 *And the waters prevailed exceedingly upon the earth; and all the high hills, that were under the whole heaven, were covered.*
	7:20 *Fifteen cubits upward did the waters prevail; and the mountains were covered.*
	7:22 *All in whose nostrils was the breath of life, of all that was in the dry land, died.*
	7:23 *And every living substance was destroyed which was upon the face of the ground, both man, and cattle, and the creeping things, and the fowl of the heaven; and they were destroyed from the earth: and*

P	J
	Noah only remained alive, and they that were with him in the ark.
8:1 *And God [Elohim] remembered Noah, and every living thing, and all the cattle that was with him in the ark: and God [Elohim] made a wind to pass over the earth, and the waters assuaged;*	8:2b *... and the rain from heaven was restrained;*
8:2a The fountains also of the deep and the windows of heaven were stopped, ...	
8:3b *... and after the end of the hundred and fifty days the waters were abated.*	8:3a *And the waters returned from off the earth continually: ...*
8:4 *And the ark rested in the seventh month, on the seventeenth day of the month, upon the mountains of Ararat.*	
8:5 *And the waters decreased continually until the tenth month: in the tenth month, on the first day of the month, were the tops of the mountains seen.*	8:6 *And it came to pass at the end of forty days, that Noah opened the window of the ark which he had made:*
8:7 *And he sent forth a raven, which went forth to and fro, until the waters were dried up from off the earth.*	8:8 *Also he sent forth a dove from him, to see if the waters were abated from off the face of the ground;*
	8:9 *But the dove found no rest for the sole of her foot, and she returned unto him into the ark, for the waters were on the face of the whole earth: then he put forth his hand, and took her, and pulled her in unto him into the ark.*
	8:10 *And he stayed yet other seven days; and again he sent forth the dove out of the ark;*

P	J
	8:11 *And the dove came in to him in the evening; and, lo, in her mouth was an olive leaf pluckt off: so Noah knew that the waters were abated from off the earth.*
8:13a *And it came to pass in the six hundredth and first year, in the first month, the first day of the month, the waters were dried up from off the earth: ...*	8:12 *And he stayed yet other seven days; and sent forth the dove; which returned not again unto him any more.*
8:14 *And in the second month, on the seven and twentieth day of the month, was the earth dried.*	8:13b *... and Noah removed the covering of the ark, And looked, and, behold, the face of the ground was dry.*
8:15 *And God* [Elohim] *spake unto Noah, saying,*	
8:16 *Go forth of the ark, thou, and thy wife, and thy sons, and thy sons' wives with thee.*	
8:17 *Bring forth with thee every living thing that is with thee, of all flesh, both of fowl, and of cattle, and of every creeping thing that creepeth upon the earth; that they may breed abundantly in the earth, and be fruitful, and multiply upon the earth.*	
8:18 *And Noah went forth, and his sons, and his wife, and his sons' wives with him:*	
8:19 *Every beast, every creeping thing, and every fowl, and whatsoever creepeth upon the earth, after their kinds, went forth out of the ark.*	
	8:20 *And Noah builded an altar unto the LORD* [Yahweh]; *and took of every clean beast, and of*

P	J
	every clean fowl, and offered burnt offerings on the altar.
	8:21 *And the LORD [Yahweh] smelled a sweet savour; and the LORD [Yahweh] said in his heart, I will not again curse the ground any more for man's sake; for the imagination of man's heart is evil from his youth; neither will I again smite any more every thing living, as I have done.*
	8:22 *While the earth remaineth, seedtime and harvest, and cold and heat, and summer and winter, and day and night shall not cease.*
9:1 *And God [Elohim] blessed Noah and his sons, and said unto them, Be fruitful, and multiply, and replenish the earth.*	

Note that J ends with Noah taking one of every kind of clean beast and fowl to sacrifice to God, which he is able to do because there are seven pairs of every kind of clean animal and there would still be six pairs left to reproduce their kinds. In P, on the other hand, only two of every kind were brought on the ark, and there is no mention of Noah's having sacrificed one of every kind of clean animal. Obviously, in the context of P, if Noah had sacrificed one of every kind of clean animal there would be no pairs of clean animals left to reproduce. The P document does not mention a sacrifice by Noah because, as a priest, the author of the P document had reserved sacrifices in his narrative until the time of the completion of the temple in Jerusalem when they could be properly done.

Again, we noted in Chapter 7 that the two creation stories in Genesis 1 and 2 are inconsistent with each other and that the inconsistency is best explained as being the result of two separate sources having been incorporated into the text. The fact that material from the same two source documents can be found in the Flood story in Genesis is evidence that the inconsistencies in Genesis 1 and 2 are, in fact, the result of the amalgamation of two source documents. And again, it is not just in the creation and Flood stories in Genesis that the various source documents can be found interspersed in the text. All of the first five books of the Bible were the constructed by an amalgamation of those source documents even though this resulted in numerous contradictions and inconsistencies through all five books. This should lead one to the conclusion that the formation of these books was the result of a purely human endeavor rather than being the "inspired" word of God.

Appendix B
Problems in the New Testament

Like the Old Testament, the New Testament also contains internal evidence of embellishment. This is particularly true of the gospels, for they contain many pious enhancements that the authors inserted in their respective texts to bolster their assertions that Jesus was the Messiah. It was not very difficult for the authors of the gospels to take such liberties. After all, as we noted in Chapter 1, the gospels were not written until many decades after the ministry of Jesus, and, in embellishing their accounts, the gospel authors had both the obscuring mists of time and a growing accretion of tales about Jesus to assist them. The results, however, were often contradictory and thus provide the evidence for the embellishments.

For example, the separate accounts of the birth and early life of Jesus that the authors of the gospels of Matthew and Luke provided are mutually incompatible—so much so that they simply cannot be reconciled with one another in any way that does not do violence to the threads of the respective narratives. By any rule of logic, this incompatibility indicates that one, or the other, or both of these accounts had been enhanced with embellishments that had no basis in reality.

We might first note that the authors of these two gospels provided genealogies of Jesus to show that he was of the lineage of David through Joseph, the husband of Mary. In Luke's gospel, the genealogy, given in Luke 3:23–38, additionally continues back to Adam, while in Matthew's gospel the genealogy, given in Matthew 1:1–17, continues back to Abraham. In comparing these two genealogies, we find that the listed names between Abraham and David are in close agreement, which is not surprising. After all, David's genealogy was well established in the Old Testament, so the authors of the gospels were able to use that material for the corresponding parts of their genealogies. When we look at the listed names between David and Joseph, however, the two genealogies cannot be reconciled with each other. For example, according to Matthew 1:16, the name of Joseph's father was Jacob:

And Jacob begat Joseph the husband of Mary, of whom was born Jesus, who is called Christ.

But, according to Luke 3:23, the name of Joseph's father was Heli:

And Jesus himself began to be about thirty years of age, being (as was supposed) the son of Joseph, which was the son of Heli,

Moreover, Matthew lists only twenty-five names between David and Joseph, while Luke lists forty names; one would think that the generations listed after David in the two genealogies would be much closer in number. In addition, with two exceptions, none of the names listed between David and Joseph are common to both genealogies.

To account for these discrepancies, some biblical apologists say that the genealogy given in Luke 3:23–38 is actually Mary's and not Joseph's. However, everything relating to the genealogy refutes that stance.

First, the wording in the genealogy given by Luke does not mention Mary at all and specifically states that Jesus was the son, "as was supposed," of Joseph—the parenthetical "as was supposed" having been inserted in the genealogy to allow for the "virgin birth" of Jesus. It is through Joseph that the lineage continues, which clearly shows that the genealogy was meant to be Joseph's and not Mary's. In providing the genealogy of Jesus through Joseph, even though Joseph was supposedly not the biological father of Jesus, Luke was presenting what would have been considered the "legal" lineage of Jesus, since Joseph was the husband of Mary.

Some apologists have said that Luke skipped over Mary's name because Hebrew genealogies did not traditionally list women. However, the author of Matthew's gospel had tried to make his gospel follow Hebrew tradition overall, and the genealogy of Jesus in his gospel *does* name women. For example, Matthew 1:3 states: "*And Judas begat Phares and Zara of <u>Thamar</u>....*" Matthew 1:5 states: "*And Salmon begat Booz of <u>Rachab</u>; and Booz begat Obed of <u>Ruth</u>....*" And in Matthew 1:6 we find: "*... and David the king begat Solomon of her that had been the <u>wife of Urias</u>.*" And, of course, in Matthew 1:16 we find: "*And Jacob begat Joseph the husband of <u>Mary</u>, of whom was born Jesus, who is called Christ.*" Moreover, the genealogies given in 1 Chronicles chapters 1 through 3 mention several women. So if the genealogy in Luke's gospel was supposed to be that of Mary, it would seem strange that Mary was not even mentioned in it.

Second, Luke 1:26–27 specifically emphasizes that Joseph was of the house of David and makes no mention at all of Mary's ancestry:

> 26 *And in the sixth month the angel Gabriel was sent from God unto a city of Galilee, named Nazareth,*
> 27 *To a virgin espoused to a man whose name was Joseph, of the house of David; and the virgin's name* was *Mary.*

Luke 2:4 goes even further and emphasizes that Joseph was not only of the house of David, but also of the lineage of David:

> *And Joseph also went up from Galilee, out of the city of Nazareth, into Judaea, unto the city of David, which is called Bethlehem; (because he was of the house and lineage of David:)*

One would therefore be justified in understanding that the genealogy that is subsequently listed in Luke 3:23–38 follows up and provides the details for those initial statements concerning Joseph's ancestry.

Third, Luke 1:36 relates that Elizabeth was Mary's cousin:

> *And, behold, thy cousin Elisabeth, she hath also conceived a son in her old age: and this is the sixth month with her, who was called barren.*

The Greek word that is translated as "cousin" in that verse is *suggenes* in the Greek, meaning a relative by blood. Luke 1:5 states the following concerning Elizabeth:

> *There was in the days of Herod, the king of Judaea, a certain priest named Zacharias, of the course of Abia: and his wife* was *of the daughters of Aaron, and her name* was *Elisabeth.*

This indicates that Elizabeth, like her husband, Zechariah, belonged to the priestly tribe of Levi (since she was "of the daughters of Aaron"), which would mean that, as a blood relative of Elizabeth's, Mary would also have been of that lineage. But that lineage is not indicated in the genealogy that Luke gave for Jesus; the lineage that is given supposedly shows that Jesus was of the tribe of Judah and was a descendant of David. Therefore, again, the genealogy given in Luke's gospel was not intended to be that of Mary.

Fourth, Luke's gospel views women with particular importance and sympathy, e.g., Luke 1:24, 7:37–48, 8:3, 10:38–42, and 23:54–56. Going along with those examples, and even more significantly, in Luke's gospel it is Mary who receives from an angel the annunciation of the forthcoming birth of Jesus (1:26–38), rather than Joseph as in Matthew's gospel. Luke also has Mary visit her cousin, Elizabeth (who is pregnant with the baby who would become John the Baptist) who blesses Mary as the one who would be the mother of the Lord (1:39–44). Luke then presents Mary's "Magnificat" soliloquy in which she rejoices in being blessed by the Lord (1:46–55). One would therefore reasonably conclude that, since Luke viewed Mary as having such importance in his gospel, he certainly would have listed her name in the genealogy he supplied if that genealogy were, in fact, hers. That he did not do so indicates that, again, the genealogy is not hers, but, rather, is Joseph's—just as the wording indicates.

It is therefore clear that the genealogies that are given in Matthew and Luke, despite the discrepancies between them, have the intent of establishing the "legal" Davidian lineage of Jesus through Joseph. The most obvious reason for the discrepancies between the two genealogies is that the authors of the two gospels independently fabricated the names that they respectively listed for the lineages between David and Joseph (though it is also possible that they used genealogies that others had contrived). It appears that, in improvising these lineages, the two gospel writers simply picked up some names from the Old Testament and made up other names to fill in the gaps. For example, Matthew appears to have used the genealogy in 1 Chronicles 3 as his source for the names from David to Zorobabel in his genealogy (though he skipped several names, apparently to make his genealogy adhere to his formula of having fourteen generations between significant points in the lineage). To complete his genealogy, Matthew, or perhaps someone who was his source for the genealogy, simply made up a list of names for the section between Zorobabel and Joseph.

In contrast, the list of names between David and Joseph in Luke's genealogy appears to be almost entirely made up. Luke did pick up some random names from the Old Testament, most notably those of Salathiel and Zorobabel (which are the KJV spellings; "Shealtiel" and "Zerubbabel" are used in most other versions of the Bible), a father-and-son pair (the son, Zerubbabel, was a governor of Judea) that is also found in Matthew's genealogy. There are, in fact, nine references to this pair of names in the Old Testament (Ezra 3:2, 3:8, and 5:2; Nehemiah 12:1; and Haggai 1:1, 1:12, 1:14, 2:2, and 2:23), so, along with the importance of Zerubbabel's position as governor, it is not very surprising that the two gospel writers coincidentally picked them for inclusion in their genealogies. But even with those names in common in the two genealogies there is a problem. According to Matthew, the father of Shealtiel was Jeconiah, whereas according to Luke the father was Neri. Since, according to 1 Chronicles 3:17, Shealtiel's father *was* named Jeconiah, it appears that Matthew did his homework. Luke, on the other hand, was not so diligent, and he simply made up a name for Salathiel's father. (It is also interesting to note that

Luke's genealogy lists seventy-seven names from Adam to Jesus, which has a correspondence in the Book of Enoch. See the Supplement to Appendix C.)

There are several more discrepancies between the gospels of Luke and Matthew relating to the birth and early life of Jesus. According to Luke's gospel, Joseph and Mary lived in Nazareth of Galilee prior to the birth of Jesus. The couple went to Bethlehem while Mary was pregnant, as Luke's story goes, because of a taxation decree by Caesar Augustus when Cyrenius (the Latin version of the name is Quirinius) was governor of Syria, and, as a consequence, Jesus was born in Bethlehem. As described in Luke 2:22, when the days that were required for Mary's purification following the birth were over (which would have been forty days, as specified by Leviticus 12:2–8), they brought the infant Jesus to Jerusalem to be presented to the Lord at the Temple. Then, after the religious requirements were accomplished, the family returned to their home in Nazareth. Luke also stated that the parents of Jesus subsequently went to Jerusalem every year for the feast of the Passover.

Matthew, however, presents a completely different story. According to Matthew's gospel, Joseph and Mary originally lived in Bethlehem of Judea, which was why Jesus was born there. At some unspecified time after the birth of Jesus, as Matthew relates the story in the second chapter of his gospel, the family fled to Egypt to escape the portended peril that King Herod had in store for the young Jesus. The family remained in Egypt until after the death of Herod, at which time it appears that Jesus would have been upwards of two years of age and perhaps older. Joseph and his family then proceeded to return to their home in Bethlehem, but, upon learning that Herod's son Archelaus ruled in Judea in place of Herod, they went to Galilee and settled in the village of Nazareth.

These two stories are completely incompatible. The events that Luke described leave no room for the events described by Matthew. In particular, in contrast to Luke's story, Matthew indicated that Joseph and his family originally lived in Bethlehem and did not settle in Nazareth until some time after the birth of Jesus.

Many believers in the Bible will doubtlessly refuse to accept that the two accounts are so incompatible. Because of that, we should analyze them in more detail.

First, the original Greek wording in the second chapter of Matthew implies that Jesus was not a new-born baby when the events described in the chapter supposedly took place. Because of a misleading translation of Matthew 2:1 in the KJV and in some other versions of the Bible, there is an erroneous perception that Jesus *was* a new-born baby. These versions of the Bible begin Matthew 2:1 with the word "when," as, for example, the KJV, *"Now when Jesus was born in Bethlehem of Judea, ... wise men from the east came to Jerusalem."* However, the word "when" does not appear in the original Greek of the verse. Several other versions of the Bible (the NEB, NIV, ESV, NASV, NRSV, NJB, and NKJV) use the word "after" instead of "when," though that word is not in the original Greek either. The NKJV, for example, states: *"Now after Jesus was born in Bethlehem of Judea ..., wise men from the East came to Jerusalem."* The original Greek word translated as "now," which is found in most of the versions of the Bible, is *de*, which, according to Friberg, in fact means "*now* (with no temporal sense)." Of the different versions of the Bible referred to in this book, the Darby version perhaps most correctly reflects the original sense of the verse:

> *Now Jesus having been born in Bethlehem of Judaea, in the days of Herod the king, behold magi from the east arrived at Jerusalem. ...*

The events that are described in the second chapter of Matthew are precipitated by the arrival of the "wise men" who came from the east. The original Greek word translated as "wise men" is *magos*, which some versions of the Bible—for example, the Darby version, above—transliterate as "magi." The word *magos* was frequently applied to astrologers, and that is the word used in Matthew 2:1–2 of the NEB: *"... After his birth astrologers from the east arrived in Jerusalem asking, 'Where is the child who is born to be king of the Jews? We observed the rising of his star. ...'"* Because the latter part of the passage makes reference to a star that the visitors took as a sign, translating *magos* as "astrologers" would appear to be appropriate. Additional evidence that the *magos* were to be understood as being astrologers is the statement that they "came from the east." The city of Babylon, in fact, was nearly due east from Judea and had been the center of astrological studies ever since the time of ancient Babylonia.[1]

Again, Matthew does not indicate the specific period of time between the birth of Jesus and the arrival of the astrologers in Jerusalem. Because of the distance involved, Matthew would have needed to allow for a certain amount of time for the astrologers to prepare for their trip to Jerusalem after they had seen the star, as well as for the trip itself. The implication would be that some time would have passed before they were supposed to have arrived in Jerusalem. Thus those versions of the Bible that begin Matthew 2:1 with the word "after" to indicate the relationship of the time between the birth of Jesus and the arrival of the astrologers would be on more firm ground than those that begin it with "when." This also implies that Jesus would not have been a newborn infant when the astrologers arrived in Bethlehem.

That conclusion is supported by the original Greek of Matthew. Depending on the version of the Bible, the young Jesus is consistently referred to in the second chapter of Matthew, not as a "baby" or an "infant," but as a "young child" (as in the KJV, NKJV, and ASV) or simply as a "child" (as in the RSV, NEB, ESV, NAB, NASV, NIV, and NJB). Here, for example, is the wording of Matthew 2:11 in the KJV:

> *And when they were come into the house, they saw the young child with Mary his mother, and worshipped him....*

In the original Greek of Matthew, the word that is translated as "young child" or "child" is *paidion*, which, though sometimes used to refer to an infant, usually implies a partially grown child.

In contrast, in Luke 2:12 and 2:16 the Greek word *brephos*, meaning "babe" or "infant" is used when an angel tells some shepherds about the new-born Jesus in Luke's quite different version of the early life of Jesus:

> 12 *And this shall be a sign unto you; Ye shall find the babe wrapped in swaddling clothes, lying in a manger.*
> 16 *So they went in haste and found Mary and Joseph, and the infant lying in the manger.*

Still, though the exact interval of time between the birth of Jesus and the arrival of the astrologers is not specified, a time frame is provided in Matthew 2:7 and 2:16. That time frame is another indication that Jesus was not a newborn infant when the events described in the second chapter of Matthew take place (here the KJV):

> 7 *Then Herod, when he had privily called the wise men, enquired of them diligently what time the star appeared....*
> 16 *Then Herod, when he saw that he was mocked of the wise men, was exceeding wroth, and sent forth, and slew all the children that were in Bethlehem, and in all the coasts thereof, from two years old and under, according to the time which he had diligently enquired of the wise men.*

Those verses indicate that Jesus could have been upwards of two years old, and not a newborn baby, when the events of the chapter were supposed to have taken place.

Again, contrary to Luke's account of the early life of Jesus, the wording of the second chapter of Matthew implies that Joseph and his family lived in Bethlehem before settling in Nazareth. Matthew 2:11 states that the wise men (or astrologers) entered the "house" where they saw the child and his mother. The word in the original Greek that is translated as "house" is *oikia*, which means "residence" or "abode" of a family. This is in contrast to Luke's story about the manger in which Mary laid the baby Jesus because there was no room at the inn (*kataluma*, "lodging place" or "inn," in the original Greek).

By itself, the wording concerning the house might be problematic, but verses 2:22–23 make it clear that, according to Matthew's gospel, Joseph and his family did *not* live in Galilee until *after* they returned to their homeland following their flight to Egypt. According to that passage, when Joseph learned that Archelaus ruled in Judea in place of Herod, he "*... turned aside into the parts of Galilee: And he came and dwelt in a city called Nazareth: that it might be fulfilled which was spoken by the prophets, He shall be called a Nazarene.*" That wording specifically indicates that Joseph and his family were making a new home for themselves in Nazareth at that time and had not previously lived there. That in turn implies that the house in Bethlehem would have been theirs.

If we accept Matthew's account for the sake of argument, Jesus may well have been upwards of two years of age, and perhaps older, by the time the family returned to Judea from their stay in Egypt. Again, this is indicated by Matthew 2:16, which states that King Herod determined that all the male children of Bethlehem "two years old and under, according to the time which he had diligently enquired of the wise men," should be killed. Still accepting Matthew's account for the sake of argument, from a few months to a year or two may have passed if we allow for a reasonable amount of time between Herod's order to kill the male children in Bethlehem and Herod's death, and also some additional time for the return of the family from Egypt. Thus, if Jesus was under two years of age at the time of Herod's order, he may have been well over two years of age by the time the family returned from Egypt and settled in Nazareth.

Now that we have investigated the background of the Mathew account, we can look at the relevant verses showing the timeline (here the NKJV):

> 1 *After Jesus was born in Bethlehem in Judea, during the time of King Herod, Magi from the east came to Jerusalem*
> 2 *saying, "Where is He who has been born King of the Jews? For we have seen His star in the East and have come to worship Him."*
> 7 *Then Herod, when he had secretly called the wise men, determined from them what time the star appeared.*
> 8 *And he sent them* [the magi] *to Bethlehem...*
> 11 *And when they had come into the house, they saw the young Child with Mary His mother....*
> 13 *Now when they had departed, behold, an angel of the Lord appeared to Joseph in a dream, saying, "Arise, take the young Child and His mother, flee to Egypt, and stay there until I bring you word; for Herod will seek the young Child to destroy Him."*

> 14 *When he arose, he took the young Child and His mother by night and departed for Egypt,*
> 16 *Then Herod, when he saw that he was deceived by the wise men, . . . sent forth and put to death all the male children who were in Bethlehem and in all its districts, from two years old and under, according to the time which he had determined from the wise men.*
> 19 *But when Herod was dead, behold, an angel of the Lord appeared in a dream to Joseph in Egypt,*
> 20 *saying, "Arise, take the young Child and His mother, and go to the land of Israel, for those who sought the young Child's life are dead."*
> 21 *Then he arose, took the young Child and His mother, and came into the land of Israel.*
> 22 *But when he learnt that Archelaus had succeeded his father Herod as ruler of Judaea he was afraid to go there, and . . . he withdrew to the region of Galilee.*
> 23 *And he came and dwelt in a city called Nazareth, that it might be fulfilled which was spoken by the prophets, "He shall be called a Nazarene."*

All this makes it clear that the events described in the second chapter of Matthew, and the resultant timeline, are completely different from those described in the second chapter of Luke. To recap the events in Luke's gospel, Joseph and Mary lived in Nazareth of Galilee and went to Bethlehem because of a taxation decree that required everyone to return to the city of their ancestors, and, as a result, Mary gave birth to Jesus in that town. Then, after the requirements of the Law of Moses were fulfilled, the family returned to Galilee.

Here are the relevant verses from the second chapter of Luke showing the timeline of the relatively uneventful (when compared with Matthew's account) trip to Bethlehem and the return to Galilee (here, the KJV):

> 4 *And Joseph also went up from Galilee, out of the city of Nazareth, into Judaea, unto the city of David, which is called Bethlehem. . .*
> 5 *To be taxed with Mary his espoused wife, being great with child.*
> 7 *And she brought forth her firstborn son;*
> 21 *And when eight days were accomplished for the circumcising of the child, his name was called JESUS. . . .*
> 22 *And when the days of her purification according to the law of Moses were accomplished, they brought him to Jerusalem, to present* him *to the Lord;*

> 39 *And when they had performed all things according to the law of the Lord, they returned into Galilee, to their own city Nazareth.*
> 40 *And the child grew, and waxed strong in spirit, filled with wisdom: and the grace of God was upon him.*
> 41 *Now his parents went to Jerusalem every year at the feast of the passover.*

Here, Joseph and his wife, along with their new-born son, return to their home in Nazareth after they had performed the ritual requirements "according to the law of Moses." Again, that would have been forty days, as that was the time required for a woman's purification after giving birth. In contrast to Matthew's account, there is no mention of the threat to Jesus from King Herod (they even take the infant Jesus to Jerusalem to present him to the Lord in the temple, with no concern about Herod); there is no escape of the family to, or of their return from, Egypt after a period of about two years; there is no wariness about Herod's son Archelaus; and there is no abandoning of their residence in Bethlehem to make a new home for themselves in Galilee. Moreover, Joseph and his family return to Jerusalem the following year (that is, less than two years after the birth of Jesus), again without any concern about Herod's threat to the life of Jesus, and each year thereafter without any wariness about Archelaus.

As we shall see, there is a very good reason for this apparent lack of concern about King Herod and Archelaus in Luke's gospel.

In addition to the incompatibilities between the narratives described in Matthew and Luke, there are also historical incompatibilities. According to Matthew's gospel, Jesus was born during the reign of King Herod, who—according to the historical record—died in 4 B.C.E.[2] The author of the Gospel of Luke likewise began his story "in the days of Herod, the king of Judea" (Luke 1:5), and related a narrative about the birth of John the Baptist and the annunciation to Mary. However, Luke went on to relate that Jesus was born following a taxation enrollment, or census, that was ordered by Caesar Augustus when Quirinius (Cyrenius in the KJV) was governor of Syria. In the KJV, Luke 2:1–2 states:

> 1 *And it came to pass in those days, that there went out a decree from Caesar Augustus, that all the world should be taxed.*
> 2 *(And this taxing was first made when Cyrenius was governor of Syria.)*

The wording in the KJV is somewhat inaccurate in that the original Greek indicates that the decree was for an *enrollment* for taxation. The NAB more correctly reflects the wording:

> 1 *In those days a decree went out from Caesar Augustus that the whole world should be enrolled.*
> 2 *This was the first enrollment, when Quirinius was governor of Syria.*

The problem is that particular taxation enrollment took place in the year 6 C.E., ten years after the death of Herod!

The historical record is clear concerning the date of the enrollment decree. When King Herod died in the year 4 B.C.E., the Romans divided his kingdom among his three sons: Archelaus, who was given Judea, Idumea, and Samaria to rule; Antipas, who was given Galilee and Perea; and Philip, who was given Trachonitis, Gaulanitis, and Batanaea. The Romans refused to allow the three sons to take on the title of king for themselves; Archelaus was to be called ethnarch, while each of the other two sons was to be called tetrarch. Nevertheless, to provide themselves with a certain amount of regality, the three sons took on the name of their father, Herod, as a title.

Herod Archelaus proved to be a particularly brutal ruler (which the author of the Gospel of Matthew undoubtedly knew, thus providing a reason for his having Joseph and his family bypass Judea and settle in Galilee following their return from Egypt), and his subjects eventually appealed to Rome for relief. As a result, the Romans banished Archelaus in 6 C.E. and brought Judea under direct Roman rule with Coponius as procurator (which was the position that Pontius Pilate would be assigned some time later).

In conjunction with the change of authority, the Romans ordered an enrollment of the citizens of Judea for the purpose of taxation, and P. Sulpicius Quirinius (again, called Cyrenius in the KJV), as the newly installed Roman governor of Syria, was charged with overseeing it. Luke exaggerated the extent of this enrollment, however, saying that Caesar Augustus decreed that "all the world should be taxed" and that everyone had to return to the city of their ancestors to be enrolled. In fact, there is no record that Augustus ordered such a world-wide decree; if there had been such a decree, the magnitude of its requirements would have certainly been so great as to have warranted a mention in the historical record. This particular enrollment, or census, applied only to the territory that was taken over by the Romans following Archelaus's deposition and did not apply to the other territories.

What this means, then, is that Luke dated the birth of Jesus some twelve years after the time that Matthew dated the birth.

That conclusion is hardly new, of course, and has been expressed by many Bible scholars and critics. In response, some apologists have tried to argue that Quirinius had also had an earlier term as governor of Syria during the reign of King Herod and that there was another enrollment at that time. But that stand is pure conjecture and has absolutely no foundation in the historical record.[3] In the first place, the historical record shows that Quintus Sentius Saturninus was the governor of Syria from 9 to 6 B.C.E., and Quintilius Varus from 6 B.C.E. until after the death of Herod. Moreover, the Romans enrolled and taxed only the inhabitants of those provinces, such as Egypt and Syria, where they ruled directly; they did not enroll and tax the inhabitants of other provinces that were ruled by "client" kings, such as Herod.[4] The very fact that the Romans ordered an enrollment for tax purposes as one of their first acts when they took direct rule over Archelaus's territory indicates that there was no previous such census in that territory.

On top of that, the Acts of the Apostles provides further evidence that there was not a previous Roman census for tax purposes in Judea during King Herod's reign. In Acts 5:36–37, Luke told of an address that Gamaliel gave to the Jewish Council when certain of the apostles were imprisoned during the early days of the Christian church. The Sadducee members of the Council wanted to execute them, but Gamaliel, who was a Pharisee, argued that God himself would see to the apostles' demise if their beliefs were false. By way of an argument for that position, Gamaliel described what had previously happened to certain rebels. He first told of the fate of a rebel named Theudas, and then stated (here from the RSV): "*After him Judas the Galilean arose in the days of the census and drew away some of the people after him; he also perished, and all who followed him were scattered.*" In Luke's narration of the address, Gamaliel did not further qualify "the census," because, for his Jewish audience, the notoriety of the referenced census made further qualification unnecessary. Luke likewise saw no need to qualify "the census" in that passage because he had already qualified it in Luke 2:2 (here again, the RSV): "*And this was the first enrollment, when Quirinius was governor of Syria.*" In the original Greek of Luke/Acts, the same word is used in both passages: *apographo*, meaning an enrollment for the purpose of taxation. (*Strong's Concordance* states it as ". . . an enrollment or registration in the public records of persons together with their income and property, as the basis of a census or valuation, i.e. that it might appear how much tax should be levied upon each one.") Luke did see a need to qualify "the census" when he first mentioned it in Luke 2:2

because he was writing for a primarily Gentile audience who likely would not have been familiar with the occasion of the first Roman census in Judea. Thus having done so, he saw no need to repeat the qualification in Acts 5:37.

But what is significant about the passage in Acts 5:37 is that a rebel named Judas did in fact historically lead a revolt against the tax enrollment imposed by the Romans when they took direct rule over Archelaus's territory in the year 6 C.E. In his *Jewish War*, Josephus described the revolt.

> And now Archelaus's part of Judea was reduced into a province [of Rome], and Coponius, one of the equestrian order among the Romans, was sent as a procurator, having the power of [life and] death put into his hands by Caesar. Under his administration it was that a certain Galilean, whose name was Judas, prevailed with his countrymen to revolt, and said they were cowards if they would endure to pay a tax to the Romans and would after God submit to mortal men as their lords.[5]

That somewhat abbreviated account of the revolt makes it clear that there was no previous enrollment for taxation instituted by the Romans. Though Josephus did not specifically mention Cyrenius in conjunction with Judas in that passage, he did do so later in the book.

> In the mean time, one Manahem, the son of Judas, that was called the Galilean, (who was a very cunning sophister, and had formerly reproached the Jews under Cyrenius, that after God they were subject to the Romans,) took some of the men of note with him, and retired to Masada....[6]

After writing his *Jewish War*, Josephus took on the task of writing a more complete history of his nation and did considerable research for that purpose. His *Antiquities of the Jews* was the result, and in it Josephus described in more detail the revolt instigated by Judas, this time including Cyrenius and the enrollment for taxes in the narrative.

> Now Cyrenius, a Roman senator ... came at this time into Syria..., being sent by Caesar to be a judge of that nation.... Coponius also ... was sent together with him, to have the supreme power over the Jews. Moreover, Cyrenius came himself into Judea, which was now added to the province of Syria, to take an account of their substance, and to dispose of Archelaus's money; but the Jews, although at the beginning they took the report of a taxation heinously, yet did they leave off

any further opposition to it, by the persuasion of Joazar, who was the ... high priest; so they, being over-persuaded by Joazar's words, gave an account of their estates, without any dispute about it. Yet was there one Judas, a Gaulonite, of a city whose name was Gamala, who, taking with him Sadduc, a Pharisee, became zealous to draw them to a revolt, who both said that this taxation was no better than an introduction to slavery, and exhorted the nation to assert their liberty. ... They also said that God would not otherwise be assisting to them, than upon their joining with one another in such councils as might be successful. ... All sorts of misfortunes also sprang from these men, and the nation was infected with this doctrine to an incredible degree; one violent war came upon us after another. ...[7]

These accounts provided by Josephus make it clear that the enrollment (the "account of their [the Judeans'] estates") for taxation—and the resultant revolt initiated by Judas—followed the deposition of Archelaus by the Romans and was overseen by Cyrenius when he became the governor of Syria. It should therefore be clear that the tax enrollment mentioned in Acts 5:36–37 is the same one that is mentioned in Luke 2:2. Since Luke associated the birth of Jesus with that tax enrollment, and since that tax enrollment took place in the year 6 C.E. after Archelaus was deposed, Luke therefore dated the birth of Jesus ten years after the death of Herod. That means, of course, that Luke dated the birth of Jesus some twelve years after Matthew dated the birth of Jesus.

The conclusion that Luke dated the birth of Jesus after the death of King Herod is supported by yet other passages in his gospel. According to Luke 3:1, John the Baptist began his baptism ministry in the fifteenth year of the reign of Tiberias Caesar. That would have been in the year 29 C.E. because Tiberias Caesar began his reign in the year 14 C.E. After describing the baptism of Jesus by John, Luke went on to state in 3:23 that (here, the RSV) "Jesus, when he began his ministry, was about thirty years of age."[8] Jesus would therefore have begun his ministry in 29 C.E. or in 30 C.E.—that is, if he began his ministry not long after John began his ministry.

Thus, if Jesus "was about thirty years of age" in 29 or 30 C.E., he would have been born about the year 2 or 1 B.C.E. at the earliest. That, in fact, is why the standard calendar was constructed centuries later so as to have its starting date of 1 A.D. at that particular point in time. That is, it was assumed that Jesus was born in the year that came to be 1 B.C.; thus, 1 A.D. was assigned as the first full calendar year in the life of Jesus (there was no year 0).

Moreover, according to Matthew and Mark, Jesus did not actually begin his ministry until after Herod Antipas had John the Baptist arrested (Luke is vague on this point and John's Gospel is silent about it).[9] Because it would likely have taken a while to gain the fame and followers he did, John the Baptist probably had been preaching for some time before he was arrested—perhaps for as long as three years. Jesus might therefore have been baptized and had begun his ministry as late as 31 or 32 C.E., in which case he would have been born in the year 1 or 2 C.E.

Therefore, if he was correct concerning both the date of the beginning of John's baptism ministry and the age of Jesus when he began his ministry, Luke, in these passages, again indicates that Jesus was born well after the death of Herod.

When he was developing his gospel, Luke apparently miscalculated the date of the Roman tax enrollment under Quirinius. That would explain the inconsistency between his placing the birth of Jesus, on the one hand, in conjunction with that enrollment, which occurred in 6 C.E., and, on the other hand, between 2 B.C.E. and 2 C.E. on the basis of the time frame in which he had Jesus begin his ministry.

That leaves Luke 1:5–80, which indicates that the birth of John the Baptist and the annunciation to Mary occurred during the reign of "Herod, the king of Judaea." However, given that the range of dates for the birth of Jesus that we have already derived from Luke's gospel is well after the death of King Herod, a solution suggests itself. It is possible that verses 1:5–80 (and also the parenthetical "as it was supposed" in 3:23) were a later addition to Luke's gospel. If those verses, which describe certain events and foreshadow the virgin birth of Jesus, were added by a later editor, the original version of Luke's gospel would not have had the apparent discrepancy of having the annunciation to Mary occurring before the death of King Herod in 4 B.C.E., while having the birth of Jesus no earlier than 2 B.C.E. (The discrepancy is "apparent" because Christians have generally assumed that Mary's conception occurred shortly after the annunciation, and that the birth of Jesus occurred only a few months after that of John the Baptist.)

Another possibility is that those verses *were* in Luke's gospel, but the original wording of verse 1:5 placed the events of the chapter during the reign of Herod Archelaus, Ethnarch of Judea, instead of during the reign of "Herod, the king of Judea." This possibility is supported by the fact that Herod (the Great) was king not only of Judea, but of the whole of Palestine. In order to bring Luke's gospel in line with Matthew's gospel, a later editor could have simply changed "Herod, ethnarch of Judea," which would refer to Herod Archelaus, to "Herod, king of Judea," which would refer to King Herod the Great. In so doing, however, he neglected to change the territory that the ruler ruled over, leaving the smaller territory ruled by Archelaus instead of the larger territory (i.e., Palestine) that was ruled over by King Herod. This possibility is also supported by Luke 2:1 which, immediately following the narrative given in Luke 1:5–80, states that the taxation enrollment under Quirinius was ordered "in those days." Thus, the first two chapters of Luke's gospel could have originally indicated that John's birth and the annunciation to Mary occurred during the last months of Herod Archelaus's reign, and that the birth of Jesus "in those days" occurred not long thereafter during the enrollment that was ordered when the Romans took over Archelaus's territory. That scenario would at least make Luke's account of the early life of Jesus historically consistent and smooth, if not factual.

We previously noted that, in contrast to Matthew's gospel, Luke's gospel had no indication that Joseph and Mary were concerned about King Herod's threat to the young Jesus or that they were wary of Archelaus. We can see now why that was the case. They had no concern about Herod or Archelaus because, according to Luke's gospel, neither Herod nor Archelaus was on the scene after the birth of Jesus: Herod was long dead and Archelaus had been deposed. Thus, any threats that those rulers may have posed for the young Jesus were not even a consideration in Luke's gospel.

Matthew and Luke, along with Mark, are called the synoptic gospels because they view the ministry of Jesus "with the same eye," which is what "synoptic" means. That term was applied to these gospels because they provide so many parallel and similar stories about the ministry of Jesus. But then, if Matthew and Luke were "synoptic," why did they not present similar stories about the early years of Jesus? As we have seen, the stories of the early years of Jesus in Matthew and Luke are so completely different that they simply cannot be meshed together without doing violence to the threads of the respective narratives.

The answer has to do with the other synoptic gospel: that of Mark. Biblical scholars, as was previously mentioned in Chapter 1, attribute the similarity of the three gospels to be the result of the authors of Matthew and Luke having used Mark as a main source of material for their own Gospels. But Mark does not relate a story about the early years of Jesus; rather, he begins his gospel with the baptism of Jesus by John. Because of that, and wanting to extend their gospels back to the early years of Jesus, Matthew and Luke had to come up with their own stories for that part of the life of Jesus. They therefore individually either made up stories that they felt were appropriate for what they were trying to get across to their readers, or they used spurious tales about the early life of Jesus that were in circulation at the time.

The reasons are quite clear as to why the authors of the two gospels felt the need of including stories of the birth and early life of Jesus in their respective gospels.

First, according to "prophecy," the Messiah was supposed to be a descendant of David. The two authors consequently felt they had to provide a genealogy to "show" that Jesus was descended from David. Each author therefore independently embellished his gospel with a fabricated genealogy.

Also according to "prophecy," the Messiah was to be born in Bethlehem. But Jesus was from Nazareth of Galilee, so the two authors needed some way to show that, though Jesus was from Nazareth, he nevertheless had been born in Bethlehem. Each author therefore came up with his own explanation. Matthew used a story relating that Joseph's family originally lived in Bethlehem but was forced to move to Nazareth because of a threat to the life of Jesus. Luke, on the other hand, used a story relating that the family had their home in Nazareth, but that the tax enrollment decree by Caesar Augustus required each male to go to the city of his ancestors, thus necessitating Joseph and Mary's trip to Bethlehem, where Jesus was born.

Both authors tried to place their stories in a historical context, but they were loose with the facts. For example, contrary to what Luke's account indicates, an enrollment ordered by the Romans as a basis for taxation certainly would not have required that everyone in the world return to the city of their ancestors. The Romans would not have cared in the least where a subject's ancestors had lived (particularly the ancestors of their Jewish subjects), and they taxed on the basis of residency, not ancestry. Though the Romans could require those who were subject to taxation to return to their place of residence for enrollment if they were away for some reason, it would have been pointless to have had everyone return to the city of their ancestors if that city were different from their place of residence. To have required such a mass movement of people would have disrupted commerce and would have been contrary to Roman practicality. The enrollment under Quirinius did not have any such provision, and that enrollment, as we noted, applied only to the residents of the territory that the Romans took over from Archelaus. As a resident of Nazareth in the territory of Galilee (according to Luke 2:4 and Luke 2:39), which was under the rule of Herod Antipas, Joseph certainly would not have been required to go to Judea to be enrolled.

Nevertheless, Luke needed a reason for having Joseph and Mary travel to Bethlehem for the birth of Jesus, and he saw in the Roman tax enrollment a means of establishing, in an "official" way, both the ancestry of Jesus and the location of his birth. All Luke needed to do was to concoct the requirement that the enrollment required everyone in the world to return to the city of their ancestors. For that purpose, he therefore expanded the enrollment decree beyond its original boundary and its original requirements, and in so doing, he "established" Joseph's Davidian ancestry by emphasizing that Bethlehem was Joseph's ancestral home.

As for Matthew's account, Herod the Great certainly would have been capable of ordering someone killed if he viewed that person as a threat to his rule (as the author of Matthew's gospel undoubtedly knew), but there is no historical record indicating that Herod ordered the killing of all of the male children up to two years of age in Bethlehem "and the coasts thereof" as stated in Matthew's gospel. If it had occurred, such an act certainly would have caused considerable bitterness among the people and surely would have been a matter for the historical record.

Rather than Herod's actually having ordered the killing of the children, it is more likely that Matthew fabricated the story in order to provide a basis for having prophecy fulfilled. By having Jesus and his parents go to Egypt to escape the supposed threat from Herod, Matthew made the birth of Jesus parallel that of Moses. He also had a "prophecy" fulfilled when Jesus came out of Egypt (as stated in Matthew 2:15). And he had another "prophecy" fulfilled when he alleged that Herod had the children in Bethlehem killed (as stated in Matthew 2:17–18), and yet another when he had Joseph's family move to Nazareth (as stated in Matthew 2:23).

But those "prophecies" invariably were either scripturally nonexistent or had nothing to do with the contexts in which Matthew placed them. For example, Matthew stated that Joseph and his family moved to Nazareth so that Jesus would fulfill that "which was spoken by the prophets, He shall be called a Nazarene." The problem is that there is no prophecy in the Hebrew Scriptures stating "He shall be called a Nazarene," and there certainly is no prophecy saying that the Messiah would be from Nazareth. For that matter, other than in Matthew 2:23, Jesus is not called a "Nazarene" in the New Testament. However, Acts 24:5 provides a clue as to why Matthew may have applied the term "Nazarene" to Jesus. In that passage, the high priest is haranguing against Paul:

> *For we have found this man a pestilent fellow, and a mover of sedition among all the Jews throughout the world, and a ringleader of the sect of the Nazarenes:*

This verse indicates that the early Jewish followers of Jesus were not called Christians, but rather Nazarenes. Now, very few, if any, of the early Jewish followers of Jesus would have been from Nazareth (the gospels are clear in indicating that Jesus had little respect in Nazareth), and, as a "sect," they certainly would not have been called Nazarenes simply because their deceased leader was originally from Nazareth. It is more likely that the word "Nazarene" had its source in what his Jewish followers *believed* about Jesus. The probable source for the term can be found in Isaiah 11:1: *"And there shall come forth a rod out of the stem of Jesse, and a Branch shall grow out of his roots."* In that passage, the word "Branch" was taken as referring to the coming Messiah. The original Hebrew word for "branch" is *netser*, which in New Testament times was apparently transliterated into *Nazoraios*, the Greek word that was in turn subsequently transliterated as "Nazarene" in the English bibles. The Greek-speaking Jewish followers of Jesus would therefore have given him the name or title of *Nazoraios*, "the Branch," since they considered him to be the prophesied Messiah who would come in the Last Days. His followers, in turn, would have been called *Nazoraion*—the term used in the original Greek of Acts 24:5—as a reflection of their belief about Jesus.

It is likely that Matthew did not understand the origin of the term *Nazoraios* (otherwise, he would have used a more credible explanation for its application to Jesus). In trying to come up with a suitable "prophecy" to explain why Jesus was called *Nazoraios* by his followers, he apparently decided that, on the basis of the similarity in sound, it must have had something to do with Nazareth. In any case, Matthew's assertion that Jesus fulfilled a prophecy because he was from the town of Nazareth, and therefore a "Nazarene," was purely a play on words on his part and had absolutely nothing to do with any actual prophecy.

The other two "prophecies" that Matthew alleged were fulfilled had no more basis than the one concerning Nazareth. In one of those "prophecies," Matthew was referring to Hosea 11:1: "When Israel was a child, then I loved him, and called my son out of Egypt." The "child" is the nation of Israel, and the passage refers to an event in the past, not the future. The passage is also hearkening to Exodus 4:22–23: "And thou shalt say unto Pharaoh, Thus saith the Lord, Israel is my son, even my firstborn: And I say unto thee, Let my son go. . . ." The other "prophecy" that Matthew referred to is found in Jeremiah 31:15. It had nothing to do with Bethlehem and also referred to a past event. The village of Ramah that is associated with Rachel in the Jeremiah passage is about five miles north of Jerusalem. Bethlehem is about five miles south of Jerusalem.

It is apparent that, insofar as these three "prophecies" are concerned, Matthew was straining to find scriptural passages in support of the story he was relating. It is also apparent that Matthew was getting carried away with having his gospel show that the birth and early life of Jesus fulfilled prophecy.

Matthew, in fact, got carried away with showing that prophecy was fulfilled in Jesus in several other parts of his gospel. One example is found in Matthew 21:1–7. The occurrence described in this passage is mentioned, with variations, in all four gospels, but only Matthew carried it to an absurd extreme. The prophecy that is referred to in that passage is found in Zechariah 9:9: "*. . . thy King cometh unto thee . . . riding upon an ass, and upon a colt the foal of an ass.*" That passage provides an example of biblical poetic reiteration, which is frequently found in the Old Testament. In poetic reiteration a passage contains two parts, the second of which restates in different words the subject of the first part. In this case, "a colt the foal of an ass" restates "an ass" of the first part. Only one animal is being referred to in the passage, and Mark, Luke, and John mention only one animal in their versions of the fulfillment of the passage. Matthew, however, erroneously took the passage in Zechariah as meaning two animals, for he had Jesus ride two animals at the same time: "*And the disciples . . . brought the ass, and the colt, and put on them their clothes, and they set him thereon.*" This again shows that Matthew was not above embellishing his gospel in order to make it conform to his understanding of Old Testament prophecy—even to the point of absurdity.

Matthew's predilection for embellishing his gospel story brings up a point. One of the primary reasons that biblical apologists try to backdate the enrollment described in Luke's gospel to the time of King Herod's reign is that Matthew's gospel is clear in stating that the birth of Jesus occurred during that reign. However, if Matthew's story of the early life of Jesus is a complete fabrication, as the evidence indicates, there is no reason to maintain unequivocally that Jesus was born during Herod's reign.

And that brings up another point. We took a lengthy and detailed look at the narratives of the early life of Jesus in the gospels of Matthew and Luke in order to determine if the authors of those two gospels provided trustworthy information about that subject. Since the evidence is clear that the information that the authors of both Matthew and Luke provided about the early years of Jesus is not trustworthy, why should those authors be considered trustworthy in anything else they say in the remainder of their books—including what they say about the death and resurrection of Jesus? Moreover, if the accounts presented by these two authors of New Testament books are untrustworthy, why should any of the accounts given by the other New Testament authors be considered trustworthy? After all, the New Testament is supposed to be a cohesive whole and all of its books are supposed to be equally "inspired."

There is another New Testament embellishment that we should perhaps examine: the "virgin birth" of Jesus.

Such an examination must be prefaced with the understanding that the Hebrew people of that era did not have a tradition that the Messiah was to be born of a virgin; rather, according to their tradition, the Messiah was supposed to be a literal descendant of David. In accordance with that tradition, as we previously noted, the gospels of both Matthew and Luke include genealogies that supposedly show that Jesus was of the lineage of David through Joseph, the father of Jesus. But both of those gospels also indicate that Jesus was born of a virgin, which, in effect, negates the genealogies.

In any case, it appears that the idea that Jesus was born of a virgin actually came about relatively late during the early years of Christianity. Aside from the gospels of Luke and Matthew, none of the New Testament books make any mention of the virgin birth and instead appear to adhere to the assumption that Jesus was a physical descendant of David. Paul, for example, stated that Jesus was *"made of the seed of David according to the flesh"* (Romans 1:3 of the KJV; the RSV passage states Jesus was "descended from David according to the flesh"). Likewise, in Acts 2:30, in speaking of David not long after the crucifixion of Jesus, Peter said: "... *knowing that God had sworn with an oath to him that of the fruit of his body, according to the flesh, He would raise up the Christ to sit on his throne.*" In saying that Jesus was born of the lineage of David "according to the flesh," these passages certainly negate the idea of a virgin birth.

Though the authors of the gospels of Luke and Matthew provided genealogies of Jesus in an attempt to show that he was of the lineage of David, they in effect negated the genealogies by bringing in the idea of the virgin birth. The author of Luke's gospel (or perhaps a later editor) apparently realized this, for he put a parenthetical statement in his genealogy of Jesus in an attempt to get around the conflict:

> *And Jesus himself began to be about thirty years of age, being (as was supposed) the son of Joseph, which was* the son *of Heli,*

In addition, the gospels of both Matthew and Luke surround their stories of the birth and first years of Jesus with events and miraculous manifestations that, if they had actually happened, would certainly have made a deep impression upon Joseph and Mary and would have unquestionably revealed to them that their son had a divine mission in life. Yet, it appears that, after those early years, the parents of Jesus had lost all recollection of the visitation of the angels; the annunciation; the meeting with Mary's cousin, Elizabeth, who was pregnant with the child who would become John the Baptist, and who acknowledged that Mary would be mother of the Lord; Mary's "Magnificat" soliloquy in which she rejoiced in being blessed as the chosen of the Lord; the virgin birth; the arrival of the shepherds in the stable; the visit of the magi; the temple testimony of Simeon that the baby Jesus would become the salvation of Israel and the light of the Gentiles; the flight to Egypt at the behest of an angel; and everything else that supposedly revealed to them that God had placed their son on earth for a special purpose.

For example, although Luke had inserted the parenthetical "as was supposed" in the genealogy of Jesus so that the genealogy would not appear to contradict the story of the virgin birth and the related manifestations, he apparently failed to notice that another passage in his gospel was damaging to that material. In Luke 2:46–50, we find that, when Jesus was twelve, he and his parents went to the temple in Jerusalem, where his parents lost track of him.

> 46 *And it came to pass, that after three days they found him in the temple, sitting in the midst of the doctors, both hearing them, and asking them questions. . . .*
> 48 *And when they saw him, they were amazed: and his mother said unto him, Son, why hast thou thus dealt with us? behold, thy father and I have sought thee sorrowing.*
> 49 *And he said unto them, How is it that ye sought me? wist ye not that I must be about my Father's business?*
> 50 *And they understood not the saying which he spake unto them.*

If they had previously experienced all of the miraculous manifestations relating to the virgin birth and the other events given earlier in the Gospel of Luke—or of Matthew, for that matter—the parents of Jesus certainly would have understood what their son was speaking about. That passage therefore does not accommodate the understanding that they would quite clearly have had about the divine mission of their son. This could perhaps be taken as evidence that the author of Luke added the story of the virgin birth and the related manifestations as an afterthought to his gospel but failed to follow through and adjust the rest of the gospel accordingly.

The next question to ask, then, is how did the idea that Jesus was born of a virgin come about?

To answer that question, we must look to the influence that Paul had on the developing church, and specifically to Paul's mission to the Gentiles. The letters that Paul wrote to the various Gentile Christian communities were the earliest of the documents that formed the New Testament canon, and he introduced many new ideas and beliefs about Jesus that certainly were foreign to what the early Jewish followers of Jesus believed and what Jesus himself would have believed. Many of the ideas that Paul introduced were, in fact, based more on Greek and other pagan mystery religions than on Jewish thought. Though Paul himself did not introduce the idea of the virgin birth of Jesus, he originated the idea that Jesus was part of the Godhead, and, once that idea was promulgated, it would not be surprising that the question of how Jesus, as a divine personage, came into this world would come up. The Gentiles who came into the church as a result of Paul's missionary activities apparently influenced the answer, for included among the pagan beliefs they would have been familiar with were stories of women who conceived and gave birth to children through the actions of some god.[10] Thus, over time, the idea developed that Jesus, as a divine personage, must also have come into this world by divine means.

In any case, Matthew wanted to provide a justification for his inclusion of the story of the virgin birth, so he tried to find something in the scriptures he could use to "foretell" it. Of course, since the idea of virgin births was foreign to Hebrew tradition and scripture, he had little success. The only scripture he could come up with was a passage that he found in Isaiah 7:14 (here, from the KJV): "... *Behold, a virgin shall conceive, and bear a son. ...*" But the word "virgin" in that passage does not properly indicate the meaning of the original Hebrew word. The original Hebrew word that was translated as "virgin" in the passage is *'almah*, which has a general meaning of "young woman." In Hebrew, the more unambiguous word for "virgin" is *bethuwlah*.

Furthermore, Matthew twisted the meaning of the Isaiah passage, for that passage was not a prophecy concerning the Messiah. The passage makes it clear that the young woman who was to give birth was someone who was living in the time of Ahaz. As Isaiah 7:14–16 states, Isaiah tells Ahaz that the young woman would "... *conceive and bear a son, and shall call his name Immanuel ..., and ... before the child shall know to refuse the evil, and choose the good, the land that thou abhorrest shall be forsaken of both her kings.*" The context thus makes it clear that the birth was to be a sign to Ahaz and was to indicate something that was supposed to occur in the near future, and not several hundred years later. The context also makes it clear that there was nothing miraculous about the birth itself: a young woman would simply conceive and bear a child.

This understanding of the verse in Isaiah is also substantiated by the use of the Hebrew words *bethuwlah* and *'almah* elsewhere in the Bible. The word "virgin" occurs twenty-six times in the Old Testament of the King James Version of the Bible. In twenty-four of those instances, the original Hebrew word is *bethuwlah*, and the contexts of those occurrences make it clear that "virgin" is the intended meaning. On the other hand, the word *'almah* is translated as "virgin" only twice: in the aforementioned verse in Isaiah and in Genesis 24:43. The context of Genesis 24:43 is such that the word could just as well mean "young woman," and that is how it is translated in the RSV (which, incidentally, also uses the term "young woman" in its translation of Isaiah 7:14), as well as in several other translations of the Bible. Elsewhere in the King James Version of the Bible, the word *'almah* is translated as "damsel" or "maid" (e.g., in Psalms 68:25 and Exodus 2:8) with no particular significance as to virginity.

Moreover, Isaiah himself used *bethuwlah* on four different occasions elsewhere in his book (in 23:12, 37:22, 47:1, and 62:5), so he certainly was no stranger to the word. For example, in Isaiah 62:5 we find: *"For as a young man marrieth a virgin* [bethuwlah]. . . . " One might therefore ask why—if Isaiah 7:14 were truly a prophecy of the virgin birth of Jesus—was not *bethuwlah* used as the all-important word for "virgin" as it was in the twenty-four other Old Testament passages—including those in Isaiah—in which "virgin" was clearly meant?

The fact of the matter is that Isaiah 7:14 had no messianic connotation, and those who lived during Old Testament times never looked upon it as being a prophecy concerning the Messiah. Certainly, if the writer of the Matthew account had not used Isaiah 7:14 as a messianic prophecy, it would never have become a part of Christian tradition concerning the birth of Jesus.

This raises the question of how the author of the Gospel of Matthew came up with that verse in Isaiah if the original Hebrew version of the passage did not imply that the young woman was a virgin.

The answer to that question can perhaps be found in the Greek Septuagint, which was the version of the scriptures that Matthew used. (As we noted in Chapter 1, the Septuagint is a Greek translation of the Hebrew Scriptures that was begun in the third century B.C.E. in Alexandria, Egypt.) In the Septuagint, the Hebrew word *'almah* in the Isaiah passage is translated into Greek as *parthenos*. That Greek word—which Matthew also used when he wrote his gospel—means "a maiden," "virgin," or "an unmarried daughter." Matthew apparently took it as meaning "virgin."

There is also one other part of the passage in Isaiah that plainly shows that it was not a prophecy concerning Jesus. Isaiah told Ahaz that the young woman would "conceive and bear a son, and shall call his name Immanuel. . . ." The son of Joseph and Mary was named "Jesus," not "Immanuel," and nowhere in the New Testament is Jesus called by the name of Immanuel.

From all this, it should be apparent that Matthew—because he had no recourse to another passage in the scriptures that would have been better suited for his purpose—appropriated the Isaiah passage and turned it into a "prophecy" concerning the "virgin birth" of Jesus. It should also be apparent that the King James translators chose to translate *'almah* as "virgin," rather than as "young woman," in the Isaiah passage because Matthew had used the passage to fulfill a "prophecy" about Jesus.

Though it is pretty much ignored today, the whole sense of the gospel when Christianity had its beginning was that the Last Days were upon the earth and that the Day of the Lord was fast approaching. In fact, the belief that God would soon overturn the existing order and establish a new world—the kingdom of heaven in which righteousness would prevail—was in the air well before Jesus came on the scene. In the Old Testament, several prophets had proclaimed that the Judgment Day was near and that the enemies of God's chosen people would soon receive their just retribution. With the repeated subjugation of the Jewish people during the period from before the end of the Old Testament era on up to the arrival of Christianity, belief in the imminent coming of the Day of Judgment intensified, and that belief was reflected in many of the writings of the time. The Book of Enoch, for example, which was written within the two centuries prior to the beginning of the Christian era, quite clearly and extensively "prophesied" the coming Day of the Lord and the kingdom of righteousness (see Appendix C).

That, then, was the milieu in which Christianity had its beginnings. Reflecting that milieu, the gospels describe John the Baptist as proclaiming, prior to the ministry of Jesus (here, Matthew 3:2): *"Repent ye: for the kingdom of heaven is at hand."* Jesus himself, of course, subsequently took up the message and also proclaimed the imminent coming of the kingdom of heaven, saying the same thing that John had said (here, Matthew 4:17):

> *From that time Jesus began to preach, and to say, Repent: for the kingdom of heaven is at hand.*

Jesus also directed his disciples to proclaim the message (here, Matthew 10:5–7):

> 5 *These twelve Jesus sent forth, and commanded them, saying, Go not into the way of the Gentiles, and into* any *city of the Samaritans enter ye not:*
> 6 *But go rather to the lost sheep of the house of Israel.*
> 7 *And as ye go, preach, saying, The kingdom of heaven is at hand.*

Even more significantly, here is what else he said to his disciples (as it is given in Matthew 16:27–28):

> 27 *For the Son of man shall come in the glory of his Father with his angels; and then he shall reward every man according to his works.*
> 28 *Verily I say unto you, There be some standing here, which shall not taste of death, till they see the Son of man coming in his kingdom.*

And here is what Jesus further had to say on the subject, (Matthew 24:3–34):

> 3 *And as he sat upon the mount of Olives, the disciples came unto him privately, saying, Tell us, when shall these things be? and what shall be the sign of thy coming, and of the end of the world?* ...
> 4 *And Jesus answered and said unto them,* ... [giving a long list of calamities that are to befall mankind during the tribulation.]
> 29 *Immediately after the tribulation of those days shall the sun be darkened, and the moon shall not give her light, and the stars shall fall from heaven, and the powers of the heavens shall be shaken:*
> 30 *And then shall appear the sign of the Son of man in heaven: and then shall all the tribes of the earth mourn, and they shall see the Son of man coming in the clouds of heaven with power and great glory.*
> 31 *And he will send out his angels with a trumpet blast, and they will gather his elect from the four winds, from one end of the heavens to the other.*
> 33 *So likewise ye, when ye shall see all these things, know that it is near,* even *at the doors.*
> 34 *Verily I say unto you, This generation shall not pass, till all these things be fulfilled.*

The term, "Son of man," which is found in both of the above passages, had the meaning of "a man"—i.e., "a human being"—in the idiom of the times. The term "son of man" was used to emphasize the human nature of an individual, and it is used in that sense in numerous places in the Old Testament (e.g., Ezekiel 26:2). The term was also used, in certain contexts, to refer to a personage sent from God, but having the appearance of a man rather than of an angel. The initial usage of the term in that sense actually occurs quite late in the Old Testament era. In fact, Daniel 7:13 is the only verse in the Old Testament that uses the term in that sense (the book of Daniel, as we saw in Appendix A, was written during the second century B.C.E). That verse states: "*I saw in the night visions, and behold, there came with the clouds of heaven one like a son of man....*" In the Book of Enoch—and in the New Testament, of course—the term "the Son of man" is frequently applied to "the anointed one" of God who would overthrow the existing order and establish the kingdom of righteousness. The Greek word meaning "the anointed one" is *christos*, from which the word "Christ" is derived. The Hebrew word meaning "the anointed one" is *mashiyach*, which is transliterated into English as "messiah." In its original more mundane usage, the Hebrew word for "messiah" referred to earthly kings, for they were anointed during their coronation. An example of that usage is found in Daniel 9:25–26, in which the "messiah," or "anointed one," is a "prince" (the original Hebrew word translated as "prince" also means "ruler").

In Matthew 24:34, above, the Greek word translated as "generation" is *genea*, which, according to the reference to this passage in *Strong's Concordance*, has the meaning of "the whole multitude of men living at the same time," and also "'an age,' i.e., the time ordinarily occupied by each successive generation." Obviously, contrary to what Jesus prophesied, the end of the world did not come to pass during the generation that was living when he spoke those words. Because of that, biblical apologists have tried to interpret the term "this generation" of that last verse as meaning something other than the generation that existed in the time of Jesus. However, Matthew 24:3–34 does not indicate that Jesus qualified the meaning of "this generation" in any way, and the context of the passage is clear. The whole sense of what Jesus said was that his listeners would "see all these things" themselves and that the generation living at that time would "*not pass, till **all** these things be fulfilled.*" What he said in Matthew 16:28, above, underscores that understanding: "*Verily I say unto you, There be some standing here, which shall not taste of death, till they see the Son of man coming in his kingdom.*"

Jesus is shown as expressing much the same thing after his arrest when he was being interrogated by the high priest. The high priest asked him if he was the Messiah, and Jesus answered (here, Matthew 26:64 in the NJB):

> *Jesus answered him, 'It is you who say it. But, I tell you that from this time onward you will see the Son of man seated at the right hand of the Power and coming on the clouds of heaven.'*

There, Jesus is telling the high priest that the time in which he himself would see the Son of man coming on the clouds of heaven was about to begin. In other words, Jesus was saying that the Day of the Lord was upon them.

The belief that the end of the world was near is also expressed in several other New Testament books. In 1 Corinthians 7, for example, Paul makes statements that make sense only if he believed the Last Days were upon the earth:

> 1 *Now concerning the things whereof ye wrote: It is good for a man not to touch a woman.*
> 6 *But I speak this by permission, and not of commandment.*
> 7 *For I would that all men were even as I myself. . . .*

> 8 *I say therefore to the unmarried and widows, It is good for them if they abide even as I.*
> 9 *But if they cannot contain, let them marry: for it is better to marry than to burn.*
> 27 *Art thou bound unto a wife? seek not to be loosed. Art thou loosed from a wife? seek not a wife.*
> 29 *But this I say, brethren, the time is short, so that from now on even those who have wives should be as though they had none,*

There, Paul is saying that he wishes, but does not command, that all men abstain from sexual relations and that those who are not married should remain unmarried unless they cannot contain themselves. The sense of what he is saying is that, since the time until the coming of the Day of the Lord and its attendant tribulation is so short, it is pointless even for those who are married to engage in marital relations and raise families. If he did not believe that the Day of the Lord was near, his wish that all men would abstain from sexual relations would mean, if followed through, that all of mankind to pass away after the generation that was then living died off. His wish that all men would abstain from sexual relations therefore makes sense only if he believed that the end of the world was at hand.

Paul also writes in 1 Thessalonians 4:13–17 (here, the NKJV):

> 13 *But I do not want you to be ignorant, brethren, concerning those who have fallen asleep, lest you sorrow as others who have no hope.*
> 14 *For if we believe that Jesus died and rose again, even so God will bring with Him those who sleep in Jesus.*
> 15 *For this we say to you by the word of the Lord, that we who are alive and remain until the coming of the Lord will by no means precede those who are asleep.*
> 16 *For the Lord Himself will descend from heaven with a shout, with the voice of an archangel, and with the trumpet of God. And the dead in Christ will rise first.*
> 17 *Then we who are alive and remain shall be caught up together with them in the clouds to meet the Lord in the air. And thus we shall always be with the Lord.*

There, Paul is telling the Thessalonians that "we who are alive"—that is, those who were then living—would not be lifted up to heaven before those who sleep (i.e., are dead) in Jesus, but that Jesus would rise up the dead first and only then would "we who are alive" be caught up with them. It is clear from Paul's words that he expected the Day of the Lord to come within the lifetimes of the believers who were alive when he was writing.

In 1 John 2:17–18 we find:

> 17 *And the world is passing away, and the lust of it; but he who does the will of God abides forever.*
> 18 *Little children, it is the last hour; and as you have heard that the Antichrist is coming, even now many antichrists have come, by which we know that it is the last hour.*

It is clear that the author of the epistle thought that it was the "last hour" before the end of the world.

In James 5:8 we find:

> *Be ye also patient; stablish your hearts: for the coming of the Lord draweth nigh.*

There, James is saying that the Lord is coming soon, meaning that the end of the world, is near.

In Hebrews 10:24–25, we find:

> 24 *And let us consider one another in order to stir up love and good works,*
> 25 *not forsaking the assembling of ourselves together, as is the manner of some, but exhorting one another, and so much the more as you see the Day approaching.*

There, the author is saying that his readers should exhort one another to do good works, all the more so as they see the Judgment Day approaching.

And in Hebrews 10:37 the author says:

> *For yet a little while, and he that shall come will come, and will not tarry.*

There, the author is saying that the Second Coming of Jesus will be in just a little while and he will not tarry.

In Hebrews 9:26 we find:

> *For then must he often have suffered since the foundation of the world: but now once in the end of the world hath he appeared to put away sin by the sacrifice of himself.*

There, the author is saying that it is now the end of the word, for Jesus has sacrificed himself.

In 1 Peter 4:7, we find;

> *But the end of all things is at hand: be ye therefore sober, and watch unto prayer.*

In saying "the end of all things is at hand," the writer obviously believed that the existing order was going to be soon overthrown with the coming Day of Judgment.

And in Acts 2:14–17 we find Peter responding to the Jews when they mocked the apostles who were speaking in tongues:

> 14 *But Peter, standing up with the eleven, lifted up his voice, and said unto them, Ye men of Judaea, and all ye that dwell at Jerusalem, be this known unto you, and hearken to my words:*
> 15 *For these are not drunken, as ye suppose, seeing it is* but *the third hour of the day.*
> 16 *But this is that which was spoken by the prophet Joel;*
> 17 *And it shall come to pass in the last days, saith God, I will pour out of my Spirit upon all flesh: and your sons and your daughters shall prophesy, and your young men shall see visions, and your old men shall dream dreams:*

There, Peter was saying that the Last Days were upon the earth; hence the actions of his fellow apostles were in accordance with the prophecies of Joel.

Likewise, Hebrews 1:1–2 makes it clear that the Last Days were considered to be upon the earth at that time:

> *God . . . [h]ath in these last days spoken unto us by* his *Son, whom he hath appointed heir of all things, by whom also he made the worlds;*

In saying "*these* last days," the writer clearly believed that the Last Days were upon the earth and the Day of Judgment was at hand.

And we find in Revelation 1:1–7 (here, the KJV):

> 1 *The Revelation of Jesus Christ, which God gave unto him, to shew unto his servants things which must shortly come to pass; and he sent and signified* it *by his angel unto his servant John:*
> 3 *Blessed* is *he that readeth, and they that hear the words of this prophecy, and keep those things which are written therein: for the time* is *at hand.*
> 7 *Behold, he cometh with clouds; and every eye shall see him, and they* also *which pierced him: and all kindreds of the earth shall wail because of him. Even so, Amen.*

There, we find that the End Times "must shortly come to pass," and "the time is at hand." The expression that Jesus "cometh with the clouds" is presented as something that is unfolding even as the passage is being written. The passage also states that "every eye shall see him," including those who "pierced him." The implication there is that those who crucified Jesus would still be alive at his second coming.

At the end of the Book of Revelation, verses 22:6–10 also clearly state that the events which had been described in the book were shortly to come to pass:

> 6 *And he said unto me, These sayings are faithful and true: and the Lord God of the holy prophets sent his angel to shew unto his servants the things which must shortly be done.*
> *7 Behold, I come quickly: blessed is he that keepeth the sayings of the prophecy of this book.*
> *10 And he saith unto me, Seal not the sayings of the prophecy of this book: for the time is at hand.*

There, according to John, the angel had shown "*the things which must shortly be done*" and that Jesus said "*Behold, I come quickly,*" and "*the time is at hand.*"

It is therefore clear that Jesus and his followers believed that they were living in the Last Days and that the Day of Judgment was at hand; it would not be long before God would overturn the existing order and establish a new world in which righteousness would prevail. As it states in 2 Peter 3:13:

> *Nevertheless we, according to his promise, look for new heavens and a new earth, wherein dwelleth righteousness.*

What is not so clear is what part Jesus himself believed he would play in the event; all we have are the accounts of his ministry that were written many years after his death, and they certainly have been colored by embellishment and expectation. At the very least, it would appear Jesus thought that, through his ministry, he was preparing mankind (or, more specifically, his fellow Jews, as indicated in Matthew 10:5–7, above, and also Matthew 15:24) for the coming Day of the Lord. He may also have thought that, in doing so, he was helping God to bring about that day. If he thought of himself as the Messiah, it would have been the Messiah of Jewish tradition—that is, a conquering Messiah who would overthrow the enemies of God. As he said in Matthew 10:34:

> *Think not that I am come to send peace on earth: I came not to send peace, but a sword.*

In that light, Matthew 27:46 perhaps provides the most telling of all of the utterances of Jesus that are ascribed to him in the New Testament. While he was hanging from the cross, and realizing that his mission had failed, he cried out:

> *My God, my God, why hast thou forsaken me?*

And that brings up an important and much-discussed question. Why was Jesus condemned to die? Even more pointedly, why was he sentenced to death by crucifixion? If Jesus had been simply preaching to his fellow Jews that they should act righteously, neither the Romans nor the Jewish establishment would have been concerned about what he was saying, nor would they have sought to have him condemned to death. But Jesus was preaching much more than simple righteousness. He was preaching that the Messiah—either as Jesus himself, or as a heaven-sent personage—would soon overthrow the existing order and establish a new kingdom, the kingdom of God. Since Jesus had gained a large following among the Jewish people, the Romans would have viewed such proclamations as a challenge to their authority, and certain of the Jewish leaders—specifically the Sadducees, who were installed as Temple priests at the whim of the Romans—would have understood quite well what the Roman response to that challenge would be. John 11:48 lays out their concerns quite clearly:

If we let him thus alone, all men will believe on him: and the Romans shall come and take away both our place and nation.

After the arrest of Jesus, he was brought before Pontius Pilate. We cannot know exactly what took place during his appearance before the Roman procurator, for the descriptions that the gospels provide of this event are highly contrived and present Pilate as being almost noble in his examination of Jesus. But, in fact, as the Roman overseer of Judea, Pontius Pilate was arrogant and brutal, and he had little regard for the sensitivities, religious or otherwise, of the Judeans that he governed. His character in this respect was made clear by Josephus, who, in his *Antiquities of the Jews*, described some of the harsh brutalities that Pilate had visited upon his Jewish subjects. One such brutality is even mentioned in Luke 13:1:

There were present at that season some that told him of the Galilaeans, whose blood Pilate had mingled with their sacrifices.

From what we know of the historical character of Pilate, then, it is highly unlikely that he would have acted in the manner that is described in the gospels. If he had sensed in any way that Jesus was preaching that the existing order was to be overthrown and a new kingdom established, he would have considered Jesus to be inciting rebellion and he would have acted accordingly. Because Jesus was crucified, a form of execution that the Romans used not only to punish criminals, but also to make an example of those who rebelled against their rule, we can take it that Pilate did act accordingly. To set an example of what would happen to those who would try to overthrow the existing order and establish another kingdom, the Romans placed an inscription above the head of the crucified Jesus—here, as given in Luke 23:36:

> *And a superscription also was written over him in letters of Greek, and Latin, and Hebrew, THIS IS THE KING OF THE JEWS.*

Decades later, the gospel writers tried to shift the responsibility for the death of Jesus by minimizing the Roman involvement and maximizing the involvement of the Jewish leaders. There was a good reason for that shift. The fledgling church had to make its way in a Roman world and the church leaders would not have wanted its believers to antagonize the Romans because of their role in the death of Jesus. Nevertheless, the fact of the matter is, again, that the Romans executed Jesus for being a perceived threat to their authority.

With his death on the cross, but still believing in his mission and in the coming Day of the Lord, his followers had to rethink the meaning of it all. If Jesus did not bring about the Day of the Lord while he was on the earth, then he would necessarily do so in a Second Coming. That belief in a Second Coming was therefore integrated into the belief that the Last Days were upon the earth.

One thing should be clear in all of this: Jesus had never intended to start a new religion, and those of his Jewish followers who remained after his death considered themselves still to be Jews. However, as we previously noted, some years after the death of Jesus, Paul came on the scene and instituted a whole new view of the life and death of Jesus. If Jesus had suffered and died on the cross, it must have been for a purpose, and Paul came up with the idea of the suffering Messiah who, with his death and "resurrection," paid for the sins of those who believed in him. Paul also came up with several other esoteric views that had more in common with Greek and other Pagan mystery religions than with Jewish thought and religious principles.[11] Paul, as the "apostle to the Gentiles," also brought in Gentile converts who were not attuned to the original message of Jesus, and, as a result, the growing church moved further away from that message. Since Paul's were the first of the writings that eventually became part of the New Testament, his views colored subsequent writings, including the gospels. In particular, Paul's concept of Jesus as a suffering Messiah who paid for the sins of those who believed in him was made an important part of the gospel stories, which were cast to make it appear that Jesus knew all along that he had been assigned that role.

As time went on, and as the early Christians expanded their literary base beyond the original Hebrew Scriptures, they wrote numerous gospels, letters, and apocalypses that were influenced by the evolving, and oftentimes inconsistent, beliefs about the mission and character of Jesus. As was noted in Part One of this book, only some of those writings ultimately made it into the New Testament canon. However, as we saw earlier in this appendix, even those writings were not always consistent with each other. Just as Matthew and Luke contradicted themselves concerning the birth and childhood of Jesus, so they likewise contradicted themselves about his death and "resurrection." According to Luke (and John as well), the resurrected Jesus showed himself to his disciples in Jerusalem (here, Luke 24:33–36):

> 33 *And they rose up the same hour, and returned to Jerusalem, and found the eleven gathered together, and them that were with them, . . .*
> 36 *And as they thus spake, Jesus himself stood in the midst of them, and saith unto them, Peace be unto you.*

However, according to Matthew, first an angel, and then the resurrected Jesus himself, instructed the two Marys to tell the disciples to meet him in Galilee, which is where the first meeting took place according to that gospel (here, Matthew 28:10–17):

> 10 *Then said Jesus unto them, Be not afraid: go tell my brethren that they go into Galilee, and there shall they see me. . . .*
> 16 *Then the eleven disciples went to Galilee, to the mountain where Jesus had told them to go.*
> 17 *And when they saw him, they worshipped him: but some doubted.*

Some early manuscripts of Mark's gospel end at Mark 16:8, just after a young man—rather than Jesus, as in Luke's gospel—in the tomb tells the two Marys, and Salome as well, to instruct the disciples to go to Galilee where Jesus would meet them. Other manuscripts of Mark's gospel add verses 9 through 20, which describe Jesus as meeting the disciples but make no mention of Galilee.

Matthew 28:11–15 adds the following story about events surrounding the resurrection:

> 11 *Now . . . behold, some of the watch came into the city, and shewed unto the chief priests all the things that were done.*
> 12 *And when they were assembled with the elders, and had taken counsel, they gave large money unto the soldiers,*
> 13 *Saying, Say ye, His disciples came by night, and stole him away while we slept. . . .*
> 15 *So they took the money, and did as they were taught: and this saying is commonly reported among the Jews until this day.*

If the guards had actually seen a resurrected Jesus as the passage implies, they certainly would have been awed by his appearance and would have realized that Jesus was indeed somehow favored of God. It is not likely that they would have thought so lightly of it that they would have risked the condemnation of God by accepting money and lying about what had happened. The same could be said of the chief priests. Upon learning that Jesus had been resurrected, they certainly would have questioned their own standing before their God and would perhaps have even fallen to their knees and begged his forgiveness for whatever part they had played in persecuting Jesus. The story therefore smacks of being an effort by the author of the gospel to explain away the "commonly reported" story that the followers of Jesus had stolen his body from the tomb.

Contrary to the expectations of the early Christians, of course, the Second Coming and the Judgment Day did not occur within the generation that had lived during the ministry of Jesus. But that did not deter the believers from believing that the Second Coming was imminent. As the years and centuries wore on, the Last Days were continually pushed into the contemporary near-term future, and always with the declaration that the time was short and that the Second Coming was imminent. So it went on, scarcely with cessation, during the past two millennia. Even today, the true believers are proclaiming in the electronic media and in best-selling books that we are in the Last Days and that the Apocalypse is fast approaching. That proclamation has remained the same for millennia and doubtlessly will remain the same for as long as there are Bible-believing Christians.

The purpose of this and the previous appendix has been to provide a limited but functional perspective on the people who wrote the Bible and on some of the problems that can be found in the books of the Bible. This perspective should help to clear away some of the widely circulated misinformation and preconceived notions that could otherwise prevent one from understanding something of how the Bible was formed and from arriving at a rational answer to the question of whether it is a revelation from God.

What this perspective shows is that the Bible was a product *of* the times and *for* the times in which it was written. We have also seen that many of the "prophecies" found in the Bible were failures as prophecies. Some of them were not fulfilled as they were supposed to have been and were thus failed prophecies. Others were written after the fact and were therefore interpretations of history rather than being true prophecies. And still others were made, through embellished narratives, to appear to have been fulfilled. We have also seen that there is considerable evidence that much of the material in the Bible is the result of fabrication by its authors, and that the Bible contains material that is mutually contradictory and historically inconsistent.

These problems—and we have barely touched on all of the problems that can be found in the Bible—are extensive enough that any rational person examining them objectively would have to seriously doubt that the Bible can be considered the infallible word of God.

When those problems are added to what we have learned about the cosmological view that permeates the Bible, the preponderance of evidence should make it clear that the Bible is a collection of myths, folklore, embellished histories, and superstitious speculations and assumptions and cannot be considered the word of God.

Appendix C
The Book of Enoch
and the Biblical Cosmos

As we previously noted, the books that can be found in the Bible were not the only "scriptural" books that were in existence when the biblical canon was being formed. Several other books and writings were influential in the development of religious thought among the ancient Hebrews and the early Christians. Some of these other books and writings were even referred to in the Bible. Jesus himself provides an example in John 7:37–38:

> 37 *In the last day, that great day of the feast, Jesus stood and cried, saying, If any man thirst, let him come unto me, and drink.*
> 38 *He that believeth on me, as the scripture hath said, out of his belly shall flow rivers of living water.*

Here Jesus is quoting a passage from scripture: "out of his belly shall flow rivers of living water." The problem is that passage is not found in the Bible, which means that Jesus must have been quoting from a book that was not included in the biblical canon. Surely, if Jesus himself considered that book to be scripture, it should have been included in the canon. If the early Christian leaders who put together the books of the New Testament canon were derelict in not including that book, even adding it to the Old Testament if that would have been appropriate, how can one say that they were inspired when they decided which books to include in the canon? The same could be said of other books that are quoted or referenced in the Bible as scripture, but which were also not included in the canon.

One such book is the Book of Enoch, which, like the Book of Daniel, is a pseudepigraphic book—that is, it was ascribed as the work of someone who was not its actual author. It was written within the first two centuries B.C.E. and ascribed to Enoch, one of the patriarchs mentioned in the Book of Genesis.

The Book of Enoch, which is also called 1 Enoch, is an apocalyptic book, and it appears to have been written by a Jew who was quite devout in his religious beliefs. That the writer of the Book of Enoch was so devoutly religious clearly demonstrates that the cosmological view he expressed in the book must have been fully compatible with his understanding of the cosmos that is found in the Hebrew Scriptures. It is certainly not likely that he would have incorporated into his writings a cosmological view that was contrary to the understanding of the biblical cosmos that he and his contemporaries had. This is important because the Book of Enoch is very clear in its biblically based cosmological descriptions, and those descriptions show that the view of the biblical cosmos that we discerned in Part Three of this book is valid.

Although the Book of Enoch did not become a part of the biblical canon, many religious figures of the time regarded it highly. It influenced the authors of many Jewish writings following the Old Testament era, including Jubilees, the Testaments of the Twelve Patriarchs, the Assumption of Moses, 2 Baruch, and 4 Ezra. Parts of several copies of the Book of Enoch were even found among the Dead Sea scrolls. It is also apparent that the authors of several New Testament books used concepts that are found in the Book of Enoch, and it also influenced the authors of many non-canonical early Christian books, such as the Epistle of Barnabas, the Apocalypse of Peter, and several apologetic works.

Soon after the early Christian church began to become more widespread, the Jewish establishment began to look with disfavor upon the Book of Enoch, perhaps because it simply had too many points of similarity with the new rival religion. Several early Christian leaders and writers, on the other hand, viewed the Book of Enoch as scripture or were familiar with it. These included Justin Martyr, Irenaeus, Origen, and Clement of Alexandria. Tertullian, in particular, had a very high regard for it. By about the fourth century, however, Christian leaders also began to look with disfavor upon the Book of Enoch, probably because some of its theological and eschatological concepts had become obsolete as a result of the evolving nature of Christianity.

Nevertheless, the acceptance of the Book of Enoch by many of the earlier Jewish and Christian leaders and writers shows that those individuals found no problem with squaring the cosmology of the Book of Enoch with that of the Bible. Again, since the Book of Enoch is very clear in its cosmological descriptions, and since those descriptions are based on the cosmology of the Bible, that acceptance by these religious figures gives substantial testimony for the validity of the view of the biblical cosmos that we discerned in Part Three of this book.

But the importance of the Book of Enoch in relation to the Bible does not end with its cosmology. A study of the Book of Enoch sheds light on the development of New Testament beliefs about the Messiah, the nature of the Messianic kingdom, the tribulation, the Day of Judgment, Sheol, the resurrection of the dead, and demonology. In fact, one could make a good case in asserting that many of those aspects of the New Testament had their beginnings in the Book of Enoch.

However, since the thrust of this appendix is to show the relationship of the cosmology of the Book of Enoch to that of the Bible, we will begin our analysis with those parts of the Book of Enoch that bear on that relationship. At the end of this appendix is a supplement that investigates some of the other aspects of the Book of Enoch that have a relation to the Bible.

As we noted in Part Three, one of the early religious figures who considered the Book of Enoch to be authoritative was the author of the Epistle of Jude in the New Testament; from his use of the Book of Enoch, he obviously was much influenced by it. The first example of that influence appears in verses 6 and 7 of the Epistle of Jude, for the eschatology of the Book of Enoch is reflected in that passage:

> 6 *And the angels which kept not their first estate, but left their own habitation, he hath reserved in everlasting chains under darkness unto the judgment of the great day.*
> 7 *Even as Sodom and Gomorrha, and the cities about them in like manner, giving themselves over to fornication, and going after strange flesh, are set forth for an example, suffering the vengeance of eternal fire.*

The passage alludes, via the Book of Enoch, to Genesis 6:2: "*That the sons of God saw the daughters of men that they were fair; and they took them wives of all which they chose.*" According to the Book of Enoch, the "sons of God" were angels who left heaven—"their own habitation"—and coupled with mortal women. The passage in Genesis does not indicate that the sons of god were punished for their actions, whereas Jude does so indicate, stating that, like the inhabitants of Sodom and Gomorrha, the wayward angels would suffer the vengeance of eternal fire on Judgment Day. And that is where the Book of Enoch comes in, for it provided Jude with the basis for that idea, as the following excerpts from the Book of Enoch show[1]:

> 6:2 *And the angels, the children of heaven, saw and lusted after [the daughters of men], and said to one another: 'Come, let us choose us wives from among the children of men . . .*
> 7:1 *And they began to go in unto them and to defile themselves with them. . . .*
> 10:4 *And again the Lord said to Raphael: 'Bind Azazel [one of the leaders of the angels] hand and foot, and cast him into the darkness. . . .*
> 10: 6 *And on the day of the great judgement he shall be cast into the fire.'*
> 54:4–6 *And I asked the angel of peace who went with me, saying: 'For whom are these chains being prepared?' And he said unto me: 'These are being prepared for the hosts of Azazel, so that they may take them and cast them into the abyss of complete condemnation. . . . And Michael, and Gabriel, and Raphael, and Phanuel shall take hold of them on that great day, and cast them on that day into the burning furnace, that the Lord of Spirits may take vengeance on them for their unrighteousness in becoming subject to Satan and leading astray those who dwell on the earth.'*

Note specifically that both Jude and Enoch mention many of the same or similar things: the angels who left heaven to defile themselves with the daughters of men (the strange flesh of the Jude passage), their being cast into darkness as punishment, their being bound with chains, their condemnation on the Judgment Day, and the "vengeance" that the condemned angels and their leader would suffer by being cast into the fire.

However, the principal and most significant part of the Epistle of Jude that relates to the Book of Enoch is found in verses 12 through 15. Jude introduced those verses by decrying how certain corrupters of the church were "spots" in the brethren's feasts of charity. He then compared the corrupters to certain aspects of nature, and it is at that point where the Enochian material begins:

> 12 *. . . clouds they are without water, carried about of winds; trees whose fruit withereth, without fruit, twice dead, plucked up by the roots;*
> 13 *Raging waves of the sea, foaming out their own shame; wandering stars, to whom is reserved the blackness of darkness forever.*
> 14 *And Enoch also, the seventh from Adam, prophesied of these, saying, Behold, the Lord cometh with ten thousand of his saints,*

> 15 *To execute judgement upon all, and to convince all that are ungodly among them of all their ungodly deeds which they have ungodly committed, and of all their hard speeches which ungodly sinners have spoken against him.*

Note that, in verse 14, Jude specifically attributed his words to Enoch. There can be no doubt that the pseudepigraphic Book of Enoch was Jude's source for this material, because the remainder of verse 14 and the whole of verse 15 is virtually a direct quote from that book. The words that Jude quoted can be found in Enoch 1:9:

> *And Behold! He cometh with ten thousand of His holy ones to execute judgement upon all, And to destroy all the ungodly: And to convict all flesh of all the works of their ungodliness which they have ungodly committed, And of all the hard things which ungodly sinners have spoken against Him.*

That passage from the Book of Enoch and the corresponding passage from the Epistle of Jude are substantially the same. The minor differences between them can be attributed either to Jude's having made a slightly loose quotation from the Book of Enoch or to the vagaries of the translation processes that brought both renditions down to us.

Note that in verse 14 Jude stated that Enoch was "the seventh from Adam." That expression of Enoch's lineage does not appear in the Bible, but it does appear in Enoch 60:8, where "Noah" describes his grandfather, Enoch, as being "the seventh from Adam": "... *on the east of the garden where my grandfather was taken up, the seventh from Adam*...." The fact that the author of the Epistle of Jude used the same expression in conjunction with the quotation that he attributed to Enoch is additional evidence that he was familiar with, and was in fact using, the pseudepigraphic Book of Enoch as he composed his epistle.

Note also that Jude stated that Enoch "prophesied of these." According to the first chapter of the Book of Enoch, some angels showed Enoch a vision. In relating the account of the vision, "Enoch" states that "...*from them I heard everything, and from them I understood as I saw, but not for this generation, but for a remote one which is for to come.*" The Book of Enoch, then, is presented as a prophecy, just as Jude indicated in his epistle.

That leaves us with the passage from verses 12 and 13 of the Epistle of Jude, in which Jude compared the corrupters of the church to certain aspects of nature. That passage, again, is as follows:

> 12 ... *clouds they are without water, carried about of winds; trees whose fruit withereth, without fruit, twice dead, plucked up by the roots;*

> 13 *Raging waves of the sea, foaming out their own shame; wandering stars, to whom is reserved the blackness of darkness forever.*

Because Jude had stated in verse 14 that Enoch "prophesied of these," we should expect to find something in the Book of Enoch that relates to the concepts in that passage. And, in fact, we find that Enoch 80:2–8 provided Jude with the sense of most of what he said in the passage. The following extract from Enoch 80:2–8 contains the relevant material:

> 2 ....*And the rain shall be kept back and the heaven shall withhold (it).*
> 3 *And in those times the fruits of the earth shall be backward, and shall not grow in their time, and the fruits of the trees shall be withheld in their time.*
> 4–5 ....
> 6 ... *And many chiefs of the stars shall transgress the order (prescribed). And these shall alter their orbits and tasks and not appear at the seasons prescribed to them.*
> 7 ....
> 8 *And evil shall be multiplied upon them, and punishment shall come upon them.* ...

Note that three of the four aspects of nature that Jude wrote about have their parallels in this passage from the Book of Enoch, and those parallels are even in the same order as the aspects of nature appearing in the Epistle of Jude. It appears, then, that this passage gave Jude his ideas about the clouds without water, the trees without fruits, and the wandering stars that would be punished for their transgressions.[2]

In addition, the first part of Enoch 80:2 states: "*And in the days of the sinners the years shall be shortened, ... and all things on the earth shall alter, and shall not appear in their time.*" Therefore, like Jude, the author of this passage from the Book of Enoch was also speaking about sinners— sinners of a future time that Jude apparently took as meaning his own.

But more importantly, in addition to the point that the Book of Enoch as a whole was presented as a prophecy, the introduction to this particular passage specifically makes the passage a prophecy. That confirms what Jude stated when he attributed the previous verses of his epistle to their source: "*And Enoch also, the seventh from Adam, prophesied of these.*"

Verses four and five, which are represented by points of ellipses in the above excerpt from the Book of Enoch, have no parallel in the passage from the Epistle of Jude. According to those verses, the moon would alter "her order" and shine more brightly than it should, and the sun would shine in the evening. Jude obviously could not say that these events were observable. As far as he would have been concerned, they would have had to await the Judgment Day for their denouement.

In place of those two verses, Jude made a reference to the "raging waves of the sea" in verse 13. He may have picked up the idea for that aspect of nature from the Book of Enoch also, for in Enoch 69:18 we find the following: *"And through that oath the sea was created, and [to limit it] He set for it the sand against the time of its anger."* The passage from Enoch, which reflects Jeremiah 5:22, refers to the anger of the sea, while the passage from Jude refers to the "raging waves of the sea, foaming out their own shame." When the "angry" waves of the sea break on the sands of the shore, they produce a foam. Thus, the passage from Enoch may have given Jude the idea for the remaining aspect of nature that he included in his epistle.

In any case, it is apparent that Enoch 80:2–8 provided Jude with most of his ideas about the aspects of nature that appear in verses 12 and 13 of his epistle. He simply put the Enochian material into his own words and used it to provide the comparisons relating to the corrupters of the church.

However, it is our investigation into biblical cosmology that bears on this passage from Enoch 80:2–8, for it provided Jude with his image of the wandering stars. Verse 6 of the Enoch passage states: *". . . And many chiefs of the stars shall transgress the order (prescribed). And these shall alter their orbits and tasks and not appear at the seasons prescribed to them."*[3]

According to the Book of Enoch, these "chiefs of the stars" were to be punished for their transgression in "wandering" from their prescribed heavenly circuits. These stars that "altered their orbits" correspond with the wandering stars of Jude verse 13: *". . . wandering stars, to whom is reserved the blackness of darkness forever."* As we shall see, the punishment to be meted out to these "wandering stars" is described elsewhere in the Book of Enoch, and it is in a "place of darkness," just as Jude stated in his epistle.

The evidence, then, is clear that Jude was using the pseudepigraphic Book of Enoch as the source for much of the content of his epistle. The evidence is also clear that Jude considered the Book of Enoch to be a genuine and inspired scripture from the hand of the biblical patriarch—as indicated by his statement that Enoch "prophesied of these."

This has formidable implications for any claims of divine inspiration for the Bible. Not only is the Book of Enoch pseudepigraphic, it was written more than three thousand years after the time when its purported author, Enoch, supposedly lived. Moreover, as we shall see, the cosmological descriptions that are found in the Book of Enoch are mythological in character and are, from a modern standpoint, absolute nonsense. Therefore, because Jude believed that the Book of Enoch was a genuine scripture written by the hand of Enoch, and because he accepted the cosmological descriptions given in the Book of Enoch, any assertion that he was writing under the inspiration of God is substantially undermined. Furthermore, the Bible is considered by its believers to be a unified whole. That being the case, if one book of the Bible is subject to question concerning its "inspiration," then the authority of the whole Bible is placed under question. Alternatively, if one believes that the whole Bible was written under the "inspiration" of God, then one must also believe that the Book of Enoch was likewise inspired, at the very least because of Jude's "inspired" acceptance of that book as scripture. That would mean that the Book of Enoch, with all of its cosmological absurdities, should have been incorporated into the Biblical canon.

With that, we can now start our investigation of the cosmology of the Book of Enoch and how it relates to the biblical cosmos.

The Book of Enoch is divided into five sections. Cosmological elements can be found throughout the book, but particularly in the third section, which is essentially a discourse in which the author views and expounds upon—by "explaining" its workings—the cosmos of the Hebrew Scriptures. The cosmos that is described in the Book of Enoch, and which is the subject of the author's discourse, in fact reflects much of what we have discerned about the biblical cosmos.

A large part of the Book of Enoch consists of a dream vision in which Enoch is escorted by angels through the cosmos and has the secrets of heaven explained to him. As we have already noted, one of Enoch's cosmological observations is referred to in the Epistle of Jude: *"wandering stars, to whom is reserved the blackness of darkness forever."* We have seen that Jude derived the idea for these wandering stars from Enoch 80:6: *"And many chiefs of the stars shall transgress the order (prescribed). And these shall alter their orbits and tasks and not appear at the seasons prescribed to them. . ."* Jude found the future punishment for these wandering stars revealed in chapters 17 and 18 of the Book of Enoch, which describe a "place of darkness" to which Enoch is brought:

> 17:1 *And they took me to a place in which those who were there were like flaming fire, and, when they wished, they appeared as men.*
> 17:2 *And they brought me to the place of darkness. . . .*
> 18:13 *I saw there seven stars like great burning mountains, and to me, when I inquired regarding them,*

> 18:14 *The angel said: "This place ... has become a prison for the stars and the host of heaven.*
> 18:15 *And the stars which roll over the fire are they which have transgressed the commandment of the Lord in the beginning of their rising, because they did not come forth at their appointed times.*
> 18:16 *And He was wroth with them, and bound them till the time when their guilt should be consummated."....*

Those who were "like flaming fire" and, when they wished, "appeared as men" were of the host of heaven—the angels who were associated with, and even were, the stars of heaven. In Enoch 80:6, the seven stars who "transgressed the commandment of the Lord" and were imprisoned in the "place of darkness" provided the basis for Jude's "wandering stars, to whom is reserved the blackness of darkness forever."

In the extract of Enoch 80:2–8 that we looked at earlier, verse 7 was represented by points of ellipses. That verse presents some further information about the "chiefs of the stars": "*... And the thoughts of those on the earth shall err concerning them, ... Yea, they shall err and take them to be gods.*"

Because some ancient societies looked upon the planets (to the naked eye, the points of light that move, or wander, among the fixed stars) as being gods or as representing gods, that verse appears to be an indication that at least some of the "chiefs of the stars" were supposed to be the "wandering stars" that we know to be the planets of our solar system.

According to the Book of Enoch, then, the "wandering stars," which we know to be planets, are the same type of entity as are the fixed stars; but because these particular stars "transgressed the commandment of the Lord" and altered their prescribed circuits across the vault of heaven, they were to be punished.

The fact that Jude wrote about wandering stars that were to be punished for failing to keep their appointed heavenly circuits puts his view of biblical cosmology squarely in the realm of mythology. And the fact that Jude got his "inspiration" from, and accepted the authority of, the Book of Enoch concerning the punishment of the wandering stars shows that he had not the slightest inkling of the true nature of the cosmos. Indeed, as already indicated, and as we shall see, the cosmology that is expounded upon in the Book of Enoch is nothing but nonsense from the modern standpoint, though it parallels the view of the cosmos that we have found in the Bible. That can be easily seen if we follow Enoch in his tour of the cosmos.

In our analysis of the relevant passages in the Bible, we found that, in the biblical cosmos, the earth is a flat disk. We also found that the base of the solid firmament, or vault, of heaven rests on the rim of—i.e., the "ends" of—the earth. Significantly, that view of the biblical cosmos is explicitly stated in the Book of Enoch. In fact, Enoch is described as seeing the places where the firmament of heaven rests on the ends of the earth. Here, for example, is the description as given in Enoch 33:1–2:

> *And from thence I went to the ends of the earth.... And ... I saw the ends of the earth whereon the heaven rests....*

The importance of that passage in relation to what we have discerned about the fundamental nature of the biblical cosmos is clear. The author of these passages in the Book of Enoch conformed the cosmology of his book to reflect the cosmology of the Bible as he and his fellow Hebrews of the time understood it to be—and they certainly would have known what constituted the cosmology of the Bible since they had a cultural connection with those who brought forth the Hebrew Scriptures. The Book of Enoch therefore confirms the validity of what we have gleaned from the Bible about the nature of its cosmological viewpoint—specifically, the Book of Enoch shows quite clearly the relationship of the solid vault of heaven to the disk of the earth in the biblical cosmos.

We also noted that the biblical cosmos contains several tiers or levels, consisting primarily of the waters above the firmament of heaven, the firmament of heaven itself, the disk of the earth beneath the firmament of heaven, and the waters of the deep, or the abyss, under the earth. That same view is also found in the Book of Enoch. An example is provided in Enoch 54:7–10, which provides a "prophecy" of the biblical Flood from the standpoint of Enoch's time:

> *And in those days shall punishment come from the Lord of Spirits, and he will open all the chambers of waters which are above the heavens, and of the fountains which are beneath the earth. And all the waters shall be joined with the waters: that which is above the heavens is the masculine, and the water which is beneath the earth is the feminine. And they shall destroy all who dwell on the earth and those who dwell under the ends of the heaven.*

That passage specifically mentions the waters above the heaven and the waters beneath the earth. In the last part of that passage, "the ends of the heaven" is a reference to the base of the vault of heaven, which, as we determined in the main part of this book, was taken to be a reference to the vault of heaven as a whole. Thus, the term "those who dwell under the ends of the heaven" is meant to be taken as those who dwell under the whole heaven all the way to its ends, and it reiterates "all who dwell on the earth" in the passage.

In Chapter 11, remember, we dealt at length in determining the meaning of Job 26:7:

> *He stretcheth out the north over the empty place, and hangeth the earth upon nothing.*

We concluded that the passage was actually a reiteration of the creation of heaven and earth, and that the statement that the earth "hangeth upon nothing" actually means that the earth hangs over the abyss of the deep. A passage in Enoch 69:16–17 makes a similar statement:

> *... And the heaven was suspended before the world was created,*
> *And for ever.*
> *And through it the earth was founded upon the water. ...*

There, upon its creation, the heaven was "suspended" before the earth was created. The firmament of heaven would have needed to be "suspended" at that time because there were no "ends of the earth" for it to rest upon. Subsequently, the earth was created "through" the heaven and was "founded upon the water" of the abyss of the deep under the earth. It was then that the firmament of heaven came to rest on "the ends of the earth."

A graphic description of a vision of the Day of Judgment in Enoch 83:3–8 also supports this view of the structure of the biblical cosmos:

> *... I saw in a vision how the heaven collapsed and was borne off and fell to the earth. And when it fell to the earth I saw how the earth was swallowed up in a great abyss. ... And ... I lifted up (my voice) to cry aloud, and said: 'The earth is destroyed.' And my grandfather Mahalalel waked me as I lay near him, and said unto me:' Why dost thou cry so, my son, and why dost thou make such lamentation?' And I recounted to him the whole vision which I had seen, and he said unto me: 'A terrible thing hast thou seen, my son, and of grave moment is thy dream- vision as to the secrets of all the sin of the earth: it must sink into the abyss and be destroyed with a great destruction.[']*

Here, at the End Times, the solid vault of heaven would collapse upon the earth and the earth would sink into the abyss beneath it. This description fits in quite well with what we have learned in Part Three about the multi-tiered nature of the biblical cosmos. Moreover, it also confirms our assessment in Chapter 10 concerning the meaning of the biblical verses that indicate that the earth does not move. Those verses, remember, actually indicate that, in the biblical cosmos, the earth is normally fixed in place and that the only movement of the earth that can occur will be during the Day of the Lord when God "removes the earth from her place." Specifically, we determined that the earth would fall into the abyss under the earth. In the Old Testament the relevant passage is found in Isaiah 24:19–20 (here from the JPS 1917):

> 19 *The earth is broken, broken down, the earth is crumbled in pieces, the earth trembleth and tottereth;*
> 20 *The earth reeleth to and fro like a drunken man, and swayeth to and fro as a lodge; and the transgression thereof is heavy upon it, and it shall fall, and not rise again.*

Enoch 34–36 provides a description that further shows the contrast between the biblical view of the cosmos and the modern view, for it describes portals in heaven from which the winds, dew, and rain emerge:

> 34 *And from thence I went towards the north to the ends of the earth, and there . . . I saw three portals of heaven open in the heaven: through each of them proceed north winds: when they blow there is cold, hail, frost, snow, dew, and rain. And out of one portal they blow for good: but when they blow through the other two portals, it is with violence and affliction on the earth, and they blow with violence.*
> 35 *And from thence I went towards the west to the ends of the earth, and saw there three portals of the heaven open such as I had seen in the east, the same number of portals, and the same number of outlets.*
> 36 *And from thence I went to the south to the ends of the earth, and saw there three open portals of the heaven: and thence there came dew, rain, and wind. And from thence I went to the east to the ends of the heaven, and saw here the three eastern portals of heaven open and small portals above them. Through each of these small portals pass the stars of heaven and run their course to the west on the path which is shown to them. . . .*

The portals of the winds are further described in Enoch 76:

> *And at the ends of the earth I saw twelve portals open to all the quarters (of the heaven), from which the winds go forth and blow over the earth. Three of them are open on the face (i.e., the east) of the heavens, and three in the west, and three on the right (i.e., the south) of the heaven, and three on the left (i.e, the north).*

In an attempt to clarify the passage, the translator parenthetically included the words "of the heaven" following "all the quarters." However, the words translated as "to all the quarters" literally mean "to all the winds." In biblical times, the four cardinal directions were referred to as the four winds, as we have already noted. The "quarters" in this translation would therefore refer to the four cardinal directions on the earth.[4]

Enoch 33, remember, described the "ends of the earth whereon the heaven rests," and Enoch 36, above, mentions "the ends of the heaven." In Chapter 10 we noted several passages that use the term "the ends of heaven" to indicate the limits of the earth. We also noted that Jesus, in Matthew 24:31, said that the elect would be gathered from "the four winds, from one end of heaven to the other." We noted that such statements would make no sense from a modern perspective of the cosmos, but that it would make perfect sense from the perspective of the biblical cosmos in which the ends, or the rim, of the solid firmament of heaven rests on the ends, or the rim, of the flat, disk-shaped earth. In the Book of Enoch, then, which was written within two centuries before the time of Jesus, we have confirming descriptions of that perspective. It would therefore appear that the biblical Jesus believed the cosmological view that is found in the Bible—as we have discerned it—*and* in the Book of Enoch.

In addition, the passage from Enoch 76, in describing the portals of heaven at the four cardinal directions from which the winds blow, provides a conceptual basis for Revelation 7:1, which we took a look at in Chapter 12:

> *And . . . I saw four angels standing on the four corners of the earth, holding up the four winds of the earth, that the wind should not blow on the earth, nor on the sea, nor on any tree.*

Enoch 76 thus provides a clarification for this passage. The four angels would have been standing by the "portals" in the solid firmament of heaven at the north, east, west, and south ends, or "corners," of the earth so they could "hold up" the four winds coming through the portals.

It should also be clear that the "portals" of heaven that are mentioned in the Book of Enoch are equivalent to the "doors" and "windows" of heaven that, as we found in Part Three of this book, are mentioned in the Bible.

As we saw above, the Book of Enoch describes portals in the vault of heaven from which the winds emerge. In Enoch 72–75, there is likewise a highly detailed and very tedious description of the portals from which the sun, moon, and stars emerge at one end of heaven and other portals into which they enter at the opposite end of heaven after making their circuits across the sky. According to Enoch's exposition, there are six portals of heaven in the east and six in the west. The sun uses the first pair of east/west portals for a month, and then the next pair for the next month, and so on for the first six months of the year, then reverses the order of the portals for the next six months. As the author of the Book of Enoch explains, this accounts for the changing position of the sun's rising and setting on the horizon over the space of a year. The following is only a part of Enoch's exposition in Enoch 72:

> ... And this is the first law of the luminaries: the luminary the Sun has its rising in the eastern portals of the heaven, and its setting in the western portals of the heaven. And I saw six portals in which the sun rises, and six portals in which the sun sets: and the moon rises and sets in these portals, and the leaders of the stars and those whom they lead: six in the east and six in the west, and all following each other in accurately corresponding order: also many windows to the right and left of these portals. And first there goes forth the great luminary, named the Sun, and his circumference is like the circumference of the heaven, and he is quite filled with illuminating and heating fire. The chariot on which he ascends, the wind drives, and the sun goes down from the heaven and returns through the north in order to reach the east, and is so guided that he comes to the appropriate (lit. 'that') portal and shines in the face of the heaven. In this way he rises in the first month in the great portal, which is the fourth [those six portals in the east]. And in that fourth portal from which the sun rises in the first month are twelve window-openings, from which proceed a flame when they are opened in their season. When the sun rises in the heaven, he comes forth through that fourth portal thirty mornings in succession, and sets accurately in the fourth portal in the west of the heaven. And during this period the day becomes daily longer and the night nightly shorter to the thirtieth morning. On that day the day is longer than the night by a ninth part, and the day amounts exactly to ten parts and the night to eight parts. And the sun rises from that fourth portal, and sets in the fourth and returns to the fifth portal of the east thirty mornings, and rises from it and sets in the fifth portal.

The author then repeats the whole process in explaining the movement of the moon in a like manner. All of this, of course, is nonsensical from the modern cosmological standpoint. However, in expounding upon the way the biblical cosmos works, the author of the Book of Enoch confirms the basic structure of that cosmos as we have discerned it from numerous biblical passages. In the excerpt from Enoch 72, above, for example, note that Enoch has this to say about the sun:

> ... the luminary the Sun has its rising in the eastern portals of the heaven, and its setting in the western portals of the heaven. ...

The eastern portals that the sun comes out of in the morning, and the western portals that it enters in the evening, would, of course be at the "ends of heaven" by the horizon. A related discourse is found in Enoch 41:5–7:

> *And I saw the chambers of the sun and moon, whence they proceed and wither they come again.... And first the sun goes forth and traverses his path according to the commandment of the Lord of Spirits.*

The chamber of the sun would likely have been behind the portals of heaven from which the sun emerges. Those passages from Enoch, above, bring to mind the following description in Psalm 19:1–6 that we looked at in Chapter 12:

> *The heavens declare the glory of God; and the firmament sheweth his handywork.... In them he set a tabernacle for the sun, which is as a bridegroom coming out of his chamber, and rejoiceth as a strong man to run a race. His going forth is from the end of the heaven, and his circuit unto the ends of it: and there is nothing hid from the heat thereof.*

According to the lengthy passage from Enoch 72 that we previously looked at, after the sun enters one of western portals of heaven in the evening, it goes around via the north (apparently on the outside of the vault of heaven) so it can emerge again from one of the eastern portals the next morning. Here is the relevant excerpt:

> *The chariot on which he ascends, the wind drives, and the sun goes down from the heaven and returns through the north in order to reach the east, and is so guided that he comes to the appropriate portal....*

What is significant about that passage is that it clarifies Ecclesiastes 1:5, which we looked at in Chapter 12:

> *The sun also ariseth, and the sun goeth down, and hasteth to his place where he arose.*

Enoch 72 therefore explains how the sun "hasteth to his place where he arose" after making its circuit across the vault of heaven. Again, from the standpoint of present-day knowledge this cosmological view is nonsense and is nothing but pure mythology. But again, this view is what is expressed in the Bible.

In Chapter 12 we noted that Job 9:7 states that "*[God] ... commandeth the sun, and it riseth not; and sealeth up the stars,*" and we also noted that the Hebrew word translated as "sealeth" in that passage is *chatham*, which means "to seal up," or "to lock up." But what does that verse mean when it says that God seals up or locks up the stars? Enoch 33:1–3 provides the context for the answer:

> *And from thence I went to the ends of the earth. . . . And . . . I saw the ends of the earth whereon the heaven rests, and the portals of the heaven open. And I saw how the stars of heaven come forth, and I counted the portals out of which they proceed. . . .*

Note that, as well as the "ends of the earth," the passage also mentions portals in the vault of heaven from which the stars emerge. A similar description is found in Enoch 36:

> *And from thence I went to the east to the ends of the heaven, and saw here the three eastern portals of heaven open and small portals above them. Through each of these small portals pass the stars of heaven and run their course to the west on the path which is shown to them. . . .*

With these passages in the Book of Enoch, we have the answer to the question of what Job 9:7 means when it says that God "seals up" the stars: He would "seal up" the portals from which the stars emerge, thus preventing them from performing their nightly circuits. Job 9:7 certainly makes more sense when we understand it in the context of the biblical view of the cosmos than it would if we try to understand it in the context of the modern view of the cosmos.

In the above passage we quoted from Enoch 72, the sun is described as ascending in a chariot:

> *And first there goes forth the great luminary, named the Sun, and his circumference is like the circumference of the heaven, and he is quite filled with illuminating and heating fire. The chariot on which he ascends, the wind drives. . . .*

That idea is also found in Enoch 75:3–4:

> *. . . the angel Uriel showed to me . . . the sun, moon, and stars and all the ministering creatures which make their revolution in all the chariots of the heaven. In the like manner twelve doors Uriel showed me, open in the circumference of the sun's chariot in the heaven, through which the rays of the sun break forth.*

The "ministering creatures" are the angels or entities in control of the chariots of the sun, moon, and stars. Significantly, these descriptions from the Book of Enoch have a bearing on Revelation 19:17:

> *And I saw an angel standing in the sun; and he cried with a loud voice, saying to all the fowls that fly in the midst of heaven, Come and gather yourselves together unto the supper of the great God.*

Therefore, in the Book of Enoch we find the basis for the idea of the angel standing in the sun that is mentioned in the Book of Revelation. In fact, there are many other similarities between the Book of Revelation and the Book of Enoch. For that matter, there are enough similarities that it is almost certain the author of the Book of Revelation, like the author of the Epistle of Jude, was quite familiar with the Book of Enoch and used it as a source of ideas.[5]

Note that the motive force for the sun's chariot is the wind: "*The chariot on which he ascends, the wind drives. . . .*" In the Book of Enoch, the wind is the primary force that maintains the biblical cosmos. For example, like the sun, the moon is also driven by the wind. Here is the relevant excerpt from Enoch 27:

> *And after this law I saw another law dealing with the smaller luminary, which is named the Moon. And her circumference is like the circumference of the heaven, and her chariot in which she rides is driven by the wind. . . .*

And here is an excerpt from Enoch 18 in which the winds are described as being the "pillars" that bear up the earth and the heaven:

> *I saw the treasuries of all the winds: I saw how He had furnished with them the whole creation and the firm foundations of the earth. And I saw the corner-stone of the earth: I saw the four winds which bear [the earth and] the firmament of the heaven. And I saw how the winds stretch out the vaults of heaven, and have their station between heaven and earth: these are the pillars of the heaven. I saw the winds of heaven which turn and bring the circumference of the sun and all the stars to their setting.*

The author apparently reasoned that, since the winds themselves need no support, they could act as the supporting pillars holding up the heaven.

In Chapter 11, we noted that it would have been reasonable to conclude that the ancient Hebrews would have located the throne of God at the point in the northern sky around which the stars circle during the course of a night. Though there is no passage in the Bible that explicitly describes that arrangement (probably because it would have been taboo to pinpoint the location of God's throne), it is suggested in Enoch 71:3–8:

> 3 *And the angel Michael seized me by my right hand....*
> 4 *And he showed me all the secrets of the ends of the heaven,*
> *And all the chambers of all the stars, and all the luminaries,*
> *Whence they proceed before the face of the holy ones.*
> 5 *And he translated my spirit into the heaven of heavens,*
> *And I saw there as it were a structure built of crystals....*
> 8 *And I saw angels who could not be counted,*
> *A thousand thousands, and ten thousand times ten thousand,*
> *Encircling that house.*

Since Enoch equated the angels with the stars of heaven, the implication of the above passage is that thousands of angels, in their guise as stars, are circling the house of God, in which, of course, God's throne would be located. Since the stars move in circles around a point in the northern night sky, the conclusion one would draw from the above passage from the Book of Enoch is that God's house is located at that same point.

As well as mentioning the chambers of the sun, moon, and stars, the Book of Enoch also mentions chambers in which the winds, rain, and dew are stored. The following passage is from Enoch 41:4:

> *And there I saw closed chambers out of which the winds are divided, the chamber of the hail and winds, the chamber of the mist, and of the clouds....*

In the Bible, Psalm 104:13 also speaks of the heavenly chambers:

> *He watereth the hills from his chambers: the earth is satisfied with the fruit of thy works.*

And Job 38:22 mentions places in which the snow and hail are stored (the Hebrew word translated as "treasures" here means "storehouse" or "treasure house"):

> *Hast thou entered into the treasures of the snow? or hast thou seen the treasures of the hail?*

Job 9:9 likewise speaks of the heavenly chambers:

> *[God] maketh Arcturus, Orion, and Pleiades, and the chambers of the south.*

The Bible thus supports the statements in the Book of Enoch concerning the chambers of heaven.

We have barely touched on the descriptions of the cosmos that are provided in the Book of Enoch—descriptions that have many parallels in the Bible. These parallel descriptions that are found in the Bible and in the Book of Enoch thus show and verify the common features of the cosmology of the ancient Hebrews that we discerned in Part Three of this book.

One can therefore conclude that, because the writers of the Bible received no guidance from God concerning an important aspect of the natural order of things—i.e., the nature and structure of the cosmos—there is no reason for believing that they received any guidance from God concerning the supernatural order of things. Consequently, one must therefore conclude that the Bible is not the word of God, and that belief in the Bible is a delusion.

Supplement to Appendix C

The Book of Enoch is an apocalyptic book, and, as such, it appears that it may have provided the basis for many of the beliefs about the Messiah, the kingdom of God, the expected tribulation, and the End Times that were prevalent in the early Christian community. The Book of Enoch, in fact, begins with the following (Enoch 1:1–2):

> *The words of the blessing of Enoch, wherewith he blessed the elect and righteous, who will be living in the day of tribulation, when all the wicked and godless are to be removed.*

The author continues by stating that the prophecies of the book are for a remote future time—that is, remote from the standpoint of Enoch, the purported author. As verses 2 and 3 continue, they state that Enoch

> *... saw the vision of the Holy One in the heavens, which the angels showed me, and from them I heard everything, and from them I understood as I saw, but not for this generation, but for a remote one which is for to come.*

The remote generation that the actual author of the passage is referring to would, of course, have been one that had a relevance to his own.

We have already seen that the author of the epistle of Jude in the New Testament used the Book of Enoch as a source for much of what is found in his epistle. It appears that several other New Testament writers also used material that is found in the Book of Enoch, though it is not clear whether they did so directly by perusing the book, or indirectly by picking up parts of it that were passed on orally and had become part of the cultural milieu of the time. For example, in 2 Peter 2:4 we find another reference to the same wayward angels that, as we noted, are mentioned in the Book of Enoch and that Jude had referred to:

> *God spared not the angels when they sinned, but cast them down to hell, and committed them to pits of darkness, to be reserved unto judgment.*

Again, that idea is not found in the Old Testament, so Peter, like Jude, apparently accepted what the Book of Enoch had to say on the matter.

The wayward angels described in the Book of Enoch may also help to clarify an obscurity in 1 Corinthians 11:10. Paul had stated that the heads of women should either be shorn or covered, and he then gave this justification (here, the NIV):

> *For this reason, and because of the angels, the woman ought to have a sign of authority on her head.*

The reference to the angels is out of place in the context of what Paul had been saying, but he added it to provide further justification for his position that women should cover their heads. The possible reason he did so was that he thought an uncovered woman would be a temptation for the angels, a concern that he could have derived from the Book of Enoch and its description of the wayward angels who had left heaven and sinned with women on the earth.

Parts of the Gospel of Luke also have some interesting parallels in the Book of Enoch. One of these is found in Luke 3:23–38, which provides the genealogy of Jesus. In that genealogy the number of generations totals seventy-seven, including that of Adam, the first, and that of Jesus, the last. Now we saw in Appendix B that Luke's genealogy of Jesus, along with Matthew's, was certainly a fabrication. One might then wonder why Luke specifically used seventy-seven as the number of generations in the genealogy of Jesus.

The answer apparently can be found in the Book of Enoch, for it indicates that there are to be seventy-seven generations from Adam until the day of judgement (and, as we saw in Appendix B, Jesus proclaimed that the day of judgement was to take place in the generation he was speaking to).

There are two passages in the Book of Enoch that give the number of generations until the day of judgement. One is Enoch 10:11–16:

> *And the Lord said unto Michael: 'Go, bind Semjaza and his associates . . . , bind them fast for seventy generations in the valleys of the earth, till the day of their judgement and of their consummation, till the judgement that is for ever and ever is consummated. In those days they shall be led off to the abyss of fire: and to the torment and the prison in which they shall be confined for ever. . . .*

That passage states that the Day of Judgment was to be seventy generations from the standpoint of the events of the passage. The question is, then, what was the starting point of the seventy generations in this passage? It must have been some time after the creation of Adam, for the human race must have had the time to propagate in order for the angels to take the daughters of men as wives for themselves (which, as we noted in the main part of this appendix, was the cause of the punishment that is being referred to in the passage). The answer is found in Enoch 91 through 93, which gives the generations from Adam until the judgement day in terms of weeks of generations—that is, each week of generations consists of seven generations. In Enoch 93:3 we find:

> *And Enoch began to recount from the books and said:*
> *'I was born the seventh in the first week,*
> *While judgement and righteousness still endured. . . .'*

There, Enoch states that he was born on the seventh day of the first week of generations; that is, he was born in the seventh generation from Adam. That of, course, reiterates Enoch 60:8–9, in which, as we saw, Noah stated that Enoch was the seventh from Adam: "... *on the east of the garden where the elect and righteous dwell, where my grandfather was taken up, the seventh from Adam....*"

In Enoch 91:12–17 and 93:4–14, Enoch recounts the events of each succeeding week of generations.[6] Then, during the seventh generation of the tenth week—that is, the seventy-seventh generation—comes the Judgement Day, as given in Enoch 91:15–17:

> 15 *And after this, in the tenth week in the seventh part,*
> *There shall be the great eternal judgement,*
> *In which He will execute vengeance amongst the angels.*
> 16 *And the first heaven shall depart and pass away,*
> *And a new heaven shall appear,*
> *And all the powers of the heavens shall give sevenfold light.*
> 17 *And after that there will be many weeks without number for ever,*
> *And all shall be in goodness and righteousness,*
> *And sin shall no more be mentioned for ever.*

This list of generations therefore indicates that the Judgement Day is to take place in the seventy-seventh generation from Adam. That in turn indicates that the seventy generations until the Judgement Day for the licentious angels as given in Enoch 10:11, above, is also to be taken as the seventieth generation from Enoch, who was the seventh from Adam, so we still have a total of seventy-seven generations.

As Enoch 91:15–17 indicates, on the Judgement Day the first heaven shall depart and pass away and a new heaven shall appear in which righteousness shall prevail forever. Though that passage does not mention the fate of the earth on the Judgement Day—the perspective in the passage, as in much of the Book of Enoch, relates to heaven—of necessity, the earth would also need to be included in the events of the Judgement Day. In fact, another passage, Enoch 45:5–6, includes the transformation of the earth along with the heaven on the Judgement Day.

> 4 *Then will I cause Mine Elect One to dwell among them.*
> *And I will transform the heaven and make it an eternal blessing and light*
> 5 *And I will transform the earth and make it a blessing:*

We can now see the logic that Luke used in creating his genealogy. According to the Book of Enoch, the Day of Judgement was to occur in the seventy-seventh generation from Adam. According to Jesus, the Day of Judgement was imminent and would take place during the generation that he was preaching to. Luke would therefore have concluded that Jesus had to be of the seventy-seventh generation from Adam, and he constructed his genealogy of Jesus to have the requisite number of generations. He was, of course, able to construct the first part of the genealogy from the genealogies provided in the Hebrew Scriptures. All he needed to do was to add a sufficient number of generations in the latter part of the genealogy to bring the total number up to seventy-seven. In Luke's mind, the fact that he had to make up the names for the ancestors of Jesus during those generations would have been irrelevant; the pious fabrication of the names in the genealogy reflected a greater "truth": the required number of generations from Adam to Jesus.

But that is not the only parallel between the Book of Enoch and Luke's gospel. In the Book of Enoch there is a lengthy passage extolling the Son of Man. That term, as we previously noted, is another name for the Messiah and was also a name that Jesus applied to himself. In the Book of Enoch, the Son of Man is also called "the Elect One" more than a dozen times. We already saw one example from Enoch 45:4, above:

> *Then will I cause Mine Elect One to dwell among them.*

In the original Greek of Luke 9:35, Jesus is called *ho eklelegmenos*, which means "the elect one" or "the chosen one." In contrast to the KJV, which uses the term "beloved son" in that passage, several other versions of the Bible—the ASV, ESV, NAB, NASV, NEB, NIV, NJB, NRS, and RSV—use the more accurate "the chosen" or "the chosen one." Here is the NASV rendition of the passage:

> *And a voice came out of the cloud, saying, "This is My Son, My Chosen One; listen to Him!"*

The same term is found in Luke 23:35:

> *And the people stood by, looking on. And even the rulers were sneering at Him, saying, "He saved others; let Him save Himself if this is the Christ of God, His Chosen One."*

In the Book of Enoch and the Gospel of Luke, we are dealing with translations from different languages, but being chosen is the same as being elected. It should be clear that the two writings are applying the same essential meaning to the original term. In fact, according to the Book of Enoch, the Son of Man was "chosen" by God (Enoch 46:2–3):

> *And I asked the angel . . . concerning that Son of Man, who he was, and whence he was, (and) why he went with the Head of*

Days? And he answered and said unto me:
This is the son of Man who hath righteousness...
Because the Lord of Spirits hath chosen him....

Significantly, as Charles points out (in his note for Enoch 40:5), apparently the only writings of that era that apply the term "the Elect one" (or "the Chosen one") to the Son of Man—i.e., the Messiah—are the Book of Enoch and the Gospel of Luke. That could indicate that Luke derived the term from his reading of the Book of Enoch.

In Enoch 48:4, which is a part of the passage extolling the Son of Man, we find another connection with the Gospel of Luke:

... And he shall be the light of the Gentiles,

In Luke 2, Simon takes the infant Jesus in his arms and prophesies, as given in verse 32 (here the RSV), that Jesus would be:

... a light for revelation to the Gentiles...

In Acts 26:23, Luke again applies the term to Jesus:

That Christ should suffer, and that he should be the first that
should rise from the dead, and should shew light unto the
people, and to the Gentiles.

As we have seen, the Greek word that means "the anointed one," or "the anointed" is *christos*, from which the word "Christ" is derived, and that the Hebrew word having the same meaning is *mashiyach*, which is transliterated into English as "messiah." Although the term "the Son of Man" is what is most commonly used in referring to the expected messiah in the Book of Enoch, in that book he is also called God's "Anointed." Here is an example in Enoch 48:10:

... And there shall be no one to take them with his hands and
raise them:
For they have denied the Lord of Spirits and His Anointed.
The name of the Lord of Spirits be blessed.

Here is another example in Enoch 52:4:

... And he said unto me: 'All these things which thou hast
seen shall serve the dominion of His Anointed that he may be
potent and mighty on the earth.'

Most of the descriptions of the Son of man that are found in the Book of Enoch are in the chapters that contain what are called the Similitudes or the Parables. Those chapters both reflected and contributed to the developing popular view of the coming messiah during the period immediately before the rise of Christianity—a view that was fully grounded in Jewish expectations.

And that brings up an important point. Though there are similarities between the Son of man, as he is described in the Book of Enoch, and the descriptions and prophecies that are applied to Jesus in the New Testament, there is one glaring difference: There is nothing in the Book of Enoch to indicate that the Son of man would need to suffer on the cross and die so that he would atone for the sins of those who believed in him. Rather, according to the Book of Enoch, the Son of man was to be the conquering Messiah of Jewish tradition, a messiah who would overthrow the existing order and establish a kingdom of righteousness on the earth; and it was through the establishment of that kingdom by the Messiah that the righteous would receive their salvation.

That, in fact, was the prevalent view of the nature of the Messiah and the coming kingdom of God that existed in the social milieu prior to the ministry of Jesus. It is also likely that Jesus himself subscribed to that same view. Of course, events turned out differently from what he thought they would be. And that turn of events, as we noted previously, was what provided the impetus for his followers to develop the idea that Jesus was a suffering Messiah who paid for the sins of those who believed in him.

Appendix D
Babylonian Myths and the Bible

Societies and their belief systems do not develop in isolation; there is usually at least some interplay with other societies during which beliefs and mythologies are exchanged and adapted. Such was also the case with ancient Israel. For a large part of their existence, the ancient Hebrews migrated over much of the land of the ancient Near East, and the peoples and cultures that they came in contact with influenced their own culture and their mythology. As a result, within the text of the Bible one can find traces of the literature and mythology of the Egyptians, the Babylonians, the Hittites, and the Phoenicians.

A study of the influences that these peoples had on the ancient Hebrews will help one to understand how the Bible developed and why some biblical accounts are similar to accounts found in the literature of other cultures. This appendix will show some of these influences, particularly those relating to the cosmology of the Bible.

We might note that the picture of the cosmos that we derived from the Bible is quite similar to the picture of the cosmos given in Babylonian mythology (actually, Mesopotamian mythology, since many of the myths were handed down through the succeeding cultures in the area). That should not be very surprising, for the ancient Hebrews were influenced by the Babylonian culture in a number of ways. This was especially so during the time that the Israelites spent in the Babylonian exile. It was during this time that the Book of Genesis reached its final form, as we previously noted. Of course, the ancient Hebrews modified over time much of what they picked up from other cultures.

Long before the Bible was written, a belief in a cosmic ocean was a part of Mesopotamian mythology, as was the belief that there was a solid heaven that divided the waters and covered the earth. In the Babylonian creation myth—which is preserved in tablets from the Akkadian culture, but which had its origins in earlier cultures—the cosmic waters were personified in the deities Apsu and Tiamat, with their "waters commingling as a single body" before the creation of the heaven and the earth. From these deities the other deities sprang, among them Marduk, the chief of the deities. As described in Tablet IV, Marduk eventually went to battle with Tiamat, who took the form a dragon:

> He brought forth Imhulla, "the Evil Wind," the Whirlwind, the Hurricane. . . .
> When Tiamat opened her mouth to consume him,
> He drove in the Evil Wind that she close not her lips.
> As the fierce winds charged her belly,
> Her body was distended and her mouth was wide open,
> He released the arrow, it tore her belly,
> It cut through her insides, splitting the heart. . . .
> He cast down her carcass to stand upon it. . . .
> And he turned back to Tiamat whom he had bound.
> The lord trod on the legs of Tiamat,
> With his unsparing mace he crushed her skull. . . .
> Then the lord paused to view her dead body,
> That he might divide the monster and do artful works.
> He split her like a shellfish into two parts:
> Half of her he set up and ceiled it as sky. . . .[1]

The tablets then go on to say that Marduk fashioned the earth from the rest of Tiamat's body. At first, this description might appear to be far from the description of the creation in Genesis, but there are echoes of the Akkadian account in the Bible. Most notably, there is the idea of the cosmic ocean that existed prior to the creation of the heaven and earth. The creation of the heaven and earth is even in the same order in the two versions.

We saw in chapter 7 that the Hebrew word that is translated as "deep" in Genesis 1:2 is *tehowm*. This word, which refers to the great deep of the cosmic ocean, may be linguistically related to the Babylonian "Tiamat." Certainly, both words refer to the cosmic waters.

Note also that Marduk drove the "Evil Wind" into Tiamat's mouth. In Genesis 1:2 we read, "*And the Spirit of God moved upon the face of the waters*" of the great deep. The Hebrew word that is translated as "spirit" is *ruwach*, which means "wind." In the RSV, there is a footnote for the word "spirit" that states, "Or wind." The NEB version of the verse reads, "*. . . and a mighty wind that swept over the surface of the waters.*" Thus both creation stories, the Babylonian and the Hebrew, have the idea of a mighty or fierce wind in conjunction with the waters of the deep before the creation began.

If the relationship between the two stories still seems rather vague, there are other passages in the Bible that are not at all so vague in having a similarity with the Babylonian myth. In Isaiah 51:9 we read, *"Art thou not it that cut Rahab and wounded the dragon?"* Also in Psalm 74:13 we read, *"Thou didst divide the sea by thy strength: thou breakest the heads of the dragons in the waters."* And in Psalm 89:10 we read, *"Thou hast broken Rahab in pieces, as one that is slain."* These passages have been secondarily applied to the Egyptians, but in them are echoes of the story of Marduk's crushing the skull of the dragon Tiamat and then cutting her body in two.

Another such passage is Job 26:12–13 (here, the JPS Tanakh):

> 12 *By His power He stilled the sea; By His skill He struck down Rahab.*
> 13 *By His wind the heavens were calmed; His hand pierced the Elusive Serpent.*

There are definite echoes of the Babylonian creation myth in that passage.

There are other similarities between the Bible and Mesopotamian mythology. The biblical story of the Flood, for example, also has clear antecedents in Mesopotamian mythology. It is found in tablets from both the Sumerian and Akkadian cultures. The Sumerian Noah was Ziusudra, while the Akkadian Noah, as described in the Epic of Gilgamesh, was Utnapishtim.

According to the Epic of Gilgamesh, Utnapishtim related the story of how the gods ordered him to build a ship and to take on board the seed of all living things. The text describes the building of the boat and its dimensions, which were ten dozen cubits on each side. Bitumen and asphalt were used to seal it. On the day it was launched, Utnapishtim brought on board "of all living things" and his family and kin. The deluge came and drowned all that were not on the boat. As the waters of the flood receded, the boat came aground. On the seventh day after, Utnapishtim sent forth a dove, which came back, for it found no resting place. Utnapishtim next sent forth a raven, which did not return. Utnapishtim then offered a sacrifice to the gods and the gods smelled the sweet savor.[2]

There is a remarkable similarity between this story and the story of the Flood in the Bible. To be sure, the Akkadian myth is full of intrigues and disputes between the gods. Also, there were several versions of the flood story in the Mesopotamian cultures. The story may have originated in a particularly bad flood in the land between the two rivers, the Tigris and the Euphrates, which was low and susceptible to flooding. A farmer may have managed to save his family and some of his flock in a makeshift boat and the story became magnified in the retelling over the years.

In any case, these stories predate the Genesis account, the final version of which, as we saw in Appendix A, actually dates from around the time of the Babylonian exile. The earliest Mesopotamian accounts even predate Moses, who traditionally was supposed to have written the Genesis account.

The creationists, of course, dispute that the story of the Flood in the Bible is mythology. According to the creationists, the Flood actually happened and the evidence for it is to be found in the fossil record, for—they say—the fossils were laid down during the Flood.

Despite the rhetoric of the creationists, the fossil record in no way supports the creation "model." According to the creation "model," all forms of life, both presently existing and presently extinct, were created during a relatively short time about 10,000 years ago. Subsequently, according to that model, there was a world-wide "hydrological" disaster—i.e., a flood—that wiped out all life except that on an ark. The fossil record, according to this model, consists of the remains of those animals that perished in the flood. Also according to this model, the animals on the ark dispersed and repopulated the earth after the flood waters receded, though some of the animals, such as the dinosaurs, subsequently became extinct because of changes in the climate that resulted from the flood.

If that scenario had actually happened, there should be no particular ordering of species in the fossil record. We should expect to find fossil specimens of the more than 4,000 presently existing mammal species, including humans, and the thousands more of presently extinct mammal species mixed in with the remains of dinosaurs. The same is also true for the thousands of species of birds that presently exist or are now extinct. Thus, we would expect to find the remains of humans, bears, mammoths, lions, horses, giant ground sloths, eagles, moas, ostriches, pigeons, etc., in the same strata in which we find fossils of dinosaurs and other early species. That, however, has never been the case in any of the fossil finds anywhere in the world. The vast majority of mammal fossils have always been found in strata above and younger than those in which dinosaur fossils have been found. The only mammal fossils that have ever been found in the same strata as dinosaur fossils have been of relatively small species that were superseded in the fossil record by new forms that evolved after the extinction of the dinosaurs. The same is true of bird fossils. No fossils of existing species of birds have ever been found in the same strata in which dinosaur fossils have been found. The only bird fossils that have been found in such strata have been of primitive and extinct species.

But what is most telling is that the fossil record distinctly shows a succession of life forms that lived over the past 3.5 billion years, from simple forms through more complex forms. The fossil record also shows a replacement of representative species over time. The order of succession follows an evolutionary sequence, and the vast majority of fossils in ancient strata are of species that do not even exist today. Fossils of most of the species that now exist do not appear in the geological record except in strata that were laid down relatively recently.

Furthermore, the layers of the geological record were laid down under an enormous variety of conditions and over an immense span of time. To say that all of these layers were laid down by a single world-wide flood is simplistic to the extreme and ignores the physical evidence. No reputable geologist adheres to the creation model for an explanation of geological deposits, for that model can provide no basis for determining where specific deposits of minerals and petroleum can be found. Geologists, rather, use the standard scientific model of ancient earth geology to predict where deposits of minerals and petroleum may be found.

What all this means is that the biblical Flood, like the biblical cosmos, is based on ancient Hebrew—and ultimately Babylonian—mythology.

Notes

Introduction

1. These individuals usually distort the words of the Bible in an attempt to make it conform to the scientific evidence, or they distort the scientific evidence in an attempt to make it conform to the words of the Bible. A few examples include Gerald Schroeder's *Genesis and the Big Bang* and his *The Science of God*; Hugh Ross's *A Matter of Days*; D. Russell Humphreys's *Starlight and Time*; John D. Morris's *The Young Earth*; Donald DeYoung's *Thousands not Billions*; and *Dismantling the Big Bang* by Alex Williams and John Harnett.

Chapter 1 The Origins of the Bible

1. In this book, the abbreviations "C.E." for "Common Era" and "B.C.E." for "Before the Common Era" are used instead of A.D. and B.C. These abbreviations have no religious connotation are often used where religious neutrality is desirable.

2. For descriptions of the many forms of early Christianity and the various writings they were based on, see Bart D. Ehrman, *Lost Christianities: The Battles for Scripture and the Faiths we Never Knew* (New York: Oxford University Press, 2003).

Chapter 2 The Consequences of Belief in the Bible

1. See Catherine Crier, *Contempt: How the Right is Wronging American Justice* (New York: Rugged Land, 2005). See also Michelle Goldberg, *Kingdom Coming: the Rise of Christian Nationalism* (New York: W.W. Norton & Company, 2006). See also Chris Heddges, *American Fascists: The Christian Right and the War on America* (New York: Free Press, 2007) See also Jan G. Linn, *What's Wrong with the Christian Right* (Boca Raton, Florida: Brown Walker Press, 2004). See also Barry W. Lynn, *Pi-*

ety and Politics (New York: Harmony Books, 2006). For a detailed examination of the Christian Coalition and its founder, Pat Robertson, see Robert Boston, *The Most Dangerous Man in America? Pat Robertson and the Rise of the Christian Coalition* (Amherst, NY: Prometheus Books, 1996).

2. See Hanna Rosen, *God's Harvard: A Christian College on a Mission to Save America* (Orlando, Fla.: Harcourt, 2007).

3. See "Boss Pat: comparing the Christian Coalition to the Tammany Hall political machine, Pat Robertson shares with top lieutenants his secret 'game plan' for taking the White House and ruling America," *Church and State*, October 1, 1997. See also "Laws of change: Christian college hopes its students will change society," *Roanoke Times* (Roanoke, Va.), August 19, 2006. See also "Justice's Holy Hires," *The Washington Post* (Outlook), April 8, 20007.

4. The architect of the Reconstructionist movement was the late Rousas John Rushdoony, Director of the Chalcedon Foundation in Vallecito, California. His massive two-volume book, *The Institutes of Biblical Law*, provides the cornerstone of Reconstructionism.

5. The speech was given by David Barton. The quoted statement was reported by Julie Schollenberger, "Concerned about Concerned Women for America," *Freedom Writer*, January 1995, p. 3.

6. The Rev. Joseph Morecraft, of the Chalcedon Presbyterian Church, Marietta, Georgia, quoted by Frederick Clarkson, "Christian Reconstructionism: Religious Right Extremism Gains Influence—Part Two," *The Public Eye*, June 1994, pp. 1–6. *The Public Eye* is a publication of the Political Research Associates.

7. For additional information about the Christian Reconstructionist movement, see Anson Shupe, "The Reconstructionist Movement on the New Christian Right," *The Christian Century*, October 4, 1989, pp. 880–882. See also Rob Boston, "Thy Kingdom Come," *Church and State*, September 1988, pp. 6–12. See also Joseph L. Conn, "The Reconstructionist Connection: Putting the 'Radical' in Radical Religious Right," *Church and State*, March 1996, p. 9. See also "Warped worldview: Christian Reconstructionists believe democracy is heresy, public schools are satanic and stoning isn't just for the Taliban anymore—and they've got more influence than you think," *Church and State*, July 1, 2006.

8. *The News-Sentinel* (Fort Wayne, Ind.), Aug. 16, 1993, as quoted in *The News & Record* (Piedmont Triad, NC) July 19, 1997.

9. Genesis 17:20.

10. Probably the foremost organization involved in promoting creation "science" is the Institute for Creation Research (ICR), which has a graduate school, as well as other divisions and activities. All of the members, faculty, and students of the organization are required to sign a "statement of faith" organized in "two parallel sets of tenets, related to God's created world and God's inspired word, respectively." The ICR "bases its educational philosophy on the foundational truth of a personal Creator-God and His authoritative and unique revelation of truth in the Bible, both Old and New testaments." The tenets specifically deny evolution and mandate adherence to the biblical account of creation.

The full text of the two parallel ICR tenets of faith, as well as an explanation of the ICR educational philosophy, was available on the ICR web site while this book was being researched.

Another prominent organization promoting creation "science" is the Creation Research Society (CRS). Applicants for membership in this organization must likewise subscribe to a statement of belief similar to that of the ICR.

11. For examples of refutations of irreducible complexity and other arguments put forth by the proponents of intelligent design, see Kenneth R. Miller, *Finding Darwin's God* (New York: Perennial, 1999). See also Eugenie C. Scott, *Evolution Vs. Creationism: An Introduction* (Berkeley: University of California Press, 2004). See also Michael Schermer, *Why Darwin Matters* (New York: Henry Holt and Company, LLC, 2006). See also Mark Perakh. *Unintelligent Design*. (Amherst, NY: Prometheus Books, 2004).

12. Several books have been published that examine and refute the arguments put forth by the creationists. See, for example, *Scientists Confront Creationism*, edited by Laurie R. Godfrey (NY: W. W. Norton & Co., 1983) pp. 99–116. See also Philip Kitcher, *Abusing Science: The Case Against Creationism* (Cambridge: The MIT Press, 1982), pp. 98–96. See also Mark Isaak, *The Counter-Creationism Handbook* (Berkeley and Los Angeles: University of California Press, 2007). See also Brian J. Alters and Sandra M. Alters, *Defending Evolution in the Classroom* (Sudbury, Mass: Jones and Bartlett Publishers, 2001).

13. See Russell Shorto, "Contra-Contraception," *The New York Times Magazine*, May 7, 2006.

14. Pat Robertson and his book, *The New World Order* (W Publishing Group: March, 1992), provide a notable example.

15. For examples, see David S. New, *Holy War: The Rise of Militant Christian, Jewish and Islamic Fundamentalism* (Jefferson, North Carolina: McFarland, 2001). See also "Apocalypse Soon; Israel's occupation

of the West Bank has no stronger supporters in the United States than the evangelicals of the Christian Coalition. But for them it's just another step in the run-up to Christ's Second Coming," *The Independent on Sunday* (London, England), June 29, 2003.

16. See *Church and State*, March 1, 2006: "Religious Right Scuttles Group's Plan to Address Global Warming." See also *The Christian Century*, March 7, 2006: "Evangelicals Split on Global Warming."

17. See "Home Schooling: Preach Your Children Well" (*New Scientist*, Nov. 11–17, 2006) for information about the home schooling movement and information about the referenced books.

Chapter 3 The Earth in Space

1. See Robert Benjamin, "How Astronomers Glimpse the Naked Galaxy," *Astronomy*, February 2007, pp. 58–63, for a graphic representation of the Milky Way Galaxy showing the location of our sun.

2. *Ibid*.

3. See Robert Irion, "The Planet Hunters," *Smithsonian*, October 2006, pp. 38–44.

Chapter 4 The Evidence for the Modern View

1. "Generally speaking" because in fact there are a few stars, such as Sirius, that are brighter than first magnitude.

2. Astronomers had once thought that Proxima Centauri was a part of the Alpha Centauri system, but it now appears that it is only passing by the system.

3. For further information about the 1987 supernova, see Ronald A. Schorn, "Happy Birthday Supernova!," *Sky and Telescope*, February 1988, pp. 134–137, and J. M. Lattimer and A. S. Burrows, "Neutrinos from Supernova 1987A," *Sky and Telescope*, October 1988, pp. 348–350. See also Richard Talcott, "Insight into Star Death," *Astronomy*, February 1988, pp. 6–23.

4. See Ron Cowen, "Supernova yields cosmic yardstick," *Science News*, January 26, 1991 (Vol. 139, No. 4), p. 59, for a report about the measurement of the distance to the supernova.

5. See "Repaired Hubble Finds Giant Black Hole," *Science News*, June 4, 1994 (Vol. 145, No. 23), pp. 356–357.

6. See Peter Goldreich, "Tides and the Earth-Moon System," *Scientific American*, April 1972, pp. 43–52. See also S. K. Runcorn, "Corals as Paleontological Clocks," *Scientific American*, October 1966, pp. 26–33.

7. See R. Monastersky, "The Moon's Tug Stretches Out the Day," *Science News*, July 6, 1996 (Vol. 150, No. 1), p.4.

8. The numbers used in this discussion were derived from Kenneth R. Miller, "Scientific Creationism Versus Evolution: The Mislabeled Debate," in *Science and Creationism*, edited by Ashley Montagu (New York: Oxford University Press, 1984) pp. 18–63.

9. See G. Brent Dalrymple, Ph.D., *The Age of the Earth* (Stanford, CA: Stanford University Press, 1991) pp. 376–387. Dalrymple analyzes the nuclides that have half lives greater than one million years, rather than those with half lives greater than 1000 years as analyzed here, but the considerations remain the same.

Dalrymple shows that hypothesis number one can be eliminated because the odds against all of the short-lived elements being absent by mere chance in our solar system are astronomical. Hypothesis number two can be eliminated for several reasons. For example, nuclides that have even numbers of both neutrons and protons tend to be three times as abundant as those with odd numbers. This is true for the stable nuclides as well as the long-lived radioactive nuclides. Several of the missing short-lived nuclides have even numbers of neutrons and protons and should be represented if they were formed relatively recently. Another reason to eliminate hypothesis number two is that all of the missing short-lived nuclides are easily made in nuclear reactors, and there is no reason to believe that they would not likewise be made in natural nuclear reactors—i.e., stars. And, in fact, some of the short-lived nuclides have been spectrographically observed in extra-solar stars, and, as we have noted, stars provide the raw material for the formation of planets. This in itself argues against hypothesis number 2. Dalrymple provides considerably more detail about the reasons for eliminating the two hypotheses than we can go into here.

10. For a more detailed explanation of this dating method, see Kenneth R. Miller (ref. note 8 above). See also Arthur N. Strahler, *Science and Earth History* (Buffalo, NY: Prometheus Books, 1987) pp. 131–134.

11. For more information on the age of the earth see Dalrymple (ref. note 9, above). See also Ivars Peterson, "A Rocky Start," *Science News*, March 20, 1993 (Vol. 143, No. 12), pp. 190–191.

12. This creationist argument was developed by Thomas G. Barnes in a series of papers and a subsequent book, *Origin and Destiny of the Earth's Magnetic Field* (El Cajon, CA: Institute for Creation Research, 1983). For a critique of Barnes's work, see G. Brent Dalrymple, Ph.D., "Origin and Destiny of the Earth's Magnetic Field," *Reviews of Creationist Books* (Berkeley, Ca: The National Center for Science Education, Inc., 1992), pp. 8–9.

13. See Jean-Pierre Valet, "From the Superchron to the Microchron: Magnetic Stratigraphy in Deep Sea Sediments" *Oceanus*, December 22, 1993. See also Maurice A. Tivey, "Paving the Seafloor--Brick by Brick: New Vehicles and Magnetic Techniques Reveal Details of Seafloor Lava Flows" *Oceanus*, September 22, 2004.

14. Patrick M. Hurley, "The Confirmation of Continental Drift," *Scientific American*, April, 1968, pp. 52–64.

15. See Miller and Strahler (ref. note 10, above) for refutations of these and other creationist "proofs."

Chapter 5 The Big Bang

1. For a more detailed explanation of the cosmological red shift, see Sten Odenwald and Richard Tresch Fienberg, "Galaxy Redshifts Reconsidered," *Sky and Telescope*, February 1993, pp. 31–35.

2. See "The Golden Age of Cosmology," *Scientific American*, July 1992, pp. 17–22. For a more complete discussion of the COBE findings, see also George Smoot and Keay Davidson, *Wrinkles in Time* (New York: William Morrow & Company, 1993).

Chapter 6 The Size of the Cosmos

1. For a report on the Hubble observation, see "Hubble scopes possible planet-forming disks," *Science News*, December 19 & 26 (combined issue), 1992 (Vol. 142, Nos. 25 & 26), p. 421. See also "Hubble eyes disks that may form planets," *Science News*, June 18, 1994 (Vol 145, No. 25) p. 391. For other observations concerning possible planet formation, see "Low-Mass Stars: Born to make Planets," *Science News*, January 16, 1993 (Vol 143, No. 3), p. 36, and "New evidence of dust rings around stars," *Science News*, October 3, 1992 (Vol 142, No. 14), p. 214. See also Ivars Peterson, "A Rocky Start," *Science News*, March 20, 1993 (Vol 143, No. 12), pp. 190–191, for information concerning the formation of our own solar system.

Chapter 8 The Firmament of Heaven

1. Ernest Klein, *A Comprehensive Etymological Dictionary of the Hebrew Language for Readers of English* (New York: Macmillan Publishing Co., 1987) p. 629. The word *raqa`* is in Hebrew characters in the definition.

2. See Claus Westermann, *Genesis 1–11: A Commentary*, translated by John J. Scullion (Minneapolis: Augsburg Publishing house, 1984) p. 117.

3. The Hebrew word *kuwn* is translated as "established" or "establish" 58 times, which is more than twice as often as the next most frequent Hebrew word, *quwm*.

4. Everett Fox, *The Five Books of Moses* (N.Y.: Schocken, 1995).

Chapter 9 The Circle of the Earth

1. Henry M. Morris, *The Biblical Basis for Modern Science* (Grand Rapids, Mich: Baker Book House, 1984), p. 246.

Chapter 10 The Pillars of the Earth

1. The Hebrew word translated as "world" in this passage is *tebel*. *Strong's Concordance* provides the following definition for *tebel*: "... the earth ...; by extens. the *globe*...." By using the term "globe" in this definition, Strong has extended ("by extens[ion]" in his words) his own modern-day understanding of the meaning of "world" onto his definition of the ancient Hebrew word for "world." Klein (*op. cit.*, p. 689) defines *tebel* only as "world," and makes no mention of "globe." In addition, none of the other lexicons referred to in this book use the term "globe" in defining *tebel*. In any case, the passage in Samuel certainly lends itself more to the idea of a flat earth—rather than a "globe"—resting on pillars.

Some biblical apologists say that the reference to pillars of the earth in this passage is a metaphor for people who bear the weight of the world on their shoulders. Even if that were so, the metaphor uses the structure of the cosmos, as the ancient Hebrews saw it, as its basis. However, what the words actually mean in the context of the verse is that God has control over the whole of creation, and it is he who establishes what happens in the world.

2. Morris, op. cit., p. 165.

3. Quoted by Andrew D. White, *A History of the Warfare of Science with Theology in Christendom* (New York: 1896. (Prometheus Books reprint; Buffalo, NY: 1993)) Vol. 1, p. 137

4. Ibid., p. 126

5. Ibid., p. 127.

6. Ibid., pp. 91–92.

7. Ibid., pp. 150–151.

8. Gerardus D. Bouw, Ph.D., *Geocentricity* (Cleveland, Oh: Association for Biblical Astronomy, 1992), p. ii.

9. Marshall Hall, *The Earth is Not Moving* (Athens, Ga: Fair Education Foundation, 1991), pp. 2–3. For a review of Hall's book, see Francis G.

Graham, "The Earth is Not Moving," *Reviews of Creationist Books* (Berkeley, Ca: The National Center for Science Education, Inc., 1992), pp. 59–61.

10. See Robert Schadewald, "The Flat Out Truth," *Science Digest*, July, 1980.

Chapter 11 The Pit of Sheol

1. See G. von Rad, *Genesis* (Philadelphia: The Westminster Press, 1961), p. 47.

Chapter 12 The Lights in the Firmament

1. Some promoters of the Bible refer to myths from various cultures around the world that describe supernaturally long days or nights. Such myths, however, can be considered a natural result of the storytelling process by primitive people who speculated on what would happen if the regular events of nature became irregular. In any case, if the solar event described in the Book of Joshua had happened when the Bible states it did, it certainly would have caused enough consternation to have been clearly and unequivocally dated and described in the written records of those civilizations mentioned in the paragraph that references this note.

2. See Joshua 11:4; Judges 7:12; 2 Samuel 17:11; 1 Kings 4:20; Jeremiah 15:8.

3. See Genesis 26:4; Exodus 32:13; Deuteronomy 10:22; Deuteronomy 28:62; Nehemiah 9:23; Exodus 32:13.

4. These various terms refer to the planet Venus, which can be so bright at certain times that it can even be visible in the light morning sky; as the passage in Isaiah indicates, Venus was sometimes called the "son of the morning." The word "Lucifer" itself is from Middle English and is derived from a Latin term meaning "light bearer." It should perhaps also be noted that this passage provides the only occurrence of the word *heylel* in the original Hebrew of the Bible. Although "Lucifer" has been equated with Satan, the passage is actually a denunciation of the king of Babylon, as Isaiah 14:4 makes clear:

> ... *thou shalt take up this proverb against the king of Babylon, and say, How hath the oppressor ceased! the golden city ceased!*

The Hebrew word translated as "proverb" in Isaiah 14:4 is *mashal*, which also means "parable," and Isaiah continues the denunciation of the king of Babylon in verses 12 through 15 in the form of a parable that is expressed in terms of the Hebrew view of the cosmos. As this passage indicates, the ancient Hebrews—or at least some of them—viewed the stars, i.e., the host of heaven, to be entities (i.e., angels); in this case Venus was viewed as the star that was called the "son of morning."

Chapter 13 The Biblical Cosmos: Metaphor or Mythology

1. For an example, see Luis Stadelmann, *The Hebrew Conception of the World* (Rome: Pontifical Biblical Institute, 1970).

2. For more detailed information about many of the discoveries of the Greek astronomers, see Sir Thomas L. Heath, *Greek Astronomy* (New York, NY: Dover Publications, Inc., 1991). It is also interesting to note that a CT scan of the so-called Antikytheria mechanism found in a second-century B.C.E. shipwreck off Greece revealed it to be a very complex system of gears that that was designed to predict eclipses from the relative positions of the earth, moon, and sun. The mechanism was more complex than any machine known to have existed for more than 1000 years after it was built. See "Enigmatic relic was eclipse calculator," *New Scientist*, December 2 – 8, 2006, p.17. One wonders how many other scientific advancements the ancient Greeks had made and have since been lost to mankind.

Chapter 14 The Implications

1. Hypatia was the daughter of Theon, who was associated with the Serapeum library in Alexandria, which was destroyed by Bishop Theophilus in 391 C.E. Hypatia worked at the palace library, which eventually was destroyed by the Moslems after their conquest of Alexandria in the year 641 C.E. By the time of the Moslem conquest most the original scrolls of the palace library were gone and had been replaced by patristic writings and "sacred" literature. When they mention the destruction of the Alexandrian "library," some present-day writers do not appear to understand that Alexandria had more than one library. For details concerning the destruction of the libraries, see Luciano Canfora, *The Vanished Library: A Wonder of the Ancient World* (Berkeley and Los Angeles: University of California Press, 1990). See *Hypatia* in the Micropedia of the Encyclopedia Britannica for further information about the societal setting of the events relating to her death. See also Ronald E. Mohar, "The Murder of Hypatia of Alexandria," *Free Inquiry*, Spring 1983 (Vol 3, No. 2), pp. 12–14.

Appendix A Problems in the Old Testament

1. For a review of the archaeological evidence relating to the Israelite invasion of Canaan, as well as detailed references to primary sources, see William H. Stiebling, Jr., *Out of the Desert? Archaeology and the Exo-*

dus/Conquest Narratives (Buffalo, NY: Prometheus Books, 1989). See also Neil Asher Silberman and Israel Finkelstein, *The Bible Unearthed: Archaeology's New Vision of Ancient Israel and the Origin of Its Sacred Texts* (New York: Free Press, 2002).

2. For a brief description of the history of Tyre, see "Tyre" in *Harper's Bible Dictionary*, pp. 1101–1102. For a more detailed description of the history of Tyre, see Nina Jidejian, *Tyre Through the Ages* (Beirut: Dar El-Mashreq Publishers, 1969). See also H. Jacob Katzenstein, *The History of Tyre* (Beer Sheva: Ben-Gurion University of the Negev Press, 1997).

3. Josh McDowell, in his book, *Evidence that Demands a Verdict* (San Bernardino, Calif: Here's Life Publishers, 1979), was a primary source for that claim about Tyre. In upcoming endnotes, we'll examine that and other erroneous material in his book.

4. Jidejian, p. 1; Katzenstein, p. 14. After the Alexandrian conquest, the Greeks called the mainland settlement *Palaetyrus*, which means "Old Tyre" in Greek. Katzenstein states (p. 14, fn 55) that the name "'Old Tyre' ... misled the latter geographers into claiming not only that the oldest settlement of Tyre was in 'Old Tyre', but that it was the first actual site of the town."

5. See Katzenstein, p. 322.

6. According to *Harper's Bible Dictionary* (p. 1102) and several other sources, Nebuchadrezzar began his siege of Tyre in the year 586 B.C.E., though some other works give the year as 585 B.C.E. (That slight discrepancy is likely the result of an overlapping year with respect to different calendars.) Quoting the *Encyclopedia Britannica*, Josh McDowell (*op. cit.*, p. 275), in fact gives 585 B.C. as the first year of the siege. However, he wrongly states that "Nebuchadnezzar laid siege to mainland Tyre three years after Ezekiel gave the prophecy." Even with the date of 585 B.C.E. instead of 586 as the beginning of the siege, three years previously would have been 588 B.C.E., which would have been the eighth year of the exile. That, of course, conflicts with Ezekiel's statement in 26:1 that he gave the prophecy in the eleventh year of the exile, which would have been 586 B.C.E, the same year that many sources give as the date of the beginning of Nebuchadrezzar's siege against Tyre. In fact, 588 B.C.E. would have been even before the Babylonian razing of Jerusalem, which precipitated the events that led up to Ezekiel's prophecy against Tyre. One might conclude that McDowell extended the time to three years because he wanted to make it appear that Ezekiel was giving a genuine prophecy rather than merely interpolating what would happen on the basis of his observation of contemporary events. It is significant that McDowell gives no basis for saying that the period was three years.

That brings up the way that McDowell quoted Ezekiel's prophecy against Tyre (*op. cit.*, p. 274). He omitted verses 1 and 2 of Ezekiel 26, and began it with verse 3. Since verse 1 indicates that Ezekiel made his prophecy against Tyre in the eleventh year of the exile, and verse 2 provides the reason why Ezekiel made the prophecy (that is, that God would cause Tyre's imminent destruction because of what the Tyrians had said about Jerusalem's destruction), including those verses would have been at cross purposes with what McDowell was trying to prove. One would therefore have reason to be cynical about McDowell's motivation in purposely omitting those important verses.

7. Some of these apologists were apparently influenced by McDowell, who states (*op. cit.*, p. 275), "When Nebuchadnezzar broke the gates down, he found the city almost empty. The majority of the people had moved by ship to an island about one-half mile off the coast and fortified a city there." Apparently in order to make the fulfillment of Ezekiel's prophecy more credible, McDowell was trying to give the impression that the city of Tyre originally consisted only of the mainland settlement. In referring to the island as merely "an island" that was used as a place of refuge during the siege, it appears that McDowell was trying to make the island out as having previously been of no real significance and that it was not the actual city of Tyre (this will be discussed further in note 10). In fact, however, as we have noted, the island actually had been the site of the city of Tyre for millennia, and it overshadowed the mainland settlement in importance because of its harbors. Examples of books by biblical apologists who have picked up on McDowell's misleading statement include *The Merciful God of Prophecy*, by Tim LaHaye; and *Did God Write the Bible?* by Dan Hayden.

8. *The New Westminster Dictionary of the Bible* (Philadelphia: The Westminster Press, 1970) states the following under the entry for Tyre: "It is not certainly known whether he [Nebuchadrezzar] actually captured any part of the 2 cities . . . ; if he did, it was probably only the one on the shore" (pp. 961–962). Jidejian (*op. cit.*, p.1) states, "Unfortunately, history has left us no account of the surrender of the city." Jidejian assumes that Nebuchadrezzar did conquer the mainland settlement, and she suggested that its inhabitants "no doubt" fled to the island city of Tyre with whatever they could carry (p. 56). However, even though Jidejian refers to and quotes several of the historical accounts concerning the siege of Tyre by Nebuchadrezzar's forces, none of those accounts specifically states that those forces took the mainland settlement. Biblical apologists usually state it as if it were a documented fact that Nebuchadrezzar did destroy the mainland settlement. McDowell, for example (*op. cit.*, p. 275), states that

the mainland city was destroyed in the year 573 B.C.E., but he gives no source for that information.

9. Josh McDowell, in his book *Evidence that Demands a Verdict*, specifically takes this approach. As previously mentioned, many other apologists took up McDowell's arguments.

10. A case in point is provided by Josh McDowell in his book, *Evidence that Demands a Verdict*. Because it was the mainland city that was eventually destroyed (some 2,000 years after Ezekiel's prophecy) and not rebuilt, and since in contradiction to Ezekiel's prophecy the island city of Tyre *was* rebuilt after its destruction, the only way that McDowell could argue that Ezekiel's prophecy was fulfilled was by subtly revising its meaning. Therefore, according to his revision, the mainland settlement was the actual city of Tyre and the object of the prophecy. Also according to his revision, instead of being the site of the actual city of Tyre when Ezekiel made his prophecy, the island was merely "*an* island" (not *the* island) of no particular significance until later when it was used as a place of refuge during Nebuchadrezzar's siege of the mainland settlement (see note 7, above). McDowell also gave a list of Ezekiel's predictions against Tyre that reflects his revision of the meaning of the prophecies (*op cit*, p. 274). In that list, McDowell states, as number 1, "Nebuchadnezzar will destroy the mainland city of Tyre," and he makes no mention of the island city at all in the remaining numbered items in the list. When he does mention the destruction of the island city in his description of the events that occurred centuries later, it is merely incidental with respect to the destruction of the mainland settlement.

It should not be surprising, then, that a critical analysis of McDowell's revisionist claims shows them to be an exercise in obfuscation, contradiction, and inconsistency. For example, McDowell says (*Op cit.* p. 277; the ellipses are mine and omit irrelevant information):

> We shall now see Tyre at present, as described by Nina Jidejian: "The 'Sidonian' port of Tyre is still in use today. Small fishing vessels lay at anchor there.... The port has become a haven today for fishing boats and a place for spreading nets."

McDowell then provides the following as a quotation (p. 277; the ellipses are mine):

> "The destiny of Tyre according to the prophet is a place where fishermen would spread their nets. The existence of a small fishing village [There is a city of Tyre today, but it is not the original city, but is built down the coast from the original site of Tyre.] upon the site of the ancient city of Tyre does not mean that the prophecy is

not fulfilled but is the final confirmation that the prophecy was fulfilled. Tyre ... passed away never to rise (rebuild) again. The fishermen drying their nets upon the rocks ... are the last link in the chain of prophecy that Ezekiel gave. ..." ... 4/47, 48

McDowell begins the next paragraph with the following: "Jidejian concludes in her excellent book that 'Tyre's stones may be found as far away as Acre and Beirut.'" The remainder of the paragraph consists of additional quotations from Jidejian's book.

Note that McDowell does not give the name of the person who was the source of the second quotation (the one beginning "The destiny of Tyre...."). Since that quotation is sandwiched between two sets of quotations attributed to Nina Jidejian, an inattentive reader might think that it is also by Jidejian, whereas, in fact, it is not. The number 4 at the end of the quote is a cross reference to the source document as listed in a bibliography at the end of McDowell's book. That document is an unpublished Master's thesis by John Clark Beck, Jr., titled *The Fall of Tyre According to Ezekiel's Prophecy*, which is located at the Dallas Theological Seminary.

I was curious about the bracketed comment, "[There is a city of Tyre today, but it is not the original city, but is built down the coast from the original site of Tyre.]," in the quotation from Beck's thesis. I checked with the library at the Dallas Theological Seminary and, in fact, was told that the bracketed comment is *not* in the original thesis. McDowell apparently inserted the bracketed comment to build on his revision of Ezekiel's prophecy. (As a matter of propriety, McDowell should have stated that he had inserted the comment.)

Though McDowell does not say so, the "city of Tyre today ... built down the coast from the original site of Tyre" in his bracketed insertion can only refer to the present-day city of Tyre on the one-time island, now peninsula. For that matter, Beck is likely also referring to that city in McDowell's quoted excerpt from his thesis, though he minimizes its size by calling it "a small fishing village." However, it would appear that McDowell is trying to make it seem that the "small fishing village" that Beck referred to is up the coast from the "city of Tyre today" and is the site of the original mainland city. In fact, however, the location of the "city of Tyre today" is *not* "down the coast" from the original site of the mainland settlement. As both Jidejian (p. 1) and Katzenstein (p. 15) state, the mainland settlement was actually nearly directly across the strait from the island city of Tyre.

It would appear that McDowell was trying place the mainland settlement some distance from the existing city of Tyre (that is, the city on the one-time island, now peninsula) in order to bolster his revision of Ezekiel's prophecy. When McDowell states in his insertion that "There is a city of Tyre today, but it is not the original city, but is built down the coast from the original site of Tyre," it is as if the "city of Tyre today" had nothing at all to do with ancient Tyre. In effect, McDowell is unequivocally stating that the "city of Tyre today" is *not* on the original site of the city of Tyre.

However, note that he had said previously, "We shall now see Tyre at present," and he quotes Jidejian's reference to the "'Sidonian' port of Tyre" as being "still in use today" and as "a place for spreading nets." In quoting the statement that "The 'Sidonian' port is still in use today," he gives the impression that the Sidonian port was a part of the ancient city of Tyre, which, in fact, it was. But that port is not located where McDowell claims Tyre to have been—that is, on his claimed site of the ancient mainland settlement up the coast. Rather, the Sidonian port is on the north side of the present-day city of Tyre on the one-time island, now peninsula (that is, the city that he claims is *not* the original city of Tyre), and it is the same port that existed in the same location in ancient times. Though it is not likely that most of his readers would be aware of that fact, McDowell does not clarify where the location of that port is. As far as a good many of his readers would know, and because of his emphasis, "Tyre at present" would refer to the site of the original mainland city "up the coast" from the present-day city of Tyre.

That impression would be reinforced by the insertion he placed in Beck's thesis. Since, according to McDowell's revised view of the geography of Tyre, "the site of the ancient city of Tyre" is up the coast from the present-day city and is where fishermen spread their nets to dry, the reader would understandably assume that Jidejian was also describing that same location when she spoke of the place where fishermen spread their nets.

It would appear that McDowell was trying to conflate the first quotation from Jidejian's book and the quotation from Beck's thesis to make it appear that Jidejian's book supports his revision of the geography of Tyre.

To top off his contradictory and confusing statements about Tyre, McDowell finished his discussion with a small map-type sketch of Tyre as a peninsula. Significantly, he provided no label or explanatory text with the map.

Another point that needs to be made is that, contrary to McDowell's assertions that Ezekiel's prophecy about Tyre has been fulfilled, Jidejian indicated in her book that Ezekiel's prophecy was *not* fulfilled, as, for example, she states in the following (*Op cit.* p.1.): "Although the city's fate was different from that prophesized by Ezekiel, nevertheless Tyre's commerce was ruined by the long siege."

To Jidejian's statement about Ezekiel's prophecy, we might add Katzenstein's (*Op cit.* p.331.): "The war was, therefore, hard for both sides, and Tyre was the actual loser, but the destruction of the city itself, prophesied by Ezekiel, did not come to pass."

One further point should be made here. As indicated above, McDowell stated that Jidejian's book is "excellent," and one would presume that McDowell read the book during his research on Tyre. However, throughout her book Jidejian emphasizes the historical importance of the island city of Tyre, and she gives its history during the centuries before, as well as after, the time when Ezekiel prophesied its destruction. Yet, McDowell ignored the information she provided about Tyre's history and he tried to convince his readers that the mainland settlement was the city of Tyre and that the island was merely "an island" of little significance until after Ezekiel made his prophecy.

11. Ecclesiasticus, chapters 44 through 50. Ezekiel mentions a Daniel (Ezekiel 14:14; 14:20; and 28:3), but it is apparent from his commentary that this Daniel lived long before the sixth century B.C.E. time of the Babylonian exile. The author of the Book of Daniel may, however, have used Ezekiel's mention of the name as the inspiration for the name of the main character of his own book.

12. See *Nebuchadrezzar* in the Encyclopedia Britannica. See also *The Interpreter's Bible* (New York: Abingdon Press, 1956), Vol. VI, p. 362.

13. For evidence concerning the date of the composition of the Book of Daniel, as well as some information about its textual problems, see the commentary on the Book of Daniel in *The Interpreter's Bible* Vol 6, pp. 341–549. See also "Daniel, Book of" in *Harper's Bible Dictionary*, pp. 205–206, and *The Interpreter's Dictionary of the Bible* (New York: Abingdon Press, 1962), Vol A-D, pp. 761–768.

14. See also Gen. 21:32, 26:1, 26:8, 26:14, 26:15, 26:18 and Ex. 13:17 and 23:31. See "Philistines" in *Harper's Bible Dictionary* for a brief history of these people.

15. For information concerning the formation of these books of the Bible, see Richard Elliot Friedman, *Who Wrote the Bible?* (New York: Harper and Row, 1989); Robin Lane Fox, *The Unauthorized Version* (New York: Alfred A. Knopf, 1992); *The Bible Unearthed: Archaeology's New Vision of Ancient Israel and the Origins of its Sacred Texts*, by Israel Finkelstein and Neil Asher Silberman (New York: The Free Press, 2001).

Appendix B Problems in the New Testament

1. *Harper's Bible Dictionary* under "Babylon" states: "At the end of the first millennium B.C., the scribal school at Babylon produced almost exclusively astronomical and astrological reports."

2. In his *Antiquities of the Jews*, Josephus placed the death of Herod shortly after an eclipse of the moon (Book 17, Ch 6, Para. 4). There were, in fact, two eclipses of the moon visible in Palestine within a reasonable time frame: one occurring in 4 B.C.E. and one in 1 B.C.E. Elsewhere in his *Antiquities*, Josephus stated that "Philip, Herod's brother [the brother of Herod Antipas, that is], departed this life, in the twentieth year of the reign of Tiberius, after he had been Tetrarch of Trachonitis, and Gaulanitis, and Batanaea also, thirty-seven years" (Book 18, Ch 4, Para. 6). The twentieth year of the reign of Tiberius would have been 33/34 C.E., and the thirty-seven years of Philip's reign would therefore have begun in 4/3 B.C.E. Since Philip began his reign as Tetrarch shortly after the death of his father, Herod the Great, this passage makes it clear that Herod's death had to have been associated with the eclipse of 4 B.C.E.

3. Some biblical apologists have maintained that two stone inscriptions found in 1912 shows that Quirinius could have been governor of Syria at an earlier date. See Sir W. M. Ramsey, *The Bearing of Recent Discovery on the Trustworthiness of the New Testament* (New York: Hodder and Stoughton, 1915), which gives Ramsey's own account of his finding of the stones, as well as his translation of the inscriptions and his speculations about their significance. However, as Ramsey's own translation shows, the inscriptions are, in fact, undated and establish only that Quirinius was chief magistrate in Antioch, Turkey, at some time. The idea that the inscription shows that Quirinius could have been governor of Syria earlier than 6 C.E. was nothing but pure conjecture on Ramsey's part.

4. Historian Robin Lane Fox is quite clear on this point. *Op. cit.*, p. 29.

5. Josephus, *The Jewish War*, Book 2, Chapter 8, Para 1.

6. Ibid. Book 2, Chapter 17 Par 8.

7. Josephus, *Antiquities of the Jews*, Book 18, Chapter 1, Par 1. Note that in his *Jewish War*, Josephus called Judas a Galilean, but here called him a Gaulonite from the city of Gamala. Galilee and Gaulanitis bordered each other in northern Palestine, and the town of Gamala was only about 15 miles from Galilee, so Judas may very well have lived in both locations at different times. As a result of his further research, Josephus might have learned that Judas would be more properly called a Gaulonite.

8. The KJV translation is as follows: "And Jesus himself began to be about thirty years of age, being (as was supposed) the son of Joseph. . . ." Now, one can "begin" to be thirty years of age, or one can be "about" thirty years of age, but to say that one "began to be about" thirty years of age is muddled. Several other versions of the Bible, such as the NIV and the NEB, translate the passage in a way similar to that of the RSV, while others translate it as the KJV does.

9. Matthew 4:12–17 and Mark 1:14.

10. One example is the Greek myth about Danae, whose father confined her in a tower because an oracle foretold that she would bear a son who would kill him. Zeus, however, passed into the tower in form of a golden shower, materialized as a young man, and impregnated her.

11. See Hyram Maccoby, *The Mythmaker: Paul and the Invention of Christianity* (Barnes and Noble Books, 1998) for the evidence concerning this.

Appendix C The Bible and the Book of Enoch

1. Unless otherwise noted, all quotations from the Book of Enoch are from R. H. Charles, *The Apocrypha and Pseudepigrapha of the Old Testament in English* (London: Oxford University Press, 1913 (1969 printing)) Vol 2, *Pseudepigrapha*. See also James H. Charlesworth, *The Old Testament Pseudepigrapha* (New York: Doubleday, 1983), *Volume I, Apocalyptic Literature and Testaments*, pp. 5–89. The numerical breakdown of the verses that Charles provided in the Book of Enoch (and followed by Charlesworth) does not lend itself to easy use in quotations. Therefore, in multi-verse passages that are quoted here the verse numbers will usually be given in the introduction to the passage rather than interspersed throughout the passage. In some cases, the verse numbering is more straightforward and is included in the passages.

2. According to the historical record, there was an extensive famine about the year 45 C.E. in the lands of the eastern Mediterranean. The famine is referred to in Acts 11:28–29. Jude may have thought that famine, or perhaps another one he was more familiar with, fulfilled Enoch's "prophecy" that the rain would be held back and the fruits would not grow in their time.

3. In translating this passage, Charles used the word "orbits" in describing the motions of the stars, which the modern reader might view with as having an intent different from that of the author of the Book of Enoch. A more appropriate word might be "circuit," because the Book of Enoch makes it clear that the stars are moving across the solid firmament of the

biblical heaven and not in orbits in the outer space of which we know. Charlesworth translates the word as "courses" (see Charlesworth, p. 59).

4. See Charlesworth, p. 55, fn 76b. Charlesworth translates it as "in all the directions."

5. See Charles, p. 180, for some examples of phrases and terms found in the Book of Revelation that are also found in the Book of Enoch. A reading of the Book of Enoch will produce many more.

6. The numbering of these chapters is not in sequence in Charles's translation; he placed the text relating the weeks of generation in sequential order, which necessitated shifting the chapters. Charlesworth, in his translation, kept the original order of the chapters, which resulted in the weeks of generation being listed out of sequence.

Appendix D The Bible and Babylonian Mythology

1. James B. Pritchard, *Ancient Near Eastern Texts Relating to the Old Testament* (Princeton, N.J.: Princeton University Press, 1969) p. 61.

2. See Pritchard, pp. 42–44 and 93–95, for translations of the tablets in which the stories are related.

Bibliography

Books

Alters, Brian J. and Sandra M. Alters, *Defending Evolution in the Classroom.* Sudbury, Mass: Jones and Bartlett Publishers, 2001

Avalos, Hector. *Fighting Words: The Origins of Religious Violence.* Amherst, NY: Prometheus Books, 2005

Barnes. Thomas G. *Origin and Destiny of the Earth's Magnetic Field.* El Cajon, CA: Institute for Creation Research, 1983

Bauer. See Danker.

Boston, Robert. *The Most Dangerous Man in America? Pat Robertson and the Rise of the Christian Coalition.* Amherst, NY: Prometheus Books, 1996

———. *Why the Religious Right is Wrong about Separation of Church and State.* Buffalo, NY: 1993

Bouw, Gerardus D. *Geocentricity.* Cleveland, Oh: Association for Biblical Astronomy, 1992

Brenton, Sir Lancelot C.L. *The English Translation of the Septuagint Version of the Old Testament.* London, 1851. Original ASCII edition copyright © 1988 by FABS International. Revised electronic version copyright © 1998-199 by Larry Nelson, Rialto Calif. The version used in this book is available in the BibleWorks™ software program.

Brown, Francis, S.R. Driver, and Charles A. Briggs. *Hebrew-Aramaic and English Lexicon of the Old Testament* (Abridged BDB-Genesius Lexicon). ASCII version Copyright © 1988-1997 by the Online Bible Foundation and Woodside Fellowship of Ontario, Canada. The version used in this book is available in the BibleWorks™ software program.

Brown, Raymond E. *The Birth of the Messiah: A Commentary on the Infancy Narratives in the Gospels of Matthew and Luke*. New York: Doubleday, 1993

Burman, Edward. *The Inquisition: Hammer of Heresy*. Dorset Press, 1984

Callahan, Tim. *Bible Prophecy: Failure or Fulfillment?* Altadena, Calif: Millennium Press, 1997

———. *Secret Origins of the Bible*. Altadena, Calif: Millennium Press, 2002

Canfora, Luciano. *The Vanished Library: A Wonder of the Ancient World*. Berkeley and Los Angeles: University of California Press, 1990

Charles, R. H. *The Apocrypha and Pseudepigrapha of the Old Testament in English*. London: Oxford University Press, 1913 (1969 printing) Vol 2, *Pseudepigrapha*

Charlesworth, James H. *The Old Testament Pseudepigrapha*. New York: Doubleday, 1983, *Volume I, Apocalyptic Literature and Testaments*

Cohn, Haim. *The Trial and Death of Jesus*. Old Saybrook, Conn: Konecky & Konecky, 2000

Crier, Catherine. *Contempt: How the Right is Wronging American Justice*. New York: Rugged Land, 2005

Dalrymple, G. Brent. *The Age of the Earth*. Stanford, CA: Stanford University Press, 1991

Danker, Frederick William (editor), *Greek-English Lexicon of the New Testament and Other Early Christian Literature*, Third Edition. Chicago: University of Chicago Press, 2000. Based on the previous English edition of Walter Bauer's *Greichesh-deutsches Wöterbuch zu den Schriften des Neuen Testaments und für frühchistlichen Literature*, sixth edition, and on the previous English editions by W.F. Arndt, F.W. Gingrich, and F.W. Danker. The version used in this book is available in the BibleWorks™ software program.

DeYoung, Donald. *Thousands not Billions*. Green Forest, AR. Master Books, 2005

Ehrman, Bart D. *Lost Christianities: The Battles for Scripture and the Faiths we Never Knew*. New York: Oxford University Press, 2003

———. *Misquoting Jesus: The Story Behind Who Changed the Bible and Why*. HarperSanFrancisco, 2005

Eldredge, Niles. *The Triumph of Evolution and the Failure of Creationism*. New York: W.H. Freeman and Company, 2000

Encyclopedia Britannica, 15th Edition. Chicago. 2005

Finkelstein, Israel, and Neil Asher Silberman. *The Bible Unearthed: Archaeology's New Vision of Ancient Israel and the Origins of its Sacred Texts.* New York: The Free Press, 2001

Fox, Everett. *The Five Books of Moses.* N.Y.: Schocken, 1995

Fox, Robin Lane. *The Unauthorized Version.* New York: Alfred A. Knopf, 1992

Friberg, Timothy, and Barbara Friberg. *Analytical Lexicon to the Greek New Testament.* Grand Rapids: Baker Academic, 2000. The version used in this book is available in the BibleWorks™ software program.

Friedman, Richard Elliot. *The Disappearance of God: A Divine Mystery.* Boston: Little, Brown and Company,1995

———. *Who Wrote the Bible?* New York: Harper and Row, 1989

Godfrey, Laurie R. (editor). *Scientists Confront Creationism.* NY: W. W. Norton & Co., 1983

Goguel, Maurice. *Jesus and the Origins of Christianity* (translated by Olive Wyon). Harper & Row, 1960

Goldberg, Michelle. *Kingdom Coming: the Rise of Christian Nationalism.* New York: W.W. Norton & Company, 2006

Gordon, Cyrus H. *The Ancient Near East.* New York: W.W. Norton & Company, 1965

Grant, Michael. *Jesus: An Historian's Review of the Gospels.* New York: Charles Scribner's Sons, 1977

Graves, Robert, and Raphael Patai. *Hebrew Myths: The Book of Genesis.* New York: McGraw-Hill, 1964

Hall, Marshall. *The Earth is Not Moving.* Athens, Ga: Fair Education Foundation, 1991

Harper's Bible Dictionary. San Francisco: Harper and Row, 1985

Hayden, Dan. *Did God Write the Bible?* Wheaton, IL: Crossway Books, 2007

Heath, Sir Thomas L. *Greek Astronomy.* New York, NY: Dover Publications, Inc., 1991

Heddges, Chris. *American Fascists: The Christian Right and the War on America.* New York: Free Press, 2007

Heidel, Alexander. *The Gilgamesh Epic and Old Testament Parallels.* Chicago: University of Chicago Press, 1949

Helms, Randel. *Gospel Fictions.* Amherst, NY: Prometheus Books, 1988

Holladay, William L. *A Concise Hebrew and Aramaic Lexicon of the Old Testament.* Leiden (The Netherlands). E.J. Brill. 1988

Hughes, Liz Rank (editor). *Reviews of Creationist Books.* Berkeley, Ca: The National Center for Science Education, Inc., 1992

Humphries, D. Russell. *Starlight and Time.* Green Forest, AR. Master Books, 1994

Isaak, Mark. *The Counter-Creationism Handbook.* Berkeley and Los Angeles: University of California Press, 2007

Jidejian, Nina. *Tyre Through the Ages.* Beirut: Dar El-Mashreq Publishers, 1969

Josephus, *Antiquities of the Jews*

———, *The Jewish War*

Kaku, Michio. *Einstein's Cosmos: How Albert Einstein's Vision Transformed Our Understanding of Space and Time.* New York: Atlas Books, LLC, 2004

Katzenstein, H. Jacob. *The History of Tyre.* Beer Sheva. Ben-Gurion University of the Negev Press, 1997

Kitcher, Philip. *Abusing Science: The Case Against Creationism.* Cambridge: The MIT Press, 1982

Klein, Ernest. *A Comprehensive Etymological Dictionary of the Hebrew Language for Readers of English.* New York: Macmillan Publishing Co., 1987

Koehler, Ludwig, and Walter Baumgartner. *The Hebrew and Aramaic Lexicon of the Old Testament.* Translated and Edited under the supervision of M.E.J. Richardson. Leiden (The Netherlands). Koninkligke Brill NV, 1994-2000. The version used in this book is available in the BibleWorks™ software program.

LaHay, Tim. *The Merciful God of Prophecy.* LaHay Publishing Group, 2002

Linn, Jan G. *What's Wrong with the Christian Right?* Boca Raton, Florida: Brown Walker Press, 2004

Lynn, Barry. *Piety and Politics.* New York: Harmony Books, 2006

Maccoby, Hyram. *The Mythmaker: Paul and the Invention of Christianity.* Barnes and Noble Books, 1998

McDowell, Josh. *Evidence that Demands a Verdict.* San Bernardino, Calif: Here's Life Publishers, 1979

Miller, Kenneth R. *Finding Darwin's God.* New York: Perennial, 1999

Montagu, Ashley (editor). *Science and Creationism*. New York: Oxford University Press, 1984

Morris, Henry M. *The Biblical Basis for Modern Science*. Grand Rapids, Mich: Baker Book House, 1984

Morris, John D. *The Young Earth*. Green Forest, AR. Master Books, 1994

New, David S. *Holy War: The Rise of Militant Christian, Jewish and Islamic Fundamentalism*. Jefferson, North Carolina, McFarland & Co., Inc., 2001

Pelikan, Jaroslav. *Whose Bible is it?* New York: Penguin Books, 2005

Perakh, Mark. *Unintelligent Design*. Amherst, NY: Prometheus Books, 2004

Plaidy, Jean. *The Spanish Inquisition: Its Rise, Growth, and End*. New York: Barnes & Noble, 1994

Pritchard, James B. (editor). *Ancient Near Eastern Texts Relating to the Old Testament*. Princeton, N.J.: Princeton University Press, 1969

Rad, G. von. *Genesis*. Philadelphia: The Westminster Press, 1961

Ramsey, Sir W. M. *The Bearing of Recent Discovery on the Trustworthiness of the New Testament*. New York: Hodder and Stoughton, 1915

Robertson, Pat. *The New World Order*. W Publishing Group: March, 1992

Rosen, Hanna. *God's Harvard: A Christian College on a Mission to Save America*. Orlando, Fla.: Harcourt, 2007

Ross, Hugh. *A Matter of Days*. Colorado Springs, CO: Navpress Publishing Group. 2004

Rushdoony, John, *The Institutes of Biblical Law*. P & R Publishing, 1973

Sagan, Carl. *Cosmos*. New York: Random House, Inc., 2002

Schermer, Michael, *Why Darwin Matters*. New York: Henry Holt and Company, LLC, 2006

Schroeder, Gerald. *Genesis and the Big Bang*. New York: Bantam Books, 1990

———. *The Science of God*. New York: Broadway, 1998

Scott, Eugenie C. *Evolution Vs. Creationism: An Introduction*. Berkeley: University of California Press, 2004.

Silk, Joseph. *The Infinite Cosmos: Questions from the Frontiers of Cosmology*. Oxford, England: Oxford University Press, 2006

Singh, Simon. *Big Bang: The Origin of the Universe*. New York: HarperCollins, 2004

Smith, Homer. *Man and His Gods*. New York: Grosset & Dunlap, 1957.

Smoot, George and Keay Davidson, *Wrinkles in Time* New York: William Morrow & Company, 1993

Stadelmann, Luis. *The Hebrew Conception of the World.* Rome: Pontifical Biblical Institute, 1970

Stiebling, Jr., William H. *Out of the Desert? Archaeology and the Exodus/Conquest Narratives*: Buffalo, NY: Prometheus Books, 1989

Strahler, Arthur N. *Science and Earth History.* Buffalo, NY: Prometheus Books, 1987

Strong, James. *Strong's Exhaustive Concordance of the Bible.* Madison, N.J. 1890. Published as *Strong's New Exhaustive Concordance of the Bible.* Iowa Falls, IA: World Bible Publishers, 1986.

———. *The New Strong's Expanded Exhaustive Concordance of the Bible.* Nashville, TN: Thomas Nelson Publishers, 2001

Thayer, Joseph Henry. *Greek-English Lexicon of the New Testament.* Divinity School of Harvard University, 1889. Electronic Version generated by International Bible Translators (IBT), Inc., 1998-2000. The version used in this book is available in the BibleWorks™ software program.

The Interpreter's Bible. New York: Abingdon Press, 1956

The Interpreter's Dictionary of the Bible. New York: Abingdon Press, 1962

The New Westminster Dictionary of the Bible. Philadelphia: The Westminster Press, 1970

Westermann, Claus. *Genesis 1–11: A Commentary*, translated by John J. Scullion. Minneapolis: Augsburg Publishing house, 1984

White, Andrew D. *A History of the Warfare of Science with Theology in Christendom.* New York: 1896. (Prometheus Books reprint; Buffalo, NY: 1993)

Williams, Alex and John Harnett. *Dismantling the Big Bang.* Green Forest AR: Master Books, 2005

Articles

"A Rocky Start." Ivars Peterson. *Science News*, March 20, 1993, Vol. 143, No. 12

"Apocalypse Soon; Israel's occupation of the West Bank has no stronger supporters in the United States than the evangelicals of the Christian Coalition. But for them it's just another step in the run-up to Christ's Second Coming." *The Independent on Sunday* (London, England), June 29, 2003.

"Boss Pat: comparing the Christian Coalition to the Tammany Hall political machine...." *Church and State*, October 1, 1997

"Christian Reconstructionism: Religious Right Extremism Gains Influence—Part Two." Frederick Clarkson. *The Public Eye*, June 1994

"Concerned about Concerned Women for America." Schollenberger, Julie. *Freedom Writer*, January 1995

"Contra-Contraception." Russell Shorto. *The New York Times Magazine*, May 7, 2006

"Corals as Paleontological Clocks." S. K. Runcorn. *Scientific American*, October 1966, pp. 26–33

"Enigmatic relic was eclipse calculator." *New Scientist*, December 2 – 8, 2006.

"Evangelicals Split on Global Warming." *The Christian Century*, March 7, 2006.

"From the Superchron to the Microchron: Magnetic Stratigraphy in Deep Sea Sediments." Jean-Pierre Valet. *Oceanus*, December 22, 1993.

"Galaxy Redshifts Reconsidered." Sten Odenwald and Richard Tresch Fienberg. *Sky and Telescope*, February 1993

"Happy Birthday Supernova!" Ronald A. Schorn. *Sky and Telescope*, February 1988

"Home Schooling: Preach Your Children Well." New Scientist, Nov. 11–17, 2006.

"How Astronomers Glimpse the Naked Galaxy." Robert Benjamin. *Astronomy*, February 2007

"Hubble eyes disks that may form planets," *Science News*, June 18, 1994 (Vol 145, No. 25)

"Hubble scopes possible planet-forming disks," *Science News*, December 19 & 26 (combined issue), 1992 (Vol. 142, Nos. 25 & 26

"Insight into Star Death." Richard Talcott. *Astronomy*, February 1988

"Justice's Holy Hires," *The Washington Post* (Outlook), April 8, 2007

"Laws of change: Christian college hopes its students will change society," *Roanoke Times* (Roanoke, Va.), August 19, 2006

"Low-Mass Stars: Born to make Planets," *Science News*, January 16, 1993 (Vol 143, No. 3)

"Neutrinos from Supernova 1987A." J. M. Lattimer and A. S. Burrows. *Sky and Telescope*, October 1988

"New evidence of dust rings around stars," *Science News*, October 3, 1992 (Vol 142, No. 14)

"Paving the Seafloor--Brick by Brick. New Vehicles and Magnetic Techniques Reveal Details of Seafloor Lava Flows." Maurice A. Tivey, *Oceanus*, September 22, 2004

"Religious Right Scuttles Group's Plan to Address Global Warming." *Church and State*, March 1, 2006

"Repaired Hubble Finds Giant Black Hole," *Science News*, June 4, 1994 (Vol. 145, No. 23)

"Supernova yields cosmic yardstick," Ron Cowen, *Science News*, January 26, 1991 (Vol. 139, No. 4)

"The Confirmation of Continental Drift." M. Hurley. *Scientific American*, April 1968

"The Flat Out Truth." Robert Schadewald. *Science Digest*, July, 1980

"The Golden Age of Cosmology," *Scientific American*, July 1992

"The Moon's Tug Stretches Out the Day." R. Monastersky. *Science News*, July 6, 1996 (Vol. 150, No. 1)

"The Murder of Hypatia of Alexandria." Ronald E. Mohar. *Free Inquiry*, Spring 1983 (Vol 3, No. 2)

"The Planet Hunters." Robert Irion. *Smithsonian*, October 2006

"The Reconstructionist Connection: Putting the 'Radical' in Radical Religious Right." Joseph L. Conn. *Church and State*, March 1996

"The Reconstructionist Movement on the New Christian Right," Anson Shupe. *The Christian Century*, October 4, 1989,

"Thy Kingdom Come." Rob Boston. *Church and State*, September 1988

"Tides and the Earth-Moon System." Peter Goldreich. *Scientific American*, April 1972

"Warped worldview: Christian Reconstructionists believe democracy is heresy, public schools are satanic and stoning isn't just for the Taliban anymore—and they've got more influence than you think," *Church and State*, July 1, 2006

Subject Index

'ab 86, 87
'abaddown 165, 166
abbreviations
   calendrical 311
   of Bible versions...................................... 5
   of lexicons 6
Abell 1689 galaxy cluster 36
Abraham 18, 177, 244
   anachronisms in Bible concerning...... 232
abussos 167, 168
abyss, the (biblical) ....67, 103, 138, 167. *See also* deep, the; *tehowm*; *abussos*.
   as home of demons 167
   as watery chaos...................157, 159, 161
   definitions of 168
   earth will fall into, in the End Times 137–38, 290–91
   in Book of Revelation................. 167, 168
   location of Sheol in... 160, 161, 162, 166, 168
   of the deep 107, 129, 164, 166, 191
   pre-existing.. 166
   Satan as king of 167
   under the earth... 135, 137, 138, 150, 161, 163, 167, 289, 290
   waters of..................... 130, 161, 163, 164
Adam 204, 284, 302
   and the genealogy of Jesus 244, 248, 300–302
'aguddah.. 121, 122
aion 180
aionas 180
akron 119
al......................................122, 132, 149, 150
   meaning of............................ 98, 132, 149
Alexander.. 225, 230
   siege and conquest of Tyre by ....222–23, 224; *See also* Tyre

Alexandria 197
   destruction of libraries of.... 198, 206, 319
   Hypatia of..................................206, 319
   Moslem conquest of 319
   Septuagint translated in 83, 101, 267
'aliyah 94, 128
'almah 266, 267
Alpha Centauri 40, 41
'amats 86
'ammuwd 133
'anan 86
Andromeda 34, 55
Antigonus 223, 225
Antikytheria mechanism......................... 319
Antiochus Epiphanes 230, 231
Antipas............................ 254, 258, 260, 326
anti-separationists .16, 17. *See also* Religious Right
Apocalypse 23, 208, 278
Apocrypha and Pseudepigrapha of the Old Testament in English, The 327
Apocrypha, the............................. 11, 12–13
   (note) 130, 139
apographo 255
Aramaic 10
Archelaus 258, 260
   deposition of......................254, 257, 259
   no mention of, in Luke's gospel 253, 259
   rule of, in Judea 248, 251, 254, 258, 259
Aristarchus 176, 197–98
Aristotle 196
Arius 12
arubbah 93, 129
'asah 83, 86
Association for Biblical Astronomy......... 146
ASV (Bible version abbreviation) 5
Athanasius 12
B.C.E. (calendrical abbreviation) 311

337

Subject Index

Babylon 233, 234, 249
Babylonia 216, 232
Babylonian Empire 232
Babylonian exile. *See* exile, Babylonian
Babylonian mythology 305
   cosmic ocean in 67, 306
   heaven in 306
Babylonians 90, 127, 216, 224
   and Ezekiel's prophecy against Tyre 218–20
   Jerusalem razed by 216, 218
BDAG (lexicon abbreviation) 6
BDB (lexicon abbreviation) 6
beliy (belii) 159, 164, 165
beliymah 159, 160, 164, 165, 166
   equated with tohuw 159
beliyyaal 164
Bethlehem 248, 249, 250, 251, 252, 253, 260, 261, 262
bethuwlah 266, 267
Bible, the, .. 44. *See also* Old Testament, the; Hebrew Scriptures, the; New Testament, the; cosmology (biblical)
   abbreviations of versions of, used in this book 5
   anachronisms in 229, 230, 232, 233
   and morality 203–5
   and the age of the earth 50
   Christian version of 12–13
   cosmological view of.... 65, 78, 143, 191, 192, 194, 281, 282, 289, 294
   effects of belief in, on society 16–26
   God of.. 65, 126, 189, 191, 192, 195, 198, 199, 204, 205
   origins of 10–14
   references to non-biblical books in 14, 175, 280
   source documents for certain books of 233–42
BibleWorks™ 6
Big Bang 55–56, 57–58
   attempts to reconcile the Bible with .. 71–72
Bill of Rights 17
birth control 22–23
black holes 42, 45
bohuw 67
boqer 141
brephos 250
C.E. (calendrical abbreviation) 311
Caesar Augustus 248
Calvin 144
canon *See* Christian scriptural canon; Hebrew scriptural canon
canopy (of water) 78
captivity, Babylonian ...... 216. *See also* exile, Babylonian

Cercle Scientifique et Historique 146
Chaldean Dynasty 232
Chaldees 232
chaqaq 102, 103, 106
chatham 173, 295
chazaq 84
cheder 94, 152
choq 106
choshek 68, 106, 141, 142
Christ 17, 269, 303
Christian Coalition 17
Christian Nationalists 17, 18, 208
Christian Reconstructionists. *See* Reconstructionists, Christian; Reconstructionism
Christian scriptural canon. ....*See also* New Testament, the
   and the Apocrypha 12–13
   development of 11–13
Christian Topology 145
Christianity 18, 208, 281
   early
      and belief in the Last Days 268
      and belief in the virgin birth 264
      and destruction of Alexandrian libraries 198
      and development of the scriptural canon 11–13
      flat earth belief in 145
      period before the rise of 304
   history of 205–8
   modern
      and belief in a flat earth 146–47
      and belief in an immovable earth 145–46
   effects of, on society 15–25
Christians 16, 17, 208
   and birth control 22
   and home schooling 24
   and morality 205
   and the Second Coming 278
   early 90
      and belief in the Last Days 186
      called Nazarenes 261
      forced conversions by 206
      scriptures of 11, 276, 280
   Old Testament edicts ignored by modern 203
christos 11, 269, 303
church and state, separation of 16–18, 207
chuwg 99, 100, 101, 102, 103, 104, 105, 107, 108
COBE 57
Concerned Women for America 17
Constantine 12, 205
continental drift 51–52
Copernicanism 145
Copernicus 143, 144, 198, 199

Subject Index

corals 46–47
cosmic ocean (biblical)...106. *See also* abyss, the (biblical); deep, the (biblical); tehowm
   as source of rain.............................. 131–32
   circle drawn of face of......................... 105
   earth and firmament of heaven located within 127–32, 137
   earth founded upon........................ 132–33
   in Babylonian myth 67, 305
   pre-existing............................ 67, 79, 103
cosmology (biblical) .186, 192, 194, 199. *See also* Enoch, Book of; cosmos, the (biblical); cosmic ocean (biblical)
   and attempts to reconcile with science 148, 195
   and credibility of the Bible.......... 200, 213
   and heaven.. 162
   and Sheol..................................... 161, 162
   and the Book of Enoch 286, 287, 288, 289, 298
   and the Epistle of Jude........................ 199
   and the Flat Earth Society................... 147
   Babylonian influence on.............. 127, 305
   different from modern cosmology 185, 201
   flat earth nature of 123, 146–47, 199
   geocentric nature of............................ 192
   the earth the only world in 179
cosmology (non-biblical) .. *See also* cosmos, the (non-biblical)
   different from biblical......................... 185
   understanding of, by ancient Greeks 195–98
cosmos, the (biblical) .....*See also* cosmology (biblical); light (of day, biblical); abyss, the
   (biblical); cosmic ocean (biblical); earth, the (biblical); firmament, the (heaven, sky)
   exists only for the earth 65
   geocentric view of 65
   initial state of.. 67
   the deep in 67
   tiers of................. . 80, 130, 131, 150, 289
cosmos, the (non-biblical) 29–62
   and the Big Bang 53–58
   earth not favored in............................. 169
   evidence for.................................... 38–52
   overview of...................................... 29–37
   size of 59–61
Crab Nebula 44
creation (biblical)..................... 169, 179, 194
   and interpretations of length of days of 72–75
   and the problem of starlight.......... 171–72
   described in Proverbs 8:26 103–4
   earth the centerpiece of.......... 65, 181, 186

   first act of 66–68
   inconsistent stories of, in Genesis. .75–77, 242
   order of.. 69
   reiterated in Job 26 .... 105–7, 290
   similarities to Akkadian account 306
   source documents of..................... 76, 234
   week 68, 72–75, 172
Creation Research Society (CRS) .... .313
creation science .... 19–20. *See also* creationism; creationists
creationism .... .21
   scientific .... .19
   updated as intelligent design 20
   young-earth.. .24
creationists 19–21, 25, 146, 172
   and debates with evolutionists............ 147
   and the Big Bang 55
   and the earth's magnetic field . .50–51, 52
   and the Flood...................................... 308
   assertions by, about age of the earth.... 44, 48, 50, 72–73, 72
   creation model promulgated by 20, 21, 308, 309
   efforts by, to foster creationism in schools 19–20
   geocentric .... .193
   old earth.. 72
   selective use of evidence by 51
   the Bible viewed as a book of science by 193
   young earth .... .72
CRS 313
Cyrenius .. 248, 256, 257. *See also* Quirinius
Cyril .... .206
Cyrus .... .231, 233
D document 233
Daniel, Book of....................... .229, 269, 281
   evidence of second century B.C.E. origin of.. 229–32
   flat earth cosmology of................. 123–24
Darby (Bible version abbreviation) 5
darkness (of night, biblical) 69, 74, 106
   cause of, different from modern view. 141
   division of, from light..... . 68, 71, 78, 106, 107, 141
   dwelling place of 142
David .... .111, 163, 164, 177, 244, 264
day (biblical) .... 71. *See also* light (of day, biblical)
   as evening and morning............ 68, 73, 75
   instituted as period of time .... .68
   interpretations of length of 72–73
   not caused by a rotating earth............. .68
day (non-biblical)
   length of, increasing 46–47
   marked by the earth's rotation............. .29

de. 249
deep, the.... 132, 201, 235, 289, 306, 307. *See also* abyss, the (biblical); tehowm
   as a tier of the biblical cosmos.... 130, 289
   as source of springs 129
   as watery chaos... 157–58, 157, 158, 159, 161
   called the great deep .... .129
   circle on face of 103–5, 105–7, 107
   earth above 134, 290
   location of Sheol in............... .160–68, 191
   pillars of the earth within..................... 190
   pre-existing......................... .67, 68, 79, 103
   tehowm translated as... 67, 103, 128, 129, 157, 161, 162, 306
   under the earth..... 129–30, 132, 158, 164, 290
Democritus 178, 196
Dialogue of the Two Great World Systems 144
Dominionists....................................... 17, 208
Doppler effect 53, 54
doulos 204
duwr .... .101
E document 233
Earth Not a Globe .... .147
earth, the (biblical) 80
   age of.................................. 72–73, 73–75
   and Job 26:7 148–68
   and light of the daytime sky .71, 72, 141, 142
   as a disk.. 135
   as a tier of the biblical cosmos... 130, 150, 289
   centerpiece of creation.......... 65, 181, 186
   circle of.... 98–100, 102–7, 108, 110, 111, 123, 124, 125, 150, 193
   corners of.............................. 119–20, 292
   cosmic ocean under 129
   creation of 65, 76, 98, 107, 158, 168, 169
   day and night created before........... .68, 69
   defined as a disk 99–105, 100, 108
   disk of.109, 110, 114, 129, 130, 133, 141, 150, 151, 166, 172, 179, 181, 201, 288, 289, 292
   ends of .109–14, 116, 117, 120, 193, 289, 290, 292, 295
   fixed in place (immovable).. 135–47, 148, 149, 170, 173–76, 181, 190, 199, 290
   flat shape of 101, 104, 108–9, 110, 111, 114, 117–19, 123–24, 127, 129, 134, 135, 137, 139, 140, 141, 146–47, 150, 151, 181, 185, 190, 192, 196, 199, 288, 292, 317
   founded upon waters 132–33
   not a sphere . 99–101, 109, 111, 114, 118, 123–24, 134, 137, 139, 147, 173, 199
   not rotating   136, 139, 140, 142, 172–73, 173–76
   on pillars..................... 133, 139, 190, 196
   only world in the cosmos.................... 179
   pillars of 133–39, 164, 190, 196, 199, 201, 317
   solid firmament above 80, 83, 92
   solid heaven above ...... .82, 86, 88, 92, 96
   state of, before creation 66–67
   vault of heaven on 114–24, 127, 135
   vault of heaven rests on .. 121–22, 288–89
   will fall into the abyss in the End Times 137–38, 290–91
earth, the (non-biblical) 22, 23, 30, 32, 38
   age of.................................. 34, 46–52, 75
   assertions by creationists about age of.. 50
   circumference of, measured by Eratosthenes 126
   decay of the magnetic field of 50–52
   description of, in space.................. .29–37
   hydrological cycle of 131
   insignificance of, in the cosmos ...59, 61, 65
   known to be a sphere by early scholars 145
   known to be a sphere by Greek scientists 101
   length of day of, increasing 46–47
   number of stars that can be seen from . 31
   rotation of...29, 46, 47, 68, 136, 139, 142
      rejected by modern geocentrists ... 146
      rejected by Protestant reformers ... 144
   studies of the magnetic field of............ 51
Ecclesiastical History of the English People 145
'eden .... .134
Eden, Garden of 76, 201, 204, 234
Egypt .... .222, 255, 267
   and the Gospel of Matthew 248, 251, 253, 254, 261, 262, 265
   Nebuchadrezzar's plan to invade........ 216
   prophecy against, by Ezekiel 226–28
Einstein, Albert .... .45, 53, 171, 172
eklelegmenos .... .302
Elizabeth (Elisabeth)............... .246, 247, 264
Elohim 233, 234
Enlightenment, the 16, 207
Enoch, Book of... 14, 96, 120, 122, 131, 138, 140, 173, 192, 280–304
   and biblical cosmology...................... .199
   and genealogy of Jesus in Luke 248, 300–302
   and idea of hell .... .160
   and Son of Man 269, 302–4
   and the Day of Judgment................... 268
   and the New Testament 299–304
   angels as stars in the .... .183

cited by Jude........ 182, 192, 199, 282–87
damaging to the authority of the New Testament 199
early Christians and cosmology of 281
heavenly host as circling stars in 155, 296–97
epi 120
Eratosthenes..................................... 126, 197
'erets 111
ESV (Bible version abbreviation) 5
Evidence that Demands a Verdict . 220, 320, 322
exile, Babylonian 216, 220, 230
  Book of Genesis and the 305, 308
  errors in the Book of Daniel about the 230
  of Ezekiel 216
  Old Testament and the 127, 233
Ezekiel
  prophecy by, about Gog, the land of Magog 228
  prophecy by, against Egypt 226–28, 227–28
  prophecy by, against Tyre .... 216–20. *See also* McDowell, Josh; Tyre
    Bible believers' claims about 226
    events taking place at time of 220
    failure of 221–24
    incorrect claims about fulfillment of 223–24
Ezra 234
Fair Education Foundation 146
Falwell, Jerry... 17
firmament, the (heaven, sky) . . 70, 78–97, 80, 102, 103, 104, 105, 112, 121, 137, 144, 156, 166, 177. *See also* heaven (biblical); heavens
  and God's abode 155
  and God's throne 88, 154, 155
  and modern geocentrists 146
  as a solid structure 79–90, 91–93, 95–96, 191
  as a tier of the biblical cosmos..... 80, 130, 150, 289
  as a vault..... . 82, 96–97, 102, 103–4, 105, 107
  attempts by apologists to identify ... 78–79
  beaten out by God.......................... 84–86
  canopy of water above 79
  creation of................................. 73, 157–59
  earth under.. 111
  ends of 115
  location of, in biblical cosmos............... 79
  made by God 79, 83–84, 108
  nature of, as defined by raqiya` 80–83
  north as a synonym for 157
  physically near the earth 124–26, 172
  placed on the disk of the earth 127

rests on the earth................... 121–22, 290
stretched over tohuw........................... 159
sun moves across................ 139, 170, 174
sun, moon, and stars, set in......... 169, 181
waters above.... 78–79, 79, 80, 82, 83, 84, 85, 88, 90, 93, 97, 98, 127, 128, 129, 130, 132, 133
waters divided by......................... 79, 127
will be rolled up in the Last Days . 92, 186
firmamentum 79, 80, 82, 83
Five Books of Moses 10
Flat Earth Society 147
flat earthers 192
Flood, the 76, 78, 79, 128, 129, 193, 194, 309
  antecedents of................................ 307–8
  claimed evidence for, by creationists.. 308
  prophecy of, in the Book of Enoch..... 289
  source documents for the biblical version 234–42
Friberg (lexicon abbreviation) 6
Friedman, Professor Richard Elliott 234
fundamentalists 19, 35, 203
galaxies 34
  and expansion of the universe 53
  formation of.. 56
  light from, red shifted..................... 53, 54
  measuring distance to 45
  movement of.................................... 54, 56
  number of, in the universe 60
  supernovas observed in........................ 42
Galilee ... 248, 251, 252, 253, 254, 260, 277, 326
Galileo 143–44, 199
Garden of Eden 201, 204, 234
geduwd 121
genea 270
genealogies of Jesus 244–48
Genesis Institute 146
Genesis, Book of 19, 76, 305
  anachronisms in............................ 232–33
  cosmological view of............................ 65
  interpretation of first verses of........ 66–67
Geocentricity.. 146
geocentrists 145, 146, 147, 192
global warming 23–25
globular clusters 32
God (non-biblical) 202
God (the biblical) 190, 191
  abode of.. 152–56
  and Islam 18
  observable universe not created by...... 66, 192
  reprehensible acts of 205
gonias 120
Google™ Maps™ 226
Greeks, ancient 101, 148, 181

cosmological views of 89, 90
scientific advancements of 319
scientific understanding of the cosmos by 99, 195–98
Green River formation 46
gyro 101
Hammurabi 203
haphak 136
har 154
heaven (biblical) ..... 67, 82, 83, 96, 182, 183, 270, 271. *See also* firmament, the (heaven, sky); sky
   as a physical part of the cosmos 126
   angels who left .... .282, 283, 300
   as a solid structure 80
   chambers of 94, 152, 297–98
   circle of .... .107, 108
   considered to be a physical place 162
   creation of 66–67, 158, 166, 168
   doors of 94
   ends of 114–19, 120, 289, 292
   God's throne in 153–54, 155
   host of 177, 182–83, 184, 288
   kingdom of 268
   lights of .... .184, 186, 194
   likened to a scroll .... .185, 186
   luminaries of 68, 70, 74
   new, to be established 186
   north as a figure of speech for. 151–58, 158
   pillars of 296
   plural nature of 90
   portals of .... .291, 292–95
   secrets of, explained to Enoch 287
   shall pass away .... .301
   vault of 82, 93, 96–97, 99, 104, 105, 107, 108, 111, 114, 115, 116, 117, 127, 172, 179, 184, 192, 201
      on disk of the earth ... 103–4, 119–22, 135, 150, 191, 288–89
      sun moves across . 172, 173, 174, 175, 176
   waters above .... .289
   waters gathered under .... .98, 103, 105
   will collapse on the earth 290
   windows of 93–94, 128, 131, 235
heavens (biblical) .. 76, 86, 89, 96, 125, 144, 154
   can be rolled up 92
   incorrectly translated as earth 117
   made by God 83–84
   measure surface of the earth .... .115
   north as a synonym for .... .157
   plural nature of 90–91
   spread out by God .... .157
   tabernacle for the sun in 115
   vault of 99, 107, 121

waters above .... .128
Hebrew scriptural canon
   development of.... 10–11. *See also* Old Testament, the
   establishment of, by Council of Jamnia .... .11
Hebrew Scriptures, the 11, 101, 233, 276, 289, 302
   and the writer of the Book of Enoch. .281, 287
   closed to new prophecies during third century B.C.E 229
   establishment of the canon of .... .11
   translated into the Septuagint . 10–11, 83, 101, 267
hell .... 160, 164, 165. *See also* Sheol
Heraclides .... .197
Herod 253, 254, 255, 258, 259, 261, 263
   and threat to Jesus ...... .248, 251, 253, 261
   claims about enrollment during reign of 255
   death of.... . 248, 251, 254, 255, 257, 258, 326
heylel 154, 183
History of Tyre, The 225
ho eklelegmenos 302
Holladay (lexicon abbreviation) 6
home schooling, Christian-based .... .24
Hubble Space Telescope .... .32, 45, 60
Hypatia 206, 319
ICR 313
Inquisition, the 143–44, 207
Institute for Creation Research (ICR) 313
intelligent design .... .19, 20–21
Islamic nations 18
J document 233, 234–42, 234, 235, 242
Jamnia, council of 11
Jasher, Book of 14, 175
Jehovah .... .233
Jerusalem 249, 253, 262, 265
   Babylonian siege of 216
   Christian crusaders in 206
   conquered by Babylonia 216
   destruction of 217, 220
   fall of, in 70 C.E .... .11
   order to rebuild .... .231
   razed by Antiochus Epiphanes .... .230
   razed by Babylonians 216, 218, 220, 320
   Tyre and 217, 224, 226, 321
Jesus
   and belief that the Last Days were on the earth 268–74
   arrest and crucifixion of 274–76
   belief in Second Coming of. 23, 205, 272, 276, 278
   believed heaven to be a solid vault 185
   believed the earth is flat 119

Subject Index

believed the stars will fall 185
believed to be Jewish Messiah 11
birth and early life of 248–59
dispute over triune nature of 12
genealogies of 244–48
quoted scripture not found in the Bible
   280
resurrection of 277–78
virgin birth of 263–67
Jesus Ben Sirach 229
Jidejian, Nina 226, 321, 324
   quoted by Josh McDowell .. 322, 323, 324
Job 26:7 148–68
   allusion to Sheol in 159–68
   and its supposed allusion to the earth in
      space 148–51
   meaning of "empty place" in 149, 151, 157–59
   meaning of "nothing" in 158–68
   meaning of "the north" in 149, 151–58
John the Baptist 96, 247, 253, 257, 258, 259, 264, 268
John, Gospel of 13, 258, 263
Joseph (father of Jesus) . 248, 250, 251, 252, 253, 254
   and the genealogies of Jesus 244–48
Josephus 256, 257, 275, 326
Joshua, Book of 204, 318
JPS 1917 (Bible version abbreviation) 5
JPS Tanakh (Bible version abbreviation) 5
Judas the Galilean 255–57
Jude, Epistle of 199
   and the Book of Enoch 182, 199
kanaph 112, 119
kataluma 251
KB (lexicon abbreviation) 6
Kiy 219
KJV (Bible version abbreviation) 5
Koran, the 18
kuwn 86, 87
   meaning of 70, 86, 132
lake bed strata 46
Large Magellanic Cloud 34, 43, 55
   distance to, verified 44
   Supernova 1987A observed in 42
Last Days 23, 25, 65, 189, 262, 276
   early Christians believed, to be upon the
      earth 268–74
   expectation of time of, changed 278
   stars will fall to earth during 201
Law, the (group of books of the Bible) 10
Lebanon 215, 225, 226
lexicons used in this book 5–6
   abbreviations of 6
light (of day, biblical) .... *See also* 'owr
   and the Big Bang 71–72

as separate entity from light of sun 68–72, 73, 140–42
creation of 68, 78, 106
division of, from darkness 68, 106, 107
dwelling place of 142
light year, definition of 30
Luke, Gospel of 13
   and the enrollment for taxation ... 253–57, 258, 259, 260–61, 263
   attempts by author of, to have Jesus fulfill
      prophecy 260–61
   birth and early life of Jesus in 248–59
   genealogy of Jesus in 244–48, 300–302
   incompatibilities between, and Gospel of
      Matthew 244–59
Luther 144
M101 .... 34, 35
M31 .... 34, 55
M87 45
ma`alah 121
ma`aseh 83
magi 249
magnitude, visual 39
magos 249
mah 159
ma'owr 70, 71
Mark, Gospel of 13, 258, 259, 263, 277
Mary (mother of Jesus) . 244, 248, 252, 253, 259, 260, 264
   and Luke's story of the annunciation .. 258
   and Luke's story of the conception of Jesus
      258
   and Luke's story of the manger 251
   and the genealogy of Jesus in Luke .. 245–47
mashal 318
mashiyach 269, 303
matsuwq 133
Matthew, Gospel of 13
   and nazarene prophecy 261–62
   attempts by author of, to have Jesus fulfill
      prophecy 260, 261–63
   birth and early life of Jesus in 248–59
   genealogy of Jesus in 244–48
   incompatibilities between, and Gospel of
      Luke 244–59
mayim 90, 106, 161
mazzalah 179
McDowell, Josh 220
   and Ezekiel's prophecy against Tyre
      (notes 6, 7, 8, and 10) 320–25
mechuwgah 102
Messiah, the 11, 262, 266, 267, 270, 274
   and the Book of Enoch 282, 299, 302, 303, 304
   as "the Branch" 262
   as a heaven-sent personage 275

beliefs about, and the gospel 244
no prophecy about, concerning Nazareth 261
Paul's view of 276
to be a literal descendant of David ... 260, 263
mezareh 152
Milky Way (as band of light in night sky) . 32
true nature of, suggested by Democritus 178, 196
Milky Way Galaxy
description of 30
distribution of stars in 31–34
location of our solar system in 31
moon dust 52
moon, the (biblical) 169, 285
angel identified with 183, 184
called the lesser light 71, 170
created on the fourth day 65
created to be timekeeper 172
God controls movement of 174–75
ma'owr incorrectly translated as, in some Bibles 71
portals of, in the Book of enoch .. 292–94
set in the firmament 78, 169, 191
moon, the (non-biblical) 30, 170–71
and the Flat Earth Society 147
distance and size of, measured by Aristarchus 197–98
eclipses of 196
morality
and Christianity 205
and the Bible 203–5, 208–9
Morris, Henry 100, 101, 102, 136
Moses 115, 117, 118, 162, 178, 203, 204, 232, 233, 234
Moslems 226
and biblical God's promise to Ishmael .. 18
and Christian crusaders 206
palace library in Alexandria destroyed by 319
Tyre conquered and razed by 225
mowt 133, 137
NAB (Bible version abbreviation) 5
Nabopolassar 232
Nabuchodonosor 230
Nabu-kudurri-usar 230
NAE 24
naphal 138
nasiy'im 112
NASV (Bible version abbreviation) 5
natah 84, 93, 156, 157
National Association of Evangelicals (NAE) 24
Nazarene 251, 261–62
Nazareth . 248, 250, 251, 252, 253, 260, 261

and "Nazarene" reference in Matthew 261–62
Nazoraion 262
Nazoraios 262
NEB (Bible version abbreviation) 5
Nebuchadnezzar ...... 123, 320, 321, 322. *See also* Nebuchadrezzar
spelling of the name of 230
Nebuchadrezzar 225, 232, 321, 322. *See also* Ezekiel, prophecy by, against Tyre; Tyre
and Ezekiel's prophecy against Tyre 218–20
and Ezekiel's prophecy against Egypt 226–28
Egyptian campaign of 216
Jerusalem razed by forces of 217
plans of, to invade Egypt 216
siege of Tyre by 220–22, 320
incorrect claims about 223–24
spelling of the name of 230
netser 262
neutron stars 42, 44
New Testament, the 204
and cosmology 199
and the Book of Enoch 280, 281, 282, 299–304
books of 13
canon of 265, 277
development of 12–13
problems in 244–78
New Westminster Dictionary of the Bible, The 321
Nicaea, Council of 12
NIV (Bible version abbreviation) 5
NJB (Bible version abbreviation) 5
NKJV (Bible version abbreviation) 5
north, the
and Ezekiel's vision 154
and Job 37:22 152–53
and Job 37:37 156
as a figure of speech for the firmament of heaven 152–56, 157
sides of, in Isaiah 14 154
NRSV (Bible version abbreviation) 5
nuclides 47–50, 315
OEC 72
oikia 250
Old Testament Pseudepigrapha, The 327
Old Testament, the 9, 73, 126, 155, 176
and cosmology 199
and Sheol 160, 166, 167
and the Apocrypha 12
formation of 127
norms of 203
order of books of 13
problems in 213–43
sources of 76

Old Tyre 320
Omega Centauri 31, 32
Operation Rescue 18
otsar 112
overpopulation 22–23
'owr 68, 70, 71, 106, 141, 142
'owtsar 95
P document 233, 234–42, 234, 235, 242
paidion 250
Palaetyrus 320
Paleomagnetism 51
paniym 91, 106
parthenos 267
Patrick Henry College 16
Paul ... 11, 91, 204, 225, 261, 264, 265, 276, 299
    and belief that the Last Days were upon the earth 270–71
    new view of Jesus instituted by 276
    letters of 13, 265
    mission to the Gentiles by 265–66
Pentateuch, the 10
peras 113
Philip (Herod's brother) 326
Philistines 177, 232
Pilate, Pontius 254, 275
pillars of the earth .....*See* earth, the (biblical), pillars of
planet .... *See also* planets
    derivation of word 181
    earth 30, 197
    formation 60
    Venus 39, 318
planetai 181
planetes 181
planets 45, 61, 171, 172
    and Greek cosmology 148, 196, 197
    and the biblical cosmos 65, 179, 181, 189
    as representing gods 288
    as wandering stars 288
    extra-solar 34, 181
    formation of 60, 181, 315
    geocentrist view of 146
    number of, in the universe 60–61, 189
    of the solar system 30, 38, 39, 181
    origins of 60
Pluto 30, 59, 181
Prophets, the (group of books of the Bible) 10, 11, 229
Proxima Centauri 39, 314
pseudepigraphic books 228
Ptolemy 198
qara' 96
qarah 128
qatsah 110, 111, 115, 118
qatseh 109, 110, 115, 116, 117

qav 126
qerach 89
quasars 34
Quirinius .... *See also* Cyrenius
    and claims about earlier governorship of 255, 326
    as governor of Syria 248, 253, 254
    tax enrollment under 258, 259, 260
quwm 317
radioactive decay 48, 49, 50
radioactive isotopes 47
raqa' 81–82, 84, 85, 108, 157
    raqiya' derived from 81, 82
    raqiya' .. 80, 81, 82, 83, 86, 88, 89, 96, 97, 121
    derived from *raqa'* 81, 82
    meaning of 80–83, 80, 82, 121
    translated as firmament 80, 84, 89, 115
    translated as stereoma in the Septuagint 83
Reconstructionism 312
Reconstructionists, Christian ... 17–19, 20, 25
    influence of, with Religious Right 19
    philosophy of 17
    theocratic goals of, for America 17–18
red dwarf stars 39, 41
red shift, cosmological 53–55, 57
Religious Right. *See also* Reconstructionists, Christian
    and global warming 25
    anti-separationist groups within 16
    efforts by, to eliminate separation of church and state 16–17
    efforts by, to limit birth control options 22
    Reconstructionist influence with 19
Revelation, Book of 12, 228
    cosmological view of 65
Robertson, Pat 17
Romans 254, 256, 260, 275
    and deposition of Archelaus 254, 257
    and division of Herod's kingdom 254
    and the enrollment for taxation.. 254, 255, 259, 260
    response of the, to Jesus 275–76
ruwach 306
Salathiel 247
science 29
    and creationism 21, 24
    and hydrological cycle 131
    and the Bible 193, 195
    attempts to reconcile the Bible with .. 71, 72, 78, 148, 193
    identified with paganism by early Christians 206
    methodology of 19
    of the ancient Greeks 196

principles of, contradicted by creationist scenarios.................... 171–72
scientific creationism.................... 19
sea floor spreading 51–52
separation of church and state...... 16–18, 207
September 11 18
Septuagint, the......................... 168
   Brenton's translation of 5
   early Jewish Christians used................. 11
   Hebrew Scriptures translated into. 10–11, 83, 101, 267
   versions of, used in the early church 11
   version of scriptures Matthew used ... 267
Serapeum library...................... 198, 206, 319
shachaq 87
shachaqim .... .86, 87
shachath 165
shakar 141
Shalmaneser V 218
shamayim 90, 116, 117
shemesh 71
Sheol 160, 191, 282
   location of..................................... 160–68
   not a place of punishment in O.T 160
she'owl...................................... 160
Sidon 215
Sidonian port (of Tyre) 215, 322, 324
skies (biblical): 87
sky, the (biblical).83, 87. *See also* firmament, the (heaven, sky); heaven (biblical)
   and the firmament...........................78, 80
   as a vault.. 85, 97
   beaten out by God...........................84–86
   can be rolled up 92
   firm nature of... 88
   horizon defines the limits of 106
   light of the daytime, not derived from sun 68–72, 140–42
   northern night, and circling stars 156
   northern part of, as God's abode 155
   solid... 89
   stars of, shall fall from........................... 92
Small Magellanic Cloud 34, 55
solar system 45, 62, 181
   age of.. .50
   and short-lived radioactive elements . 315
   formation of... 48
   location of, in Milky Way Galaxy 31
   nearest star outside of............................ .39
   planets of .. 30, 34, 38, 39, 170, 179, 288
   size of, in relation to the universe 59
   view of, by Aristarchus............... .197, 198
   view of, by Heraclides 197
Son of man...................... 269, 270, 302–304
source documents
   and the Flood story 234–42

and the two accounts of creation in Genesis 76, 234
sphaira 101
stars (biblical)
   as angels 297
   as host of heaven 92, 93, 154
   circle throne of God............................. 155
   set in the firmament of heaven 78, 154
   will fall to the earth 93, 184–85, 191, 201
stars (non-biblical)
   Bible believers and the problem of light from.. 171–72
   circular movement of, in northern sky 155
   distribution of, in Milky Way Galaxy. 31–34
   evidence for being other suns 39–46
   geocentrist view of 146
   life cycles of 40–45
   measurement of distance of 38–39, 43–44
   number of, in the universe 60
   red dwarf 41
   supernova 41–42
   white dwarf..................................40, 41, 61
stellar evolution 40, 41, 42, 43, 44, 45
stereoma 82, 83
Strong's Concordance (lexicon abbreviation) 5
Strong's Exhaustive Concordance of the Bible 5
Strong's Expanded Exhaustive Concordance of the Bible, The New 6
suggenes 246
sun, the (biblical)..................... 169
   angel identified with.................. .183, 184
   called the greater light 71
   chariot of, in Book of Enoch 295–96
   circuit of 115
   created on the fourth day 65, 73, 74
   is small and near the earth 176, 184
   light of daytime sky not derived from 68–72, 140–42
   moves and not the earth....... 139–42, 170, 172–73, 173–76
   portals of, in the Book of Enoch ..292–94
   set in the firmament...... 78, 169, 191, 196
   statement about, in the Apocrypha (note) 139
   tabernacle for....................................... 115
sun, the (non-biblical) .... .30, 149, 172
   and Galileo's troubles with the Church 143–44
   and rotation of the earth...................... 139
   and the ancient Greek scientists . 148, 197
   end of life cycle of................................. 61
   few other stars near by.......................... 32
   life cycle of... .41
   physical description of.......................... 30

Supernova 1987A 42–43, 42–44
supernovas .... 42, 45. *See also* Supernova 1987A
   observed in other galaxies 42, 44
   process leading to explosion of 41–42
Sur 216, 225
Sûr 225
synoptic gospels 259
tabernacle (for the sun) 115
taklith 106
talah .... .149
tavek 79
tavek 79
tebel 317
tectonic plates 51
tehowm ..... 67, 103, 128, 129, 130, 157, 158, 166, 167. *See also* abyss, the (biblical); deep, the
   and Babylonian Tiamat .... .306
   as abyss under the earth 161
   as primaeval ocean 67, 103, 104
   as Sheol 162
   equivalent to Greek *abussos* 168
   meaning of 67, 129
   translated as *abussos* in Septuagint 168
Ten Commandments .... .203
Terry, Randall .... .18
Thayer (lexicon abbreviation 6
The Earth is Not Moving 146
Theophilus .... .198, 319
thermonuclear reactions .. 30, 41, 42, 45, 56, 57
Tiamat 306, 307
tidal rhythmites 47
tides, ocean 46, 47
tohuw .... .157, 158, 159
   as watery chaos... 157–58, 158, 159, 161, 166
   equated with *beliymah* 159
   equated with *tehowm* 157–58
   meaning of 67, 157
Treaty with Tripoli 17
tribulation, the .... .25, 271, 282, 299
tsaphown 151, 152, 154, 156
Tsor 216, 225
Tyre.....*See also* Ezekiel, prophecy by, against Tyre

ancient 215
Christian council at .... .12
History of, after Alexander's conquest 223–24
present-day city of 225–26
Sidonian port of 215, 322, 324
siege and conquest of, by Alexander 222–23
siege of, by Nebuchadrezzar 220–22
Tyre Through the Ages 226
universe, the. .... *See also* Big Bang
age of 34, 46
   expansion of 54–56, 55–56
   galaxies in 34
   looking back in time in 56–57
   number of planets in 60–61
   number of stars in 60
   size of 59–60
Ur of the Chaldees .... .232
Ushu 216, 218
Vanished Library, The 319
Venus 39, 197, 318
   called son of the morning .... .318, 319
virgin birth 245, 263–67
Wesley 144
white dwarf stars 40, 41
White House, the .... .17
Who Wrote the Bible? 234
Wisdom (personified) 67, 86, 103
Writings, the (group of books of the Bible) 10, 11, 229
yaal 164
yacad 122, 132, 134
Yahweh 10, 186, 189, 202, 233, 235
yam 103, 104
yamim 104
yareach 71
yashab 98
yatsaq .... .84
YEC 72
yowm 68, 71, 73
zahab 152
zebul .... .174
Zechariah 246
Zetetic Astronomy 146
Zorobabel 247

Bible Verse Index

Note: Those verse numbers marked with an asterisk (*) denote that the verse is referred to in the text rather than quoted. Otherwise the verse numbers denote that the verse is quoted in the text, either in whole or in part. In some cases, multiple versions of a verse or verses may be sequentially quoted from different versions of the Bible.

1 Chronicles
1–3*..................245
3*,......................247
3:17*..................247
16:30..................136

1 Corinthians
7:1–29............270–71
11:10..................299

1 John
2:17–18..............272

1 Peter
4:7......................272

1 Samuel
2:8......................133
2:10....................110
13:5....................177

1 Thessalonians
4:13–17..............271

2 Peter
3:10–12..............180

2 Chronicles
4:3........................84
18:18..................182

2 Corinthians
6:15....................164
12:2......................91

2 Kings
2:11....................125
23:5....................179

2 Peter
2:4......................299
3:8........................72
3:10–13..............185
3:10–13*............186
3:13..........186, 274

2 Samuel
22:5–6*..............164
22:5–7..........163–64
22:8....................135
22:10....................93
22:16–17......163–64

Acts
2:14–17..............273
2:30....................264
5:36–37*......255, 257
5:37....................255
5:37*..................256
13:47..................111

24:5....................261
24:5*..................262
26:23..................303

Amos
8:9......................174
9:1–2..................162
9:2......................125
9:6............121, 122
9:6*..........121, 122

Colossians
3:22*..................204

Daniel
2:28....................124
4:4......................230
4:10–11..............123
7:13....................269
8:10............184, 185
9:24–25*............231
9:25–26..............270
11:31*................230
12:3..............70, 141

Deuteronomy
1:10....................178
1:10*..................178
4:32....................117

348

5:21 204	26:4 223	1:9–13 98
5:8 130	26:4–6 223	1:10* 103, 105
8:7 129	26:5 223	1:11–13* 74
10:14 91	26:6–8* 224	1:14–17* 78
13:7 109	26:7 219	1:14–19 169
15:12 204	26:7–12 218–19	1:14–19* 176, 177
28:64 109	26:8 .... 223, 224	1:15–19* 74
30:3–4 115–16	26:14–17 219	1:16* 69, 71
30:4 116	26:17 223	1:16–19 170
30:4* 117	26:19–21 219	1:16–19* 170
33:13 130	28:3* 325	1:20 91
	29:1–13 226–27	1:20–31* 74
Ecclesiastes	29:17–18 .... 222, 227	2* 76, 242
1:4 138	29:19–20 227	2:1–2* 74
1:5 .... 139, 294	31:15 161	2:4–22 75–76
1:5* 139	32:4 91	2:4–22* 76
1:7 131		2:4–25* 234
1:7* 131	**Ezra**	6:2 282
1:15 173	3:2* 247	6:9–9:1 236–42
1:15* 173	3:8* 247	7:11 .... 93, 129
12:2 .... 70, 71, 141	5:2* 247	7:11* 129
		7:11–12 128
Ecclesiasticus	**Genesis**	7:11–12* 129
16:18 130	1* 73, 76, 242	8:2 93
Ch 44–50 325	1:1–2 66	8:13b* 235
	1:1–2:3 73–74	11:1–9* 125
Esdras	1:1–2:3* 75, 76, 234	11:28* 232
4:34 139	1:1–10* 106	11:31* 232
	1:2 ..67, 103, 106, 157,	15:7* 232
Exodus	306	15:15 178
2:8* 267	1:2* ..67, 79, 157, 158,	22:17 177
4:22–23 262	167, 168, 306	24:43* 266
20:4 130	1:3 68	28:12 125
20:11 75	1:3* 72	37:35 160
24:10 89	1:3–4 106	
25:12 84	1:3–5 140	**Habakkuk**
39:3 81	1:3–5* 68, 70, 71	3:11 174
	1:4–5 68	3:11* 175
Ezekiel	1:5 71	
1* 155	1:5* 68	**Haggai**
1:1–26 153	1:6 .... 81, 97	1:1* 247
1:4* 154	1:6–7 ... 79, 97, 127–28	1:12* 247
1:22* 90	1:6–8 ... 78, 96, 97, 103	1:14* 247
1:22–26 88	1:6–8* ... 79, 81, 89, 96	2:2* 247
1:22–27* 104	1:6–10* 103	2:23* 247
14:14* 325	1:7 .... 83, 157	
14:20* 325	1:7* 78	**Hebrews**
15:3 149	1:7–8 88	1:1–2 ..... 179, 180, 273
24:1–2 216	1:9 124	1:1–2* 180
26:1–2 217	1:9* 105, 170	9:26 272
26:1–2* 321	1:9–10 80	10:24–25 272
26:2 .... 217, 269	1:9–10* 103	10:37 272
26:3 220		11:3 179
26:3–6 217		

11:3* 180

Hosea
11:1 262

Isaiah
7:14 266
7:14* 266, 267
7:14–16 266
11:1* 262
11:12 119
13:4–5 118
13:10 173
13:13 .... 135, 136
13:13* 139
14:12 183
14:12–15. ..... 154, 183, 318
14:12–15* 155, 183
14:13* 152
14:4 318
14:4* 318
22:18 101
23:14* 267
24:10* 157
24:18 93
24:19–20 291
24:20 138
24:20* 138, 139
34:4 .... 92, 185
34:11* 157
37:22* 267
38:8 173
38:17 165
38:17–18 160
38:18* 160
40:22... ..... 98, 99, 111, 125
40:22* ... 99, 100, 101, 102, 104, 105, 107, 125
40:28 109–10
40:28* 110
41:29* 157
44:13 102
44:24 108
45:12 .... 83, 156
45:22 110
47:1* 267
51:9 307
55:10 131
55:10* 131, 132

62:5 267
62:5* 267
64:1 95
65:17 180
66:1 153
66:22 180

James
5:8 272

Jeremiah
5:22* 286
10:9 81
10:13 .... 94, 95, 112
25:32* 109
25:32–33 109
31:15* 262
31:37 135
33:22 177
49:36 .... 118, 119

Job
6:18 159
9:6 .... 133, 135
9:6* 134
9:7 .... 173, 294
9:7* 173, 295
9:8 156
9:9 .... 94, 297
9:9* 94
10:20 160
10:21–22* 161
19:12 121
19:12* 121
22:14 .... 107, 108
22:14* 99, 104, 105
26:5–6 .... 161, 165
26:5–7 166
26:6 .... 165, 166
26:6* 166, 168
26:7 ..... 148, 156, 157, 158, 159, 166, 168, 290
26:7* ... 148, 149, 150, 151, 154, 157, 158, 159, 164, 165, 168
26:10 105
26:10* ... 105, 106, 107
26:12–13 307
28:23 111
37:2–3 112
37:9 .... 94, 151
37:9* 151

37:18 ....84, 85, 87, 88, 157
37:18* 84, 85, 90, 154
37:22 ..... 152, 155, 156
37:22* . 151, 152, 153, 154, 155, 156
38* 95, 114, 141
38:4–5 126
38:4–6 .... 133, 134
38:6 134
38:7 182
38:12 .... 141, 142
38:12* ... 114, 142, 143
38:12–13 113
38:13* 114
38:14 136
38:19–20 142
38:19–20* 142, 143
38:22 .... 95, 297
38:22–23 95

Joel
2:10 95

John
7:37–38 280
11:48 275

Jonah
2:2–6 161–62
2:5 162

Joshua
10:12 175
10:12–13 174
10:12–13* 175, 176

Jude
6–7 282
12–13 284–85
12–13* 286
12–15 283–84
13 .... 182, 286
13–14 181
13–14* 182
14* 284
14 .... 182, 285

Judges
5:20 182

Leviticus
12:2–8* 248

Bible Verse Index

Luke
1:5 .... 246, 253
1:5* 258
1:5–80* 258, 259
1:24* 246
1:26–27 246
1:26–38* 246
1:36 246
1:39–44* 247
1:46–55* 247
2:1* 259
2:1–2 253
2:1–2 254
2:2 255
2:2* 257
2:4 246
2:4* 260
2:4–41 252–53
2:12 250
2:16 250
2:22* 248
2:32 303
2:39* 260
2:46–50 265
3:21–22* 96
3:23 .... 245, 264
3:23* 258
3:23–38* 244, 245,
246, 300
7:37–48* 246
8:3* 246
8:30–31 167
8:31 167
9:35 302
9:35* 302
10:38–42* 246
13:1 275
23:35 302
23:36 276
23:54–56* 246
24:33–36 277

Malachi
3:10 93

Mark
16:8* 277
16:9–20 277

Matthew
1:1–17* 244
1:3 245
1:5 245

1:6 245
1:16 245
3:2 268
4:8 123
4:17 268
2:1 249
2:1* 248, 249
2:1–2 249
2:1–23 251–52
2:7 250
2:11 250
2:11* 250
2:15* 261
2:16 250
2:16* 251
2:17–18* 261
2:22–23 251
2:23* 261
10:5–7 268
10:5–7* 274
10:34 274
15:24* 274
16:27–28 268
16:28 270
21:1–7* 262
21:7 263
24:3–34 269
24:3–34* 270
24:29 184
24:29* 185
24:31 .... 119, 292
24:31* 119, 185
24:34* 270
26:64 270
27:46 274
28:10–17 277
28:11–15 277

Micah
6:2 133

Nahum
1:11 164

Nehemiah
1:9 117
1:9* 117
12:1* 247

Numbers
5:12–31* 203
16:31–33 162
16:39 81–82

Proverbs
8:23–24 67
8:26–29 102
8:26–29* 103, 106
8:27 .... 102, 105
8:27* 86, 99, 100,
102, 103, 104, 105,
106, 107
8:27–28 .... 86, 87, 88
8:27–28* 102, 103,
104
8:27–29* 106, 107
8:28 103
8:28* .86, 87, 102, 103
8:28–29 103
8:29* 103, 105
11:4 164

Psalms
2:8 111
11:4 153
14:2 125
18:4–5 164
18:11 87
19:1–6 ..115, 126, 140,
294
24:1–2 .... 132, 158
24:1–2* 132
24:2 .... 122, 132
33:14 125
48:10 113
55:23 165
57:10 87
67:7 110
68:25* 267
69:14* 161
69:14–15 161
69:15 161
74:13 307
74:16 .... 70, 141
74:16* 71
75:3 .... 134, 135, 137
77:17 87
77:17* 87
78:23–24 94
78:69 138
82:5 133
88:3–7 163
89:10 307
93:1 .... 136, 138
93:1* 144
96:10 136

98:3 110
102:25 83
103:4 165
103:12 124
103:19 153
104:1–3 128
104:5 138
104:5* 138
104:13 .... 94, 297
104:19 .... 140, 173
104:19* 173
105:29 136
115:16 126
135:7 112
136:6 132

140:10 137
144:5 93
148:4 128

Revelation
1:1–7 273
1:20 183
4:1–6 89
5:13 130
6:13 92
6:13–14 185
7:1 .... 120, 292
7:1* 120
9:1 .... 167, 183
9:11 .... 166, 167

19:17 .... 184, 296
20:1–3 168
20:3* 168
20:7–8 120
21:1 181
22:6–10 274

Romans
1:3 264
10:7 167
10:7* 168
10:18 113

Zechariah
9:9 263